Titles in This Series

W9-CMQ-077

Titles in This Series

Titles in This Series

Titles in This Series

Mathematical Developments
Arising from Linear Programming

CONTEMPORARY MATHEMATICS

114

Mathematical Developments Arising from Linear Programming

Proceedings of a Joint Summer Research Conference
held at Bowdoin College, June 25–July 1, 1988

Jeffrey C. Lagarias and
Michael J. Todd, Editors

AMERICAN MATHEMATICAL SOCIETY • PROVIDENCE, RHODE ISLAND

The AMS-IMS-SIAM Joint Summer Research Conference on Mathematical Developments Arising from Linear Programming was held at Bowdoin College, Brunswick, Maine, on June 25–July 1, 1988, with support from the National Science Foundation, Grant DMS-8746804, the U.S. Army Research Office, Grant DMS-8845058, and the Office of Naval Research, Grant 1111/87/A0544.

This work relates to Department of Navy, Grant N00014-88-J-1019 issued by the Office of Naval Research. The United States Government has a royalty-free license throughout the world in all copyrightable material contained herein.

1980 *Mathematics Subject Classification* (1985 *Revision*). Primary 90C, 65K, 49D; Secondary 52A25, 58F07.

Library of Congress Cataloging-in-Publication Data

AMS-IMS-SIAM Joint Summer Research Conference on Mathematical Developments Arising from Linear Programming (1988: Bowdoin College)
 Mathematical developments arising from linear programming: proceedings of the AMS-IMS-SIAM joint summer research conference held June 25–July 1, 1988, with support from the National Science Foundation, the U.S. Army Research Office, and the Office of Naval Research/Jeffrey C. Lagarias and Michael J. Todd, editors.
  p. cm.—(Contemporary mathematics, ISSN 0271-4132; 114)
 Includes bibliographical references.
 ISBN 0-8218-5121-7
 1. Mathematical programming—Congresses. 2. Linear programming–Congresses.
I. Lagarias, Jeffrey C., 1949– . II. Todd, Michael J., 1947– . III. Title. IV. Series:
Contemporary mathematics (American Mathematical Society); v. 114.
QA402.5.A454   1988   90-22942
519.7'2—dc20   CIP

Contents

Preface

This volume contains the proceedings of the AMS Summer Research Conference on Mathematical Developments Arising from Linear Programming held at Bowdoin College, June 25–July 1, 1988. This conference presented current research in linear and nonlinear programming and related areas of mathematics. There has been intense work in this area, much of it in extending and understanding the ideas underlying N. Karmarkar's interior-point linear programming algorithm, which was proposed in 1984. This research effort is interdisciplinary, and the conference brought together mathematicians, computer scientists, and operations researchers.

The state of the field in 1987 is illustrated in *Progress in Mathematical Programming, Interior-Point and Related Methods* (N. Megiddo, ed.), Springer-Verlag, Berlin and New York, 1989. To put the results presented in this volume in perspective, we first review the state of knowledge at that time.

Karmarkar's algorithm is an interior-point method for solving linear programs. It requires as input a linear program provided with a special initial starting point in the interior of the polytope of feasible solutions, called the *center*. There are two related types of algorithms, the *projective scaling algorithm*, which uses projective transformations, that Karmarkar proved to be a polynomial time algorithm, and the *affine scaling algorithm*, which uses affine transformations, and which has not, in general, been proved to converge in polynomial time (and probably does not). The affine scaling algorithm, however, has computational advantages in practice, and many of the computer implementations of "Karmarkar's algorithm" actually use affine scaling ideas. For both algorithms there is a vector field on the polytope of feasible solutions, which yields differential equations giving trajectories of feasible solutions inside the polytope all of which go to an optimal solution. The affine and projective scaling methods have different trajectories in general, but they have one trajectory in common, the *central trajectory* or *central path*, which turns out also to be a logarithmic barrier function trajectory previously studied in connection with nonlinear programming algorithms. Karmarkar's method, and many subsequent algorithms, approximately follow the central trajectory. Karmarkar originally proved that a linear program (in *equality form*) in n dimensions and with input size L takes $O(nL)$ iterations to converge and $O(n^{3.5}L)$ arithmetic operations in total. J. Renegar, using path-following ideas, found an algorithm requiring at most $O(\sqrt{n}L)$ iterations. By early 1987 P. Vaidya and, independently, C. Gonzaga had obtained methods that followed the central trajectory requiring $O(n^3L)$ arithmetic operations in total. Karmarkar's algorithm also used nonlinear programming ideas involving minimizing a "potential function." Such ideas carry over to give a polynomial time algorithm for convex quadratic programming as was shown by S. Kapoor and P. Vaidya in 1987, and for a class of linear complementarity problems by M. Kojima, S. Mizuno, and A. Yoshise. Finally, much work was done on developing computationally efficient versions of interior-point methods, including variants of the projective scaling algorithm that can handle problems in standard form and which do not require advance knowledge of the optimal objective function value.

Now we describe the results presented in this volume. The conference had ten invited talks which were intended to provide broad views of recent work in various areas related to linear programming. Section 1 presents papers based on such talks. The remaining sections present contributed papers classified by subject area, most of which use interior-point ideas.

The papers in Section 1 reflect the wide range of areas of mathematics on which linear programming impinges. Convex polytopes are the basic mathematical objects underlying linear programming problems and the simplex method. The paper of C. L. Lee describes recent results on the combinatorial structure of convex polytopes. K. H. Borgwardt surveys "average-case" polynomial running time bounds obtained for variants of the simplex method under various probability models. The paper of N. Megiddo analyzes several different notions of "approximate solution" of a linear programming problem. N. Karmarkar presents new results that use Riemannian geometric methods to study the behavior of trajectories underlying interior-point methods. His results yield insights concerning the running time of such algorithms. Finally the paper of A. Bloch surveys gradient-like flows arising from several algorithms, including interior-point linear programming algorithms, the QR-algorithm for diagonalizing symmetric matrices, and methods for the total least squares problem, and matching problems. It shows that these flows are related to completely integrable Hamiltonian systems by suitable nonlinear transformations.

Section 2 presents results on interior-point methods for linear programming. The paper of Y. Ye gives an $O(\sqrt{n}L)$ iteration algorithm using a new class of potential functions that apparently does not require staying close to the central trajectory. This is a significant advance, because the previously known $O(\sqrt{n}L)$ algorithms take "small" steps, so that such algorithms must take on the order of $\sqrt{n}L$ iterations to get close to an optimal solution. The idea of Ye allows algorithms that can greedily take bigger steps and still the worst-case analysis applies. (In practical implementations one takes bigger steps than the complexity analyses allow.) Ye has since shown that those ideas extend to "projective" algorithms. In another direction, in early 1988 the linear programming community in the West discovered that the affine scaling algorithm was proposed in 1967 by the Soviet mathematician I. I. Dikin, and that he published a proof of convergence for it in 1974. R. J. Vanderbei and J. C. Lagarias give an exposé of Dikin's proof of convergence, which applies under the assumption of primal nondegeneracy. One of the difficulties of interior-point methods is that they require linear programs to be transformed to a form that comes with an initial interior feasible solution. The paper of I. Lustig presents a new method for obtaining an initial feasible point for a linear program and its dual. E. Barnes discusses another method for obtaining a feasible starting point for the affine scaling algorithm and gives a convergence result for this algorithm. The paper of K. Anstreicher considers ellipsoids containing dual optimal solutions for a projective scaling interior point method in the case of primal degeneracy. M. Ašić, V. Kovačević-Vujčić, and M. Radosavljević-Nikolić analyze the asymptotic behavior of Karmarkar's algorithm, obtaining results valid for degenerate linear programs. They also propose a rounding method to go from an interior feasible point to the exact optimal solution.

Section 3 presents results on the trajetories determined by "infinitesimal" versions of the affine scaling and projective scaling algorithms. The paper of C. Witzgall, P. Boggs, and P. Domich and the paper of I. Adler and R. Monteiro both prove that limiting behavior exists for affine scaling trajectories. Their results apply even to degenerate linear programs, including cases where the convergence of the affine scaling algorithm has not yet been proved. The paper of R. C. Monteiro analyzes the boundary behavior of projective scaling trajectories.

Section 4 presents results for nonlinear programming problems. F. Jarre, G. Sonnevend, and J. Stoer give a polynomial-time interior point method for solving convex quadratic programming problems having convex quadratic constraints. B. Kalantari studies the problem of finding a zero of a quadratic form over a simplex, a problem which is NP-complete in general. He gives a "projective" algorithm for finding a local minimum of a certain potential function, which gives a polynomial-time algorithm for solving certain convex quadratic programs including linear programming. S. Mehrotra and J. Sun present an interior-point method for smooth convex programming. A. Goldstein gives a criterion specifying a quadratic convergence region when using Newton's method to find a zero of a nonlinear function.

Section 5 presents results for integer programming and multi-objective programming. The paper of N. Karmarkar gives an interior-point approach to solving 0–1 integer programming problems. Such problems, which are NP-complete in general, are converted to nonconvex quadratic programs on a hypercube. Karmarkar isolates a subclass of such problems for which the set of optimal solutions is connected. He suggests that this approach will solve a large class of 0–1 integer programs not previously considered tractable, including many set covering problems. J. Mitchell and M. Todd study perfect matching problems, which are a class of integer programs known to be solvable in polynomial time. They describe a cutting plane method for such problems that uses interior-point methods to solve linear programming relaxations of the problem and present computational data. Finally, the paper of S. Abhyankar, T. Morin, and T. Trafalis outlines two methods for solving multi-objective linear programs, including an interior-point method to find a single efficient solution, and a method of circumscribed algebraic sets to find the entire set of efficient solutions.

We would like to take this opportunity to thank the anonymous referees. We also thank the National Science Foundation and Office of Naval Research for their support of the conference. The breadth of topics covered owed much to the valuable advice of the organizing committee, consisting of Victor Klee and Steve Smale. Finally, the success of the meeting owed much to the excellent local arrangements and support of the AMS staff, in particular Ms. Carole Kohanski.

The papers in this volume are in final form and no version will be submitted for publication elsewhere, except for the paper by Irvin J. Lustig and the paper by John E. Mitchell and Michael J. Todd.

Jeffrey C. Lagarias
Michael J. Todd

- 1 -
Recent Progress
and New Directions

Contemporary Mathematics
Volume **114**, 1990

Some Recent Results on Convex Polytopes

CARL W. LEE

ABSTRACT. We sample a few results on the combinatorial structure of convex polytopes, including Lawrence's volume formula, f-vectors and h-vectors, associated algebraic structures, shellability, bistellar operations and p.l.-spheres, connections with stress and rigidity, triangulations, winding numbers, the moment map, and canonical convex combinations.

1. Introduction

The study of polyhedra has enjoyed rapid growth, stimulated partly by the development of mathematical programming in the last few decades, and partly by more recently discovered connections with commutative algebra and algebraic geometry. We informally survey a few results on the combinatorial structure of convex polytopes, beginning with Lawrence's volume formula. This leads naturally into the notions of the f-vector and the h-vector. These, in turn, have algebraic significance in associated algebraic structures. Examining these structures in the context of two inductive methods for constructing polytopes, shellings and bistellar operations, reveals an interplay with stress and rigidity. One consequence is a new proof that p.l.-spheres are Cohen–Macaulay. Gale transforms play a role here and can be used to define a class of triangulations of a convex polytope. They also provide a geometric interpretation of the h-vector in terms of winding numbers. We conclude with a brief discussion of a toric variety associated with a rational simplicial convex polytope. The components of the h-vector appear as the dimensions of its homology groups, and its moment map suggests a canonical way to express a point of the polytope as a convex combination of the vertices.

2. Lawrence's volume formula

Let us start by considering a d-dimensional convex polyhedron P of the form $P = \{x \in \mathbf{R}^d : Ax \leq b, x \geq 0\}$, where A is an $m \times d$ matrix and

1980 *Mathematics Subject Classification* (1985 *Revision*). Primary 52A25.
Supported, in part, by grant DMS-8802933 of the National Science Foundation.

$0 < b \in \mathbf{R}^m$. Further, assume that P is bounded (hence a *polytope*) and *simple* (i.e., nondegenerate). Select any linear function $z = c^T x$ that is nonconstant on every edge of P. Introduce m slack variables, one for each constraint. At every vertex (basic feasible solution) v of P we record the (necessarily nonzero) reduced costs $\bar{c}_{i_1}, \ldots, \bar{c}_{i_d}$ of the d nonbasic variables, the current value $\bar{z}$ of z, and the determinant $|B|$ of the current basis matrix. If we have arrived at v from a sequence of simplex pivots, $|B|$ is the product of the pivot elements. Lawrence [20] proves that the volume of P equals

$$(1) \qquad \sum_v \frac{\bar{z}^d}{d!\,|B|\,\bar{c}_{i_1} \cdots \bar{c}_{i_d}}$$

where the sum is taken over all vertices of P. The formula can be modified to handle polytopes that are not simple.

Thus we can theoretically compute the volume of P without triangulating it first, although as Lawrence points out there may be some difficulties in practice because the sum involves terms of differing sign that can be quite large compared to the volume of P.

As a trivial example, let us calculate the volume of the unit 3-cube $\{x \in \mathbf{R}^3 : 0 \le x_i \le 1, i = 1, 2, 3\}$ using the function $z = x_1 + 2x_2 + 4x_3$. With slack variables the description becomes

$$x_1 + x_4 = 1,$$
$$x_2 + x_5 = 1,$$
$$x_3 + x_6 = 1,$$
$$x_1, \ldots, x_6 \ge 0.$$

The desired numbers are then as follows.

v	reduced costs	$\bar{z}$
$(0,0,0)$	$(\bar{c}_4, \bar{c}_5, \bar{c}_6) = (-1, -2, -4)$	0
$(1,0,0)$	$(\bar{c}_1, \bar{c}_5, \bar{c}_6) = (+1, -2, -4)$	1
$(0,1,0)$	$(\bar{c}_4, \bar{c}_2, \bar{c}_6) = (-1, +2, -4)$	2
$(1,1,0)$	$(\bar{c}_1, \bar{c}_2, \bar{c}_6) = (+1, +2, -4)$	3
$(0,0,1)$	$(\bar{c}_4, \bar{c}_5, \bar{c}_3) = (-1, -2, +4)$	4
$(1,0,1)$	$(\bar{c}_1, \bar{c}_5, \bar{c}_3) = (+1, -2, +4)$	5
$(0,1,1)$	$(\bar{c}_4, \bar{c}_2, \bar{c}_3) = (-1, +2, +4)$	6
$(1,1,1)$	$(\bar{c}_1, \bar{c}_2, \bar{c}_3) = (+1, +2, +4)$	7

In every case $|B| = 1$, so the volume equals

$$\frac{0^3}{6(-1)(-2)(-4)} + \frac{1^3}{6(1)(-2)(-4)} + \frac{2^3}{6(-1)(2)(-4)} + \frac{3^3}{6(1)(2)(-4)}$$

$$+ \frac{4^3}{6(-1)(-2)(4)} + \frac{5^3}{6(1)(-2)(4)} + \frac{6^3}{6(-1)(2)(4)} + \frac{7^3}{6(1)(2)(4)} = 1.$$

3. h-vectors, f-vectors, and faces

It may seem surprising enough that the different sign patterns of the reduced costs contribute to the signs of the terms of (1) in just the right way, but in fact the reduced cost signs tell us even more. Let h_i be the number of vertices for which there are exactly i positive reduced costs (and hence exactly $d - i$ negative reduced costs). Define the h-vector $h(P)$ to be $(h_0, \ldots, h_d)$. So, from the previous example, the h-vector of the 3-cube is $(1, 3, 3, 1)$. From the h-vector it is possible to determine the number of faces of P of all dimensions:

$$(2) \qquad f_j = \sum_{i=j}^{d} \binom{i}{j} h_i, \qquad i = 0, \ldots, d,$$

where f_j equals the number of j-dimensional faces of P.

The vector $f(P) \equiv (f_0, \ldots, f_d)$ is called the f-vector of P. For the 3-cube it is $(8, 12, 6, 1)$. Formula (2) implies that the f-vector can be derived from the h-vector by constructing a triangle in a manner similar to Pascal's triangle, but replacing the right-hand side of the triangle by the h-vector. The f-vector emerges in reverse at the bottom. For example,

$$
\begin{array}{ccccccccc}
 & & & & \mathbf{1} & & & & \\
 & & & \mathbf{1} & & \mathbf{3} & & & \\
 & & \mathbf{1} & & \mathbf{4} & & \mathbf{3} & & \\
 & \mathbf{1} & & \mathbf{5} & & \mathbf{7} & & \mathbf{1} & \\
\mathbf{1} & & \mathbf{6} & & \mathbf{12} & & \mathbf{8} & &
\end{array}
$$

One way to verify (2) is to use the one-to-one correspondence between the $d + m$ variables and the $d + m$ constraints. The original d variables correspond to the nonnegativity constraints, and the m slack variables correspond to the m explicit constraints. A constraint is enforced when the corresponding variable is set equal to zero. At a vertex v, let $S(v)$ be the set of nonbasic variables with positive reduced cost. For every subset T of $S(v)$, consider the set of all points in P for which the constraints corresponding to $S(v) \backslash T$ hold with equality. If $\operatorname{card}(T) = j$ then we obtain a face of dimension j. Moreover, if we carry out this process for every vertex of P, we will encounter every face of P once and only once [7, §18]. So, the number of faces of dimension j equals the number of ways of finding an $S(v)$ of cardinality $i \geq j$ and selecting a subset of size j.

The above argument works as long as P is simple and z is nonconstant on every edge. Thus an enumeration of the vertices of P (by the simplex method, for example) provides us with enough information to enumerate *all* of the faces of P efficiently.

Inverting (2) yields

$$(3) \qquad h_i = \sum_{j=i}^{d} (-1)^{i+j} \binom{j}{i} f_j, \qquad i = 0, \ldots, d.$$

This implies that $h(P)$ is independent of the choice of linear function $z = c^T x$. In particular, $z = -c^T x$ would reverse the signs of all reduced costs but yield the same h-vector, so the h-vector must be symmetric; i.e., $h_i = h_{d-i}$, $i = 0, \ldots, d$. This system of equations is equivalent to the *Dehn–Sommerville equations* [7, §17].

4. Algebraic significance of the h-vector

Now let P be a simplicial (rather than simple) d-dimensional polytope and $h(P)$ be the h-vector of a simple polytope dual to P. By replacing f_j by f_{d-j-1} and using the Dehn–Sommerville equations, one can obtain formulas analogous to (2) and (3) directly in terms of the f-vector of P.

$$(4) \qquad h_i = \sum_{j=0}^{i} (-1)^{i-j} \binom{d-j}{d-i} f_{j-1}, \qquad i = 0, \ldots, d$$

$$(5) \qquad f_j = \sum_{i=0}^{j+1} \binom{d-i}{d-j-1} h_i, \qquad j = -1, \ldots, d-1$$

So, for example, we say that the h-vector of the octahedron (Figure 1) is $(1, 3, 3, 1)$.

The boundary complex of a simplicial polytope is an example of a simplicial complex. Let V be a finite set, say $V = \{1, \ldots, n\}$. A *simplicial complex* Δ on V is a nonempty collection of subsets of V that is closed under inclusion. For $F \in \Delta$, F is called a *face* of Δ, and its *dimension*, $\dim(F)$, is taken to be $\mathrm{card}(F) - 1$. The *dimension* of Δ itself, $\dim(\Delta)$, is $\max_{F \in \Delta} \dim(F)$.

Given $F \in \Delta$, define $\overline{F}$ to be 2^F, the collection of all subsets of F, and $\partial \overline{F}$ to be $2^F \backslash \{F\}$. The *link* in Δ of F is $\mathrm{lk}_\Delta F \equiv \{G \in \Delta \colon G \cap F = \varnothing, G \cup F \in \Delta\}$. If F is not the empty set, we also define $\Delta \backslash F \equiv \{G \in \Delta \colon F \nsubseteq G\}$ to be the *deletion* of F from Δ. Given two simplicial complexes Δ_1 and

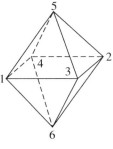

$$f(P) = (1, 6, 12, 8)$$
$$h(P) = (1, 3, 3, 1)$$

FIGURE 1. The octahedron.

Δ_2 on disjoint sets V_1 and V_2, respectively, the *join* of Δ_1 and Δ_2 is $\Delta_1 \cdot \Delta_2 \equiv \{F_1 \cup F_2 : F_1 \in \Delta_1, F_2 \in \Delta_2\}$.

For $(d-1)$-dimensional simplicial complex Δ, its *f-vector* is $f(\Delta) \equiv (f_{-1}, f_0, \ldots, f_{d-1})$ where f_j is the number of j-dimensional faces of Δ. The *h-vector* of Δ is then defined by formula (4).

Taking Δ to be any $(d-1)$-dimensional simplicial complex on $V = \{1, \ldots, n\}$, form the polynomial ring $R \equiv \mathbf{C}[x_1, \ldots, x_n]$, which has a natural grading by degree. For monomial $m = x_{i_1}^{a_{i_1}} \cdots x_{i_k}^{a_{i_k}}$ where each $a_{i_j} > 0$, the *support* of m, supp(m), is the set $\{i_1, \ldots, i_k\}$. We now let I be the ideal of all nonfaces; i.e., $I \equiv \langle m : \text{supp}(m) \notin \Delta \rangle$. Factoring out I from R yields the ring $A \equiv A_0 \oplus A_1 \oplus A_2 \oplus \cdots$, which inherits the grading by degree. This is known as the *Stanley–Reisner ring* of Δ [33], [38]. If Δ is the boundary complex of the octahedron given in Figure 1, for example, we take $R = \mathbf{C}[x_1, \ldots, x_6]$ and $I = \langle x_1 x_2, x_3 x_4, x_5 x_6 \rangle$.

The ring A is *Cohen–Macaulay* if and only if it contains elements $\theta_1, \ldots, \theta_d$ of degree one with the following property: $B \equiv A/\langle \theta_1, \ldots, \theta_d \rangle = B_0 \oplus B_1 \oplus \cdots \oplus B_d$ has only finitely many nonzero components, graded by degree, and $\dim(B_i) = h_i$, $i = 0, \ldots, d$ as vector spaces over $\mathbf{C}$. When this happens, Δ is called a *Cohen–Macaulay complex* [38]. For example, it can be shown for the octahedron that $\theta_1 = x_1 - x_2$, $\theta_2 = x_3 - x_4$, and $\theta_3 = x_5 - x_6$ have the desired property, so the boundary complex of an octahedron is Cohen–Macaulay.

Reisner [33] proved that boundary complexes of simplicial polytopes are Cohen–Macaulay, as are the properly larger classes of shellable complexes, p.l.-spheres, and homological spheres. He did this by providing a homological characterization of the class of all Cohen–Macaulay complexes. One corollary of this result is Stanley's new proof of the Upper Bound Theorem for convex polytopes and its extension to homological spheres [37].

Further, for simplicial polytopes Stanley [39] proved that for suitable θ_i there exists an element $\omega \in B_1$ with the following property: After the ideal generated by ω is factored out of B, the result is an algebra $C \equiv C_0 \oplus \cdots \oplus C_{\lfloor \frac{d}{2} \rfloor}$, graded by degree, such that $\dim(C_i) = g_i \equiv h_i - h_{i-1}$, $i = 1, \ldots, \lfloor \frac{d}{2} \rfloor$. His proof exploits a connection between rational convex polytopes and certain complex projective toric varieties. This far-reaching result leads to a complete characterization of f-vectors of simplicial and simple polytopes, as well as tight upper and lower bounds on the numbers of faces of unbounded, simple polyhedra with a given number of bounded and unbounded facets and recession cone of specified dimension [3, 9, 10, 21].

One particular consequence is that the h-vector is unimodal, which was one part of the Generalized Lower-Bound Conjecture [31]. The fact that $h_2 - h_1 \geq 0$ provides a new proof of the Lower-Bound Theorem [1], [2]. Kalai [17] has yet another proof of this inequality based upon rigidity considerations. Let E denote the set of edges of a polytope. Kalai considers the *stress space* for a convex polytope that is defined to be all functions $\lambda: E \to \mathbf{R}$ such that

$$\sum_{u \; : \; uv \in E} \lambda_{uv}(u - v) = 0$$

for all vertices v. Since the 1-skeleton of a simplicial d-polytope is generically d-rigid, the stress space is a vector space of dimension $h_2 - h_1$, and hence is nonnegative.

5. Shellability

A $(d-1)$-dimensional simplicial complex is *shellable* if its maximal faces (*facets*) are each of dimension $d - 1$ and can be ordered $F_1, \ldots, F_m$ such that for $i = 2, \ldots, m$ there is a unique minimal face G_i of F_i that is not in the union of $\overline{F}_1, \ldots, \overline{F}_{i-1}$. Suppose when adding F_i in the shelling process that the minimal face G_i contains exactly k vertices. Then from formula (4) one readily sees that h_k increases by one, while the other h_j remain unchanged. A shelling of the octahedron (Figure 2) illustrates this.

Bruggesser and Mani [8] proved that the boundary complexes of all convex polytopes are shellable. An appropriate ordering of facets can be obtained by translating P so that the origin is in its interior, taking the polar of P, choosing a linear function $c^T x$ as in Section 2 that is nonconstant on every edge of the polar, and ordering the vertices of the polar $v_1, \ldots, v_m$ such that $c^T v_1 \leq \cdots \leq c^T v_m$. The corresponding ordering of the facets of P is a shelling order. Shellings constructed in this manner are called *line shellings*. In terms of P itself, we can take a line in general position through the origin

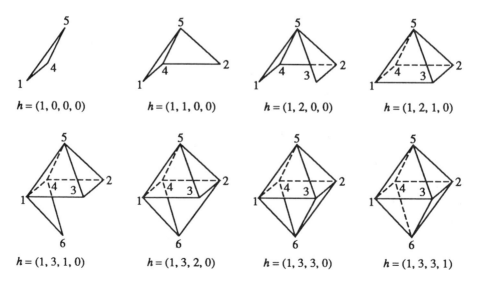

$h = (1, 0, 0, 0)$ $h = (1, 1, 0, 0)$ $h = (1, 2, 0, 0)$ $h = (1, 2, 1, 0)$

$h = (1, 3, 1, 0)$ $h = (1, 3, 2, 0)$ $h = (1, 3, 3, 0)$ $h = (1, 3, 3, 1)$

FIGURE 2. Building an octahedron by shelling.

(which is assumed to be in the interior of P) and list the facets according to the order in which their supporting hyperplanes are pierced as we travel along the line from the origin to infinity and back to the origin again from the opposite direction.

One way to prove that A is Cohen–Macaulay for a shellable complex is to show that the dimensions of the B_j change in exactly the same way as the h-vector during the shelling. Kind and Kleinschmidt [18] found an inductive proof that shellable complexes are Cohen–Macaulay in this manner.

6. Bistellar operations and p.l.-spheres

So far the methods in the previous section have not led to a more elementary proof of the existence of the element ω. One obstacle might be that the intermediate complexes during the shelling are not themselves polytopes. However, there are other ways to construct a polytope inductively using elementary operations such that one has a polytope at every intermediate stage.

Given a nonempty face F in a simplicial complex Δ and an element $v \notin V$, the *stellar subdivision* of F is $\operatorname{st}(v, F)[\Delta] \equiv (\Delta \backslash F) \cup (\overline{\{v\}} \cdot \partial \overline{F} \cdot \operatorname{lk}_\Delta F)$. The opposite of a stellar subdivision is an *inverse stellar subdivision*. A simplicial complex Δ is a p.l. (*piecewise linear*) $(d-1)$-*sphere* if Δ is obtainable from the boundary of a d-dimensional simplex by a sequence of stellar and inverse stellar subdivisions [16].

Now suppose that F is a nonempty face of a $(d-1)$-dimensional complex Δ and G is a nonface of Δ such that $\operatorname{lk}_\Delta F = \partial \overline{G}$. In this case assume $k + l = d + 1$ where $k = \operatorname{card}(F)$ and $l = \operatorname{card}(G)$. A certain combination of a stellar subdivision of F and an inverse stellar subdivision at the same location is called a *bistellar operation* and results in the simplicial complex $\operatorname{bist}(G, F)[\Delta] \equiv (\Delta \backslash F) \cup (\partial \overline{F} \cdot \overline{G})$.

Formula (4) shows that such a bistellar operation increases $g_l = h_l - h_{l-1}$ by one, decreases $g_k = h_k - h_{k-1}$ by one, and leaves all other differences $g_i = h_i - h_{i-1}$ unchanged. Figure 3 shows how the octahedron can be constructed using three bistellar operations.

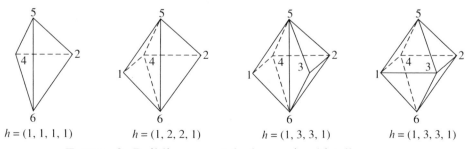

$h = (1, 1, 1, 1)$ $h = (1, 2, 2, 1)$ $h = (1, 3, 3, 1)$ $h = (1, 3, 3, 1)$

FIGURE 3. Building an octahedron using bistellar operations.

Ewald [12] showed that starting with a d-simplex, any simplicial d-polytope can be obtained by a sequence of bistellar operations such that one has a simplicial d-polytope at each intermediate stage of the sequence. Pachner [32] then proved that every $(d-1)$-dimensional p.l.-sphere can be constructed from the boundary of a d-simplex using bistellar operations.

Kind and Kleinschmidt's proof for shellable complexes suggests trying to do something analogous for p.l.-spheres: Show that the dimensions of B_i change in exactly the same way as the h-vector during a bistellar operation. This is possible, and the result is a new, inductive proof that p.l.-spheres are Cohen–Macaulay [27].

7. Connections with stress

To carry out the above proof, it suffices to work with $\mathbf{R}$ instead of $\mathbf{C}$. Given an element θ_1 in A_1, multiplication by θ_1 is a linear map from the vector space A_i to the vector space A_{i+1}, $i = 0, 1, 2, \ldots$

$$A_0 \overset{\cdot\theta_1}{\to} A_1 \overset{\cdot\theta_1}{\to} A_2 \overset{\cdot\theta_1}{\to} \cdots.$$

Now dualize the vector spaces and the maps:

$$\overline{A}_0 \overset{\cdot\overline{\theta}_1}{\leftarrow} \overline{A}_1 \overset{\cdot\overline{\theta}_1}{\leftarrow} \overline{A}_2 \overset{\cdot\overline{\theta}_1}{\leftarrow} \cdots.$$

Since we are interested in factoring out the images of θ_1, we wish to keep the kernels of $\overline{\theta}_1$. Repeating this process with the other θ_i, we get vector spaces $\overline{B}_0, \overline{B}_1, \overline{B}_2, \ldots$ of common kernels.

To describe these kernels explicitly, let $\theta_i = \sum_{j=1}^n a_{ij} x_j$ and put $V_j = [a_{1j}, \ldots, a_{dj}]^T$. Let M_i be all monomials in the variables $x_1, \ldots, x_n$ of degree i. The space $\overline{B}_i$ is isomorphic to the set of vectors $(c_m)_{m \in M_i}$ indexed by elements of M_i with the property that

(6) $c_m = 0$ if $\mathrm{supp}(m) \notin \Delta$,

and

(7) $\sum_{j=1}^n c_{x_j m} V_j = 0$ for all $m \in M_{i-1}$.

It is easy to see that $\overline{B}_1$ is the vector space of all linear relations on the V_j, and so has dimension $h_1 = n - d$ if the matrix whose columns are the V_j has full row rank. This would be the case, for example, if Δ is the boundary complex of a simplicial d-polytope containing the origin in its interior and the V_j are chosen to be its vertices.

When the simplicial complex is the boundary complex of the octahedron in Figure 1 and the V_j are taken to be its vertices, elements of $\overline{B}_2$ are of the form $c_{x_1 x_3} = c_{x_1 x_4} = c_{x_2 x_3} = c_{x_2 x_4} = p$, $c_{x_1 x_5} = c_{x_1 x_6} = c_{x_2 x_5} = c_{x_2 x_6} = q$, $c_{x_3 x_5} = c_{x_3 x_6} = c_{x_4 x_5} = c_{x_4 x_6} = r$, and $c_{x^2_i} = 0$, $i = 1, \ldots, 6$.

Now assume that the V_j are chosen in linearly general position, i.e., that every subset of size at most d is linearly independent. Consider a bistellar operation with F, G and k, l as before. Then there is a unique linear relation among $\{V_j : j \in F \cup G\}$, say, $\sum_{j \in F \cup G} a_j V_j = 0$, and $a_j \neq 0$ for all $j \in F \cup G$. Set $a_j = 0$ for all $j \notin F \cup G$. For $1 \leq i \leq d$ define $(c_m)_{m \in M_i}$ by $c_m = m(a)$; i.e., substitute a_j for x_j in m. During the bistellar operation, such elements are gained in $\overline{B}_i$ for $i \geq l$ when G is added, and lost in $\overline{B}_i$ for $i \geq k$ when F is removed. The structure of the links of F and G is enough to guarantee that the dimensions of these $\overline{B}_i$ change by ± 1 in exactly the right way. The proof is completed by demonstrating that $\dim(\overline{B}_i) = 1$, $i = 0, \ldots, d$ for the boundary of a d-simplex.

Now let us consider the special case when Δ is the boundary complex of a simplicial d-polytope P. Translate P, if necessary, so that its vertices v_j are in linearly general position. Then one can choose V_j equal to v_j. There is some evidence to support the proposal that ω can be taken to be $x_1 + \cdots + x_n$, for suppose this ω is factored out of B and this operation is viewed in a dual fashion as we did with the θ_i. The vector spaces $\overline{C}_i$ of kernels are described by

$$(8) \qquad\qquad c_m = 0 \quad \text{if supp}(m) \notin \Delta$$

and

$$(9) \qquad \sum_{j=1}^{n} c_{x_j m} V_j = 0 \quad \text{and} \quad \sum_{j=1}^{n} c_{x_j m} = 0 \quad \text{for all } m \in M_{i-1}.$$

This is the affine analogue to condition (7). One easily sees that $\overline{C}_1$ is the vector space of all affine relations on the V_j and so its dimension equals $g_1 = n - d - 1$. But $\overline{C}_2$ also is a familiar object: It is isomorphic to the stress space used by Kalai and mentioned in Section 4. The correspondence is given by taking $\lambda_{v_i v_j}$ to be $c_{x_i x_j}$. So the dimension of $\overline{C}_2$ equals $g_2 = h_2 - h_1$ and the dimensions of $\overline{C}_1$ and $\overline{C}_2$ are both correct. It remains to be seen whether the dimensions of the other $\overline{C}_i$ for $i = 3, \ldots, \lfloor d/2 \rfloor$ equal g_i for this choice of ω. The spaces $\overline{C}_i$ suggest a natural way to extend the notion of stress space to higher dimensional faces, which might prove useful [27].

8. Triangulations of polytopes

One way to see how a simplicial polytope can be built up by bistellar operations is to look at a *Gale transform*, which can be defined for any convex d-polytope P, whether simplicial or not [30]. Let $v_1, \ldots, v_n \in \mathbf{R}^d$ be its vertices and consider the matrix A given by

$$(10) \qquad\qquad A = \begin{bmatrix} v_1 & \cdots & v_n \\ 1 & \cdots & 1 \end{bmatrix}.$$

Let $y_1, \ldots, y_{n-d-1}$ be a basis for the nullspace of A (the space of all affine relations on the vertices) and list these vectors as the rows of a matrix

$$\begin{bmatrix} y_1^T \\ \vdots \\ y_{n-d-1}^T \end{bmatrix}.$$

The set V' of the n columns of this matrix, $v_1', \ldots, v_n' \in \mathbf{R}^{n-d-1}$, are the points of a Gale transform of P. There is a one-to-one correspondence between the vertices v_i of P and the points v_i' of the transform, and hence between subsets X of V and subsets X' of V'. One key property is that conv(X) is a face of P if and only if the origin is contained in the relative interior of the convex hull of $V'\backslash X'$ in the Gale transform. This property is maintained even if the points in V' are independently scaled by positive numbers. In this case we say we have a *scaled Gale transform*.

Returning to the example of the octahedron in Figure 1, we see that the matrix A is

$$\begin{bmatrix} 1 & -1 & 0 & 0 & 0 & 0 \\ 0 & 0 & 1 & -1 & 0 & 0 \\ 0 & 0 & 0 & 0 & 1 & -1 \\ 1 & 1 & 1 & 1 & 1 & 1 \end{bmatrix}.$$

The two rows of the following matrix form a basis for the nullspace of A:

$$\begin{bmatrix} 1 & 1 & 0 & 0 & -1 & -1 \\ 0 & 0 & 1 & 1 & -1 & -1 \end{bmatrix}.$$

The result is the Gale transform given in Figure 4. Note that the points of a Gale transform need not be distinct.

Given a Gale transform of any convex d-polytope P, scale the points by positive numbers so that there is no hyperplane missing the origin that contains more than $d' \equiv n - d - 1$ points of V'. Choose any halfline L in $\mathbf{R}^{d'}$ starting at infinity and ending at the origin, but otherwise in general position. As you travel along this line, you will pass through the relative interior of various simplices of the form conv(X') where card(X') = d'.

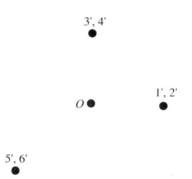

FIGURE 4. A Gale transform of the octahedron.

The complement of each such X' corresponds to a d-simplex in a triangulation of the polytope P, and the order induced by the halfline is a shelling order for the triangulation [24], [25], [30]. (That this procedure resembles line shellings is no coincidence.) Such a triangulation will be called a *Gale triangulation* of P. During the shelling process the boundary of the triangulation changes by bistellar operations, so if P is simplicial this induces a construction of the boundary complex of P by a sequence of bistellar operations.

In Figure 5, the halfline in the scaled Gale transform of the octahedron induces a triangulation of the octahedron into four simplices, each sharing the common interior edge 56. The corresponding bistellar operations are precisely those depicted in Figure 3.

If $d = 2$ or $n \le d + 3$ then every triangulation of P is a Gale triangulation. In fact, Gale transforms can be used to prove that in these cases the collection of all subdivisions of P, ordered by refinement, is isomorphic to the face lattice of some $(n - d - 1)$-dimensional polytope [23], [24], [25].

In general, not all triangulations of convex polytopes are Gale triangulations. For example, there exist 3-polytopes with seven vertices that have triangulations unobtainable in this way [24]. However, any triangulation induced by *pulling* or *placing* the vertices in any order is a Gale triangulation. In fact, if a triangulation T is determined by pulling the vertices in the order $v_1, \ldots, v_n$ and the triangulation T' is determined by placing the vertices in the opposite order, then in the Gale transform there exists an oppositely directed pair of halflines L and L' inducing T and T', respectively [25].

There are several equivalent ways of defining Gale triangulations. One way is to take the vertices $v_1, \ldots, v_n$, lift them into general position in $\mathbf{R}^{d+1}$ yielding $(v_1, t_1), \ldots, (v_n, t_n)$, determine their convex hull, and project the facets of the "lower half" of the resulting $(d+1)$-polytope back into $\mathbf{R}^d$. See, for example, [14]. Another way is to take the matrix A given in (10), choose a vector $c \equiv (c_1, \ldots, c_n)$ in general position, and form the polyhedron

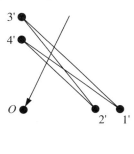

FIGURE 5. Inducing a Gale triangulation of the octahedron.

$Q = \{x : x^T A \le c^T\}$. Then the vertices of Q are in one-to-one correspondence with simplices in a triangulation of P [15].

This latter perspective can be used to prove that any simple d-polytope P with n facets can be realized as a facet of a simple $(d+1)$-polytope P' with $n+1$ facets that has an edge-path diameter not exceeding $2n - 2d$. If P is the feasible region for a linear program, this implies that we can solve the linear program from any starting point with at most $2n - 2d$ pivots, if we somehow know which pivots should be made [25]. It has been conjectured that the edge-path diameter of any simple d-polytope with n facets is at most $n - d$, but this has not been settled and related examples have suggested that it might be false. The establishment of a good upper bound is one of the significant open problems in the theory of convex polytopes [19].

One interesting d-polytope to try to triangulate efficiently is the d-cube. The smallest triangulation of the 4-cube has 16 four-dimensional simplices, but for higher dimensions the minimum number is not known [11], [22], [34], [35], [36]. There is a close connection between Hadamard matrices and Gale transforms of the d-cube that might be exploited to shed some light on this problem and the more general task of finding new, interesting triangulations of the d-cube.

9. Winding numbers

The relationship between bistellar operations, Gale transforms, and triangulations leads fairly easily to the following result [28], also known to Lawrence. Let W be a collection of at least $e+1$ points in $\mathbf{R}^e$ such that no hyperplane contains more than e points of $W \cup \{O\}$, where O is the origin. Choose an integer $0 \le k < (n-e)/2$. For X a subset of W of cardinality e, let us say that X (or $\mathrm{conv}(X)$) is of *type* k if the hyperplane $H = \mathrm{aff}(X)$ partitions the remaining $n - e$ points into two sets, one of which, say F, has cardinality k. Figure 6 shows the subsets of type 0, 1, and 2 in a set of seven points. (The origin is not marked.)

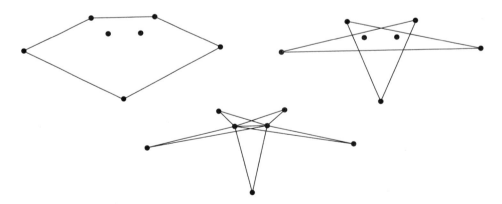

FIGURE 6. Subsets of type 0, 1, and 2.

For such a subset X, define the *sign* of X, $\mathrm{sg}(X)$, to be *positive* if F and O lie on opposite sides of H, and *negative* if F and O lie on the same side of H. Finally, define $\alpha(X)$ to be the measure of the solid angle with vertex O determined by X. The value $\alpha(X)$ is normalized to equal the fraction of the surface area of a unit sphere centered at the origin that is intersected by the cone determined by X. Set

$$(11) \qquad\qquad w_k \doteq \sum_X \mathrm{sg}(X)\alpha(X)$$

where the sum is taken over all X of type k. Then one can prove that this kth *winding number* w_k is in fact a nonnegative integer.

In the case that W is the scaled Gale transform of some simplicial $n - e - 1$ polytope P, the result follows by proving that $w_k = h_k - h_{k-1}$. The result for general W then follows readily. This suggests trying to prove the nonnegativity of w_k directly, which would yield a new proof of the unimodality of the h-vector. Such a direct proof has already been found for $e \leq 2$ [28].

The unimodality of the h-vector of simplicial d-polytopes with n vertices was first conjectured as a part of the Generalized Lower-Bound Conjecture. The second part of the conjecture is that $h_k = h_{k+1}$ for some $k < \lfloor d/2 \rfloor$ if and only if P admits a triangulation with no simplex of dimension less than $d - k$ in the interior. This part of the conjecture is still unresolved in the general case, but has been confirmed when $n \leq d + 3$ as a part of the winding number proof for $e \leq 2$.

10. The moment map

We conclude with a brief discussion of a connection between convex polytopes and algebraic varieties. See also [41]. Let Q be a rational convex d-polytope with vertices $v_1, \ldots, v_n$. Consider all nontrivial affine relations $a = (a_1, \ldots, a_n)$ on the vertices: $D = \{a : \sum_{i=1}^n a_i v_i = 0, \sum_{i=1}^n a_i = 0, a_i \in \mathbf{R}$, not all zero$\}$. For any a, define $A_+ = \{i : a_i > 0\}$, $A_- = \{i : a_i < 0\}$, and $A_0 = \{i : a_i = 0\}$. We say that a *conforms* to a' if $A_+ \subseteq A'_+$ and $A_- \subseteq A'_-$. It is not difficult to see that there exists a finite set $\{a^1, \ldots, a^m\}$ of integer $a^j \in D$ such that every integer $a \in D$ is a nonnegative integer combination of a^j conforming to a.

Fine [13] constructs a variety in the following way. Let $u_1, \ldots, u_n$ be indeterminates. For each integer $a = (a_1, \ldots, a_n) \in D$ associate the relation

$$(12) \qquad\qquad \prod_{i \in A_+} u_i^{a_i} = \prod_{i \in A_-} u_i^{-a_i}.$$

Let $A_Q = \{u \in \mathbf{C}^n : u$ satisfies (12) for all integer $a \in D\}$. One can show that $A_Q = \{u \in \mathbf{C}^n : u$ satisfies (12) for all $a \in \{a^1, \ldots, a^m\}\}$. So, A_Q is a variety in $\mathbf{C}^n$.

Note that $u \in A_Q$, $k \in \mathbf{C}$ implies that $ku \in A_Q$ since $\sum_{i \in A_+} a_i = \sum_{i \in A_-}(-a_i)$ for each $a \in D$. For $u, v \in A_Q$, let $u \sim v$ if $u = kv$ for some $0 \neq k \in \mathbf{C}$. Let $P_Q = (A_Q \backslash \{0\})/ \sim$, a projective variety in $\mathbf{CP}^{n-1}$. Note that $u, v \in P_Q$ implies that $uv \in P_Q$ under component-wise multiplication; $u, v \in P_Q$ satisfying $v_i \neq 0$ whenever $u_i \neq 0$ implies $u/v \in P_Q$ under component-wise division (taking $0/0 = 0$); and $u \in P_Q$ satisfying $u \in \mathbf{R}^n_+$ implies $\sqrt{u} \in P_Q$, where $\sqrt{u} = (\sqrt{u_1}, \dots, \sqrt{u_n})$.

For example, let Q be a d-simplex. Then there are no affine relations among the vertices, so $A_Q = \mathbf{C}^{d+1}$ and $P_Q = \mathbf{CP}^d$.

As a second example, take Q to be a square with vertices labeled v_1, v_2, v_3, and v_4, consecutively around the perimeter. Then the unique affine relation $v_1 + v_3 = v_2 + v_4$ yields the relation $u_1 u_3 = u_2 u_4$.

For F a nonempty face of Q ($F = Q$ allowed), define $\operatorname{supp}(F) = \{i: v_i \in F\}$. For $u \in P_Q$, define $\operatorname{supp}(u) = \{i: u_i \neq 0\}$. For any $u \in P_Q$ it can be shown that there is a face F of Q such that $\operatorname{supp}(u) = \operatorname{supp}(F)$, so P_Q is the disjoint union of sets of the form B_F, where $B_F = \{u \in P_Q: \operatorname{supp}(u) = \operatorname{supp}(F)\}$ for F a face of Q.

In fact there is at least one "canonical" element of P_Q associated with each face F of Q: Take $u = (u_1, \dots, u_n)$ where

$$u_i = \begin{cases} 1 & \text{if } i \in \operatorname{supp}(F); \\ 0 & \text{otherwise}. \end{cases}$$

The above results can be used to show a direct connection between the homology groups of the variety and the graded components of the ring B, and hence that the dimensions of the homology groups are equal to the h_i when Q is simple.

Define the *moment map* $\phi: P_Q \to Q$ by

$$\phi(u) = \frac{\sum_{i=1}^n |u_i|^2 v_i}{\sum_{i=1}^n |u_i|^2}.$$

Note that if $u \in P_Q$ then $|u| = (|u_1|, \dots, |u_n|) \in P_Q$. Let $R_Q^+ = \{u \in P_Q: u_i \in \mathbf{R}, u_i \geq 0 \text{ for all } i\}$. We have the commutative diagram:

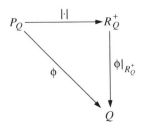

For example, let F be a face of Q and consider $u = (u_1, \ldots, u_n)$ such that

$$u_i = \begin{cases} 1 & i \in \mathrm{supp}(F); \\ 0 & \text{otherwise.} \end{cases}$$

Then

$$\phi(u) = \sum_{i \in \mathrm{supp}(F)} \frac{1}{\mathrm{card}(\mathrm{supp}(F))} v_i,$$

which is the centroid of the vertices of F.

As another example, suppose Q is a d-simplex. For $u \in R_Q^+$, scale u so that $\sum_{i=1}^n u_i^2 = 1$. Then $\phi(u) = \sum_{i=1}^n u_i^2 v_i = \sum_{i=1}^n \lambda_i v_i$ where $u_i = \sqrt{\lambda_i}$. Hence there is a one-to-one correspondence between the elements of R_Q^+ and the points of Q.

Fine [13] proved that for any Q, $\phi|_{R_Q^+}$ is a bijection between R_Q^+ and Q. See also [41]. Another way to show this is to describe $\phi|_{R_Q^+}^{-1}$ explicitly [26]. For any point $x \in Q$, consider all $\lambda = (\lambda_1, \ldots, \lambda_n)$ such that

$$\sum_{i=1}^n \lambda_i x^i = x, \qquad \sum_{i=1}^n \lambda_i = 1, \qquad 0 \le \lambda_i \le 1.$$

Call this set L. There is a unique λ^* that minimizes $\sum_{i=1}^n \lambda_i \log \lambda_i$ over L. It turns out that $\sqrt{\lambda^*} = (\sqrt{\lambda_1^*}, \ldots, \sqrt{\lambda_n^*}) \in R_Q^+$ equals $\phi|_{R_Q^+}^{-1}(x)$.

The inverse of the moment map offers a canonical way of expressing any point $x \in Q$ as a convex combination of the vertices that is a natural generalization of barycentric coordinates for a simplex. The function $-\sum_{i=1}^n \lambda_i \log \lambda_i$ is the familiar entropy function and suggests the following whimsical interpretation of the canonical convex expression: Suppose two individuals A and B are playing a game on a polytope Q. A referee chooses a point x in Q, which is known to both A and B, and A chooses a way of expressing x as a convex combination $\sum_{i=1}^n \lambda_i v_i$ of the vertices of Q. Interpreting the λ_i as probabilities assigned to the vertices, A then randomly chooses a vertex of Q using this probability distribution. B now attempts to guess the vertex A has chosen by asking questions of the form "Is the vertex in the set S?" where S is a subset of the vertices. The object of B is to guess the vertex using as few questions as possible, so the object of A is to choose the λ_i that keep B guessing as long as possible, even if B should happen to discover which λ_i his opponent has chosen. The inverse of the moment map provides the best choice of λ_i.

11. Nonsimplicial polytopes

There is considerable interest in extending some of the results for simplicial polytopes to general convex polytopes. The discovery of the generalized Dehn–Sommerville equations by Bayer and Billera [4], [5], [6], the notion of the generalized h-vector, and the connections with intersection homology

[**40**] are very encouraging and suggest that there is still much to be done to understand fully the interplay between geometry and algebra that has evolved with the study of convex polyhedra.

REFERENCES

1. D. Barnette, *The minimum number of vertices of a simple polytope*, Israel J. Math. **10** (1971), 121–125.
2. ____, *A proof of the lower-bound conjecture for convex polytopes*, Pacific J. Math. **46** (1973), 349–354.
3. D. Barnette, P. Kleinschmidt, and C. W. Lee, *An upper bound theorem for polytope pairs*, Math. Oper. Res. **11** (1986), 451–464.
4. M. M. Bayer, *The generalized Dehn–Sommerville equations revisited*, preprint.
5. M. M. Bayer and L. J. Billera, *Generalized Dehn–Sommerville relations for polytopes, spheres and Eulerian partially ordered sets*, Invent. Math. **79** (1985), 143–157.
6. ____, *Counting faces and chains in polytopes and posets*, Combinatorics and Algebra (Boulder, Colo., 1983), Contemp. Math., vol. 34, Amer. Math. Soc., Providence, R.I., 1984, pp. 207–252.
7. A. Brøndsted, *An introduction to convex polytopes*, Graduate Texts in Math., vol. 90, Springer-Verlag, Berlin and New York, 1983.
8. H. Bruggesser and P. Mani, *Shellable decompositions of cells and spheres*, Math. Scand. **29** (1971), 197–205.
9. L. J. Billera and C. W. Lee, *A proof of the sufficiency of McMullen's conditions for f-vectors of simplicial convex polytopes*, J. Combin. Theory Ser. A **31** (1981), 237–255.
10. ____, *The numbers of faces of polytope pairs and unbounded polyhedra*, European J. Combin. **2** (1981), 307–322.
11. R. W. Cottle, *Minimal triangulation of the 4-cube*, Discrete Math. **40** (1982), 25–29.
12. G. Ewald, *Über stellare Äquivalenz konvexer Polytope*, Resultate Math. **1** (1978), 54–60.
13. J. Fine, *Geometric progressions, convex polytopes, and torus embeddings*, preprint.
14. M. Haiman, *Constructing the associahedron*, preprint.
15. A. J. Hoffman, personal communication.
16. J. F. P. Hudson, *Piecewise linear topology*, W. A. Benjamin, New York–Amsterdam, 1969.
17. G. Kalai, *Rigidity and the lower bound theorem* I, Invent. Math. **88** (1987), 125–151.
18. B. Kind and P. Kleinschmidt, *Schälbare Cohen–Macauley-Komplexe und ihre Parametrisierung*, Math. Z. **167** (1979), 173–179.
19. V. Klee and P. Kleinschmidt, *The d-step conjecture and its relatives*, Math. Oper. Res. **12** (1987), 718–755.
20. J. Lawrence, *Polytope volume computation*, preprint.
21. C. W. Lee, *Bounding the numbers of faces of polytope pairs and simple polyhedra*, Ann. Discrete Math. **20** (1984), 215–232.
22. ____, *Triangulating the d-cube*, Discrete Geometry and Convexity (New York, 1982), Ann. New York Acad. Sci., New York Acad. Sci., vol. 440, New York, 1985, pp. 205–211.
23. ____, *The associahedron and triangulations of the n-gon*, European J. Combin. **10** (1989), 551–560.
24. ____, *Some notes on triangulating polytopes*, proceedings, **3**. Kolloquium über Diskrete Geometrie, Institut für Mathematik, Universität Salzburg, Salzburg, May 1985, pp. 173–181.
25. ____, *Regular triangulations of convex polytopes*, preprint.
26. ____, *A note on convex polytopes, the moment map, and canonical convex combinations*, in preparation.
27. ____, *p.l.-spheres and convex polytopes*, in preparation.
28. ____, *Winding numbers and the generalized lower-bound conjecture*, preprint.
29. P. McMullen, *The maximum numbers of faces of a convex polytope*, Mathematika **17** (1971), 179–184.

30. ____, *Transforms, diagrams, and representations*, Contributions to Geometry, (Proc. Geom. Sympos., Siegen, 1978), Birkhäuser, Basel, 1979, pp. 92–130.

31. P. McMullen and D. W. Walkup, *A generalized lower-bound conjecture for simplicial polytopes*, Mathematika **18** (1971), 264–273.

32. U. Pachner, *Shellings of simplicial balls and p.l. manifolds with boundary*, Bericht Nr. 90, Ruhr-Universität Bochum, Bochum, 1987.

33. G. A. Reisner, *Cohen–Macaulay quotients of polynomial rings*, Adv. in Math. **21** (1976), 30–49.

34. J. F. Sallee, *A note on minimal triangulations of an n-cube*, Discrete Appl. Math. **4** (1982), 211–215.

35. ____, *A triangulation of the n-cube*, Discrete Math. **40** (1982), 81–86.

36. ____, *The middle-cut triangulations of the n-cube*, SIAM J. Algebraic Discrete Methods **5** (1984), 407–419.

37. R. P. Stanley, *The upper-bound conjecture and Cohen–Macaulay rings*, Stud. Appl. Math. **54** (1975), 135–142.

38. ____, *Cohen–Macaulay complexes*, Higher Combinatorics, (Proc. NATO Advanced Study Inst., Berlin, 1976), NATO Adv. Study Inst. Ser., Ser. C: Math. and Phys. Sci. **31**, Reidel, Dordrecht, 1977, 51–62.

39. ____, *The number of faces of a simplicial convex polytope*, Adv. in Math. **35** (1980), 236–238.

40. ____, *Generalized h-vectors, intersection cohomology of toric varieties, and related results*, Commutative Algebra and Combinatorics (Kyoto, 1985), Adv. Stud. Pure Math., vol. 11, North-Holland, Amsterdam–New York, pp. 187–213.

41. T. Oda, *Convex bodies and algebraic geometry: An introduction to the theory of toric varieties*, Springer-Verlag, Berlin-Heidelberg, 1985.

DEPARTMENT OF MATHEMATICS, UNIVERSITY OF KENTUCKY, LEXINGTON, KENTUCKY 40506
E-mail addresses: lee@s.ms.uky.edu   and   lee@ukma.bitnet

Contemporary Mathematics
Volume **114**, 1990

Probabilistic Analysis of the Simplex Method

KARL HEINZ BORGWARDT

ABSTRACT. We are interested in the average number of pivot steps required for solving LPs. It is a common experience shared by almost all operations research experts that the Simplex Method is a very good tool for solving linear programming problems occuring in practice. On the other hand we know that all the usual variants of the method have an exponential worst-case behavior. To understand and to explain that discrepancy has provided a great challenge for mathematicians for over three decades.

The best way to close that gap and to give a formal description of practical efficiency seemed to be a probabilistic analysis of the average behavior. For that purpose one has to specify a stochastic model defining a (fictive) distribution of linear programming problems occurring in practice. Then a chosen fixed variant of the Simplex Method has to be analyzed under that model.

This paper is a survey report on three major streams and successful approaches for such an analysis. It is interesting that all these approaches are based on parametric variants.

1. Introduction

What I am telling here is a rather old story. Most of the mathematical results mentioned here were derived during the years 1981–1984.

Throughout the talk we consider the following type of linear programming problems:

(1.1)
$$\text{maximize} \quad v^T x$$
$$\text{subject to} \quad a_1^T x \le b^1, \ldots, a_m^T x \le b^m$$
$$\text{where} \quad v, x, a_1, \ldots, a_m \in \mathbf{R}^n, \, b \in \mathbf{R}^m \quad \text{and} \quad m \ge n.$$

The matrix

$$A = \begin{pmatrix} a_1^T \\ \vdots \\ a_m^T \end{pmatrix} \in \mathbf{R}^{m \times n}$$

1980 *Mathematics Subject Classification* (1985 *Revision*). Primary 90C05; Secondary 68C25.
Key words and phrases. Linear programming, complexity of linear programming, probabilistic analysis of algorithms, stochastic geometry.

21

consists of the row vectors

$$a_i^T \quad \text{and} \quad b = \begin{pmatrix} b_1 \\ \vdots \\ b^m \end{pmatrix}.$$

So, we can describe the above inequality system by $Ax \leq b$. The feasible set $X = \{x \in \mathbf{R}^n | Ax \leq b\}$ is a convex polyhedron. We prefer that type because we do not need slack variables, which would increase the dimension of the space and would destroy our ability to imagine what is going on geometrically. The Simplex Method solves such problems by running two phases.

(1.2)

        Phase I: Calculation of a vertex $x_0 \in X$. If this is impossible, then STOP.

        Phase II: Construction of a sequence $x_0, \ldots, x_s$ of vertices of X such that successive vertices are adjacent and the objective strictly increases from vertex to vertex. STOP as soon as the optimal vertex x_s is reached or STOP as soon as the nonexistence of an optimal solution becomes obvious at a vertex x_s.

Since we want to study the complexity of that procedure, we are mainly interested in the number s. This is the crucial figure of complexity because it is easy to verify that the single pivot step requires $0(mn)$ arithmetic operations for all usual variants and because Phase I works in a very similar way to Phase II—only a modification of the problem is required. In order to simplify our considerations, it makes sense to concentrate on Phase II for the moment. We will return to questions concerning the complete algorithm later. So, we are going to study the value of s (the number of vertex exchanges in Phase II), which turns out to coincide with the number of pivot steps in the case of nondegenerate problems. Nondegeneracy will be assumed throughout our analysis.

Our definition of Phase II is not yet complete, because we have to give a rule for choosing the successor vertex if there are more than one adjacent and improving vertices. Since we are interested in a probabilistic analysis, we want to answer the following question:

How great is s "on the average" for fixed dimensions m and n?

In order to state clearly what we mean by "on the average", we have to introduce a stochastic model for the real-world distribution of the problems.

It is clear now that the average number of steps will depend highly on

    the variant used,
    the chosen stochastic model.

Throughout this paper we shall rely on the following *nondegeneracy assumptions* under the stochastic models discussed here, which do not influence our calculated expectation values:

(1.3)   Every submatrix of (A, b) and of $\begin{pmatrix} A \\ v^T \end{pmatrix}$ is of full rank.

One of the consequences of (1.3) is the fact that every feasible X is pointed, because there cannot be a straight line that is feasible everywhere. For understanding the geometry of the problems, we introduce the following two terms.

Basic solution x_Δ. This is a solution of a system of equations

$$a_{\Delta^1}^T x = b^{\Delta^1}$$
$$\vdots$$
$$a_{\Delta^n}^T x = b^{\Delta^n}$$

with $\Delta = \{\Delta^1, \ldots, \Delta^n\} \subset \{1, \ldots, m\}$ and $\Delta^i < \Delta^j$ for $i < j$. A basic solution becomes a vertex if the remaining restrictions $a_i^T x \leq b^i$ ($i = 1, \ldots, m$; $i \notin \Delta$) are satisfied at x_Δ.

Line. This is a solution set of a system of equations

$$a_{\Delta^1}^T x = b^{\Delta^1}$$
$$\vdots$$
$$a_{\Delta^{n-1}}^T x = b^{\Delta^{n-1}}.$$

A line may contain an edge of X if there is a section satisfying all the restrictions.

It is a conseqence of nondegeneracy that these systems have unique solutions, that every vertex of X corresponds to a set of exactly n active restrictions, that every pivot step leads to a new vertex, and that an existing optimal point will be unique.

In the following section, we will describe three successful approaches that are based on different stochastic models. We shall emphasize the reasons why the analysis could be done rather than showing how it was done.

2. Smale's work

In 1982 Steve Smale [10] did an analysis where the linear programming problem was imbedded into the more general linear complementarity problem. The linear programming problem he dealt with was of the type

(2.1)
$$\text{maximize} \quad v^T x$$
$$\text{subject to} \quad a_1^T x \leq b^1, \ldots, a_m^T x \leq b^m \text{ and } x \geq 0$$
$$\text{where} \quad x, v, a_1, \ldots, a_m \in \mathbf{R}^n, b \in \mathbf{R}^m.$$

This problem was regarded as a special case of the linear complementarity problem

find w, z such that for given q, M

(2.2) $w - Mz = q$, $w^T z = 0$, $w \geq 0$, $z \geq 0$

where w, z, $q \in \mathbf{R}^{m+n}$, $M \in \mathbf{R}^{(m+n)\times(m+n)}$.

When we specialize M to

$$\begin{pmatrix} 0 & -A \\ A^T & 0 \end{pmatrix}$$

and q to

$$\begin{pmatrix} b \\ -v \end{pmatrix},$$

a solution of (2.2) will give the solution of (2.1) and of the corresponding dual problem.

Geometrically, the task is to represent q as a conical combination of $m+n$ columns (alternatively taken out from $-M$ or the identity matrix I), i.e.,

$$q = Iw - M \quad \text{and} \quad w^T z = 0, \qquad w \geq 0, z \geq 0.$$

We solve the problem by Lemke's algorithm. Usually, we do not know the representation for q, but we know it for $e = 1$ (the vector of $m + n$ ones). There we have $e = Ie - M0$.

Now imagine a straight line between e and q and parametrize a movement from e to q by setting

(2.3) $q_\lambda = \lambda q + (1 - \lambda)e \quad \text{for } \lambda \in [0, 1].$

We know the conical representation for $\lambda = 0$, but we want it for $\lambda = 1$. When we move with increasing λ, we try to maintain $q_\lambda = w_\lambda - Mz_\lambda$, $w_\lambda \geq 0$, $z_\lambda \geq 0$ by adapting w_λ and z_λ. Every time the set of positive w-entries has to change, we enter a new cone (generated by $m + n$ column vectors out of $-M$ or I). We can calculate the corresponding change of representation by a pivot step in Lemke's algorithm. Consequently, we get an upper bound for the number of pivot steps in Lemke's algorithm by counting the number of cones that had been intersected by $[e, q]$.

The stochastic assumptions used in the analysis of Lemke's algorithm were as follows:

(2.4)

(1) Let the distribution of (A, b, v) be absolutely continuous.

(2) Let A, b, v be distributed independently, and let particularly the columns of A be distributed independently.

(3) Let the probability measure of (A, b, v) be invariant under coordinate permutations (in columns of A resp. b).

Smale's analysis came to the following result about $E_{m,n}(s^L)$, the expected number of pivot steps in Lemke's algorithm for (m, n)-problems.

THEOREM 2.5. $E_{m,n}(s^L) \leq \mathscr{C}(n)(1 + \ln(m + 1))^{n(n+1)}$ *under conditions* (1)–(3) *where* $\mathscr{C}(n)$ *is a constant depending on* n *but not on* m.

This result showed that $E_{m,n}(s^L)$ is polynomial in m, but the dependence in n could still be exponential.

3. The shadow-vertex algorithm and other parametric variants

Lemke's algorithm can be interpreted as an application of the principle of parametric programming. Even more direct is the use of parametric variants under the next two approaches.

For that reason we explain how the variants work. As stated before, we consider Phase I to be done and a start vertex x_0 to be given. Now x_0 itself is an optimal vertex with respect to a certain objective direction $u \in \mathbf{R}^n$.

Having the optimal vertex with respect to $u^T x$, we desire to have the same for $v^T x$. Now we project the polyhedron X onto the two-dimensional plane spanned by u and v and we obtain a two-dimensional polyhedron $\Gamma(X, u, v)$. In $\Gamma(X, u, v)$, some of the vertices of X have disappeared in the interior; the others have become vertices of $\Gamma(X, u, v)$ (the shadow of X). The latter vertices of X will be called shadow vertices. As a result of their definition, the vertices x_0 and x_s are such shadow vertices. And there is a Simplex Path connecting both, touching only shadow vertices. This path is even unique as a consequence of nondegeneracy if we run through the smaller angle between u and v. Hence the number of shadow vertices S is an upper bound for the number of pivot steps for realizing the **shadow-vertex path**. This realization is done with little effort, we only need a representation for a second objective u in the Simplex Tableau, as we had for v before. Running along the Simplex Path from x_0 to x_s can now be regarded as walking through the vertices maximizing an objective $(1 - \lambda)u + \lambda v$ with $\lambda \in [0, 1]$ while λ increases.

And also very useful for the following is the concept of **cooptimality**. A point $y \in X$ is called (u, v)-cooptimal if y is u-optimal ($u^T y$ is maximal) among all points of $X \cap \{x | v^T x = v^T y\}$. That means that $u^T y$ cannot be improved without changing $v^T x$.

Obviously, every cooptimal vertex is optimal with respect to an objective $(1 - \lambda)u + \lambda v$ with $\lambda \in [0, 1]$, and it is clear that every cooptimal edge lies on the boundary of the shadow of X and that every cooptimal vertex is a shadow vertex. Hence we have

(3.1) $$s \leq \mathscr{C} \leq S.$$

($\mathscr{C}$ = number of (u, v)-cooptimal vertices; S = number of shadow vertices.)

4. Results under the Sign-Invariance Model

The following stochastic model—the Sign-Invariance Model—could be analyzed and it lead to a much more optimistic judgment on the average behavior.

(4.1) Let the distribution of (A, b, v) and of $(S_1 A S_2, S_1 b, S_2 v)$ be identical for all sign matrices $S_1 \in \mathbf{R}^{m \times m}$ and $S_2 \in \mathbf{R}^{n \times n}$. Here a sign matrix is a diagonal matrix with $+1$ or -1 in the diagonal entries.

In order to get an impression of the quality of a Phase II algorithm under that model, it suffices to consider a relaxed version of sign invariance, the so-called flipping model. Here we consider again our data set for the linear programming problem

(4.2)
maximize $v^T x$
subject to $a_1^T x \, b^1, \ldots, a_m^T x \, b^m$
but the omitted directions of the inequalities ($\leq$ or $\geq$) are regarded as random variables, which are chosen independently for the m inequalities. $\leq$ and $\geq$ shall have probability $\frac{1}{2}$.

By the way, we generate 2^m problems out of one data set. The idea is now to solve all these problems, to count the pivot steps, and to divide by the number of problems. The feasible regions of these problems will be called cells. Note that cells may be empty.

The following result of Haimovich (1983) [7] (and Adler [12]) gives an impression of why Phase II runs very quickly.

THEOREM 4.3. $E_{m,n}(\mathscr{C} \mid$ a cooptimal path exists$) \leq n(m - n + 2)/(m + 1)$ under the flipping model.

SKETCH OF A PROOF.

Only $\binom{m}{0} + \binom{m}{1} + \cdots + \binom{m}{n}$ problems out of the 2^m generated have nonempty cells (partition of space). There are $\binom{m}{n}$ basic solution and $\binom{m}{n-1}$ lines.
Every cooptimal path ends up in a basic solution or in a ray that is a part of a line.
Every basic solution is v-optimal in exactly one cell.
Every edge (line segment) is cooptimal in exactly one cell.
Every cell has at most one cooptimal path.
Every line is divided by the remaining $m - n + 1$ restrictions into $m - n + 2$ segments. One of them gives the ray mentioned above.

Hence the number of cells with cooptimal path is $\binom{m}{n} + \binom{m}{n-1}$, and the number of traversed segments (edges) for running through all cooptimal paths is $\binom{m}{n-1}(m - n + 2)$. $\square$

In 1983 three different and independently written papers showed that the whole work (including Phase I and Phase II) could be done in quadratic time (on the average). The authors were (1) Todd, (2) Adler/Megiddo, and (3) Adler/Karp/Shamir [11], [2], [1].

THEOREM 4.4. $E_{m,n}(s^C) \leq 2(n+1)^2$ *under the Sign-Invariance Model* (4.1) *where* s^C *is the number of pivot steps required for the complete method.*

This result was derived in the papers of Todd and Adler/Megiddo by a very skillful choice of the start vector in Lemke's algorithm (compare Section 2). They chose a vector $(\delta, \delta^2, \delta^3, \ldots, \delta^{m+n}) \in \mathbf{R}^{m+n}$ with $\delta > 0$ arbitrarily small to have an initial representation. They realized the so-called Lexicographic-Lemke Algorithm. The advantage of this algorithm comes from the fact that we have good control over which sides a cone is entered and left.

Closer to our geometrical interpretation is the method used in the paper of Adler/Karp/Shamir, the so-called **Lexicographic Constraint-by-Constraint Method**. This algorithm works directly in the original space as follows:

Initialization: Determine a basic solution x_Δ with $\Delta = \{1, \ldots, n\}$ and choose u (the objective making x_Δ optimal) as $u = \delta a_1 + \delta^2 a_2 + \cdots + \delta^n a_n$ with $\delta > 0$ arbitrarily small. Then x_Δ is an optimal vertex on $X^{(n)} = \{x | a_1^T x \leq b^1, \ldots, a_n^T x \leq b^n\}$.

Typical steps: Start at $\overline{x}$, the optimal vertex for $u^T x$ on $X^{(n+k-1)} = \{x | a_1^T x \leq b^1, \ldots, a_{n+k-1}^T x \leq b^{n+k-1}\}$.

If $a_{n+k}^T \overline{x} \leq b^{n+k}$ then $\overline{x}$ is also u-optimal on $X^{(n+k)} = \{x | a_1^T x \leq b^1, \ldots, a_{n+k}^T x \leq b^{n+k}\}$. We proceed to the next step.

If $a_{n+k}^T \overline{x} > b^{n+k}$, then use the shadow vertex algorithm to minimize $a_{n+k}^T x$ on $X^{(n+k-1)}$ until $a_{n+k}^T x \leq b^{n+k}$ is achieved. All path points are $(u, -a_{n+k})$ cooptimal, hence the first point on our path with $a_{n+k}^T x = b^{n+k}$ maximizes $u^T x$ on $X^{(k+n)}$. If we achieve $a_{n+k}^T x \leq b^{n+k}$ we proceed to the next step. If it is impossible to achieve $a_{k+n}^T x \leq b^{k+n}$, then the original problem was infeasible. We STOP.

Final step: We start at $\overline{x}$, which is u-maximal on $X^{(m)} = X$. We apply the shadow-vertex algorithm to maximize $v^T x$ on X.

This amounts to an $(m - n + 1)$-fold application of the shadow-vertex-algorithm. But the very tricky choice of u at the initial step enabled Alder, Karp, and Shamir to show that the expected number of pivot steps is not only $0(mn)$ but $0(n^2)$.

In the paper by Adler/Megiddo, it was shown that slight additional conditions on the distributions of the entries of A suffice to establish also a lower bound on the average behavior such as $\mathscr{C}n^2$ with $\mathscr{C} > 0$.

Although the algorithms used in [11], [2], [1] were formally different, they realized exactly the same simplex paths [13] on nondegenerated problems.

As observed often in probabilistic analysis of algorithms, one must be very careful in the interpretation of the results and be aware of the influence of the stochastic assumptions.

This is also a very important point for the Sign-Invariance Model.
We have already mentioned that many cells will be empty. The quotient

$$\frac{\text{Number of feasible problems}}{\text{Number of generated problems}} = \frac{\binom{m}{0} + \binom{m}{1} + \cdots + \binom{m}{n}}{2^m}$$

tends to 0 for $m \to \infty$ and n fixed (asymptotically).

Also, the expected number of vertices per generate problem

$$\frac{2^n \binom{m}{n}}{2^m}$$

tends to 0 asymptotically.

Only *conditioning on feasible problems* can avoid averaging over a lot of
infeasible and easy problems. Here the expected number of vertices per
nonempty cell is

$$\frac{2^n \binom{m}{n}}{\binom{m}{0} + \cdots + \binom{m}{n}} \to 2^n \quad \text{asymptotically}$$

But the most important influence comes from the **average redundancy rate**
(share of redundant constraints). Here we have (conditioned on *nonempty
cells*).

$$\text{Average redundance rate} = \frac{\binom{m-1}{n}}{\binom{m}{0} + \cdots + \binom{m}{n}} \to 1 \quad \text{asymptotically}$$

Adler and Megiddo showed in [2] that for $m = 0(2n)$ we know that
$E_{m,n}(s^C | \text{problem is feasible and has an optimal solution}) = 0(n^{2.5})$. (Compare also [1] and [9].) The average redundancy rate for $m \leq 2n$ is $\leq \frac{1}{2}$.
This shows that the results are reliable and realistic for $m = 0(2n)$, but that
the model makes problems easy when $m \gg n$. And this holds even in the
absence of infeasible problems.

So we must be aware of the danger that the results reflect the model rather
than the quality of the algorithm.

5. Results under the Rotation-Symmetry Model

The danger of having a dramatic share of infeasible problems or of redundant
constraints could be avoided in the stochastic model analyzed by the author
in several papers ([3], [4], [5]). We deal with problems of the type

$$\text{maximize} \quad v^T x$$

(5.1) $\qquad \text{subject to} \quad a_1^T x \leq 1, \ldots, a_m^T x \leq 1$

$\qquad \text{where} \quad v, x, a_1, \ldots, a_m \in \mathbf{R}^n, m \geq n.$

Here the origin is guaranteed to be feasible in any case.

The stochastic model is

> Let $a_1, \ldots, a_m$, v and u (the direction determining the start vertex)
> be distributed

(5.2)   independently,

> identically,

> symmetrically under rotations of $\mathbf{R}^n$.

Note that (5.1) can easily be generalized to having positive right sides $b^i > 0$ when these right sides are independently and identically distributed over $(0, \infty)$ (b^i independent of all $a_1, \ldots, a_m$). Normalization of the inequalities will then lead back to (5.2).

Now remember our inequality $s \leq S$ when we decide to apply the shadow-vertex algorithm in Phase II. Candidates for becoming shadow vertices are only the $\binom{m}{n}$ basic solutions x_Δ solving

$$a_{\Delta^1}^T x = 1$$
$$\vdots$$
$$a_{\Delta^n}^T x = 1.$$

We observe the following one-to-one correspondence.

(5.3) $x_\Delta \longleftrightarrow \Delta = \{\Delta^1, \ldots, \Delta^n\} \longleftrightarrow \mathrm{CH}(a_{\Delta^1}, \ldots, a_{\Delta^n})$(convex hull).

In addition to the primal polyhedron $X = \{x \mid Ax \leq b\}$ we introduce the "polar polyhedron" $Y = \mathrm{CH}(0, a_1, \ldots, a_m)$.

The following equivalencies enable us to derive the average number of shadow vertices directly from the input data.

LEMMA 5.4.

(1) x_Δ is a vertex of $X \iff \mathrm{CH}(a_{\Delta^1}, \ldots, a_{\Delta^n})$ is a facet of Y.

(2) Let x_Δ be a vertex of X. Then

$$x_\Delta \text{ is a shadow vertex} \iff \mathrm{CH}(a_{\Delta^1}, \ldots, a_{\Delta^n}) \cap \mathrm{span}(u, v) \neq \varnothing.$$

By the addition theorem for expectation values we get the following integral expression for $E_{m,n}(S)$.

(5.5) $E_{m,n}(S) = \binom{m}{n} \int_{\mathbf{R}^n} \cdots \int_{\mathbf{R}^n} P(\mathrm{CH}(a_1, \ldots, a_n)$ defines a facet of Y and is intersected by $\mathrm{span}(u, v)) \, dF(a_1) \cdots dF(a_n)$.

Here F denotes the distribution function under consideration.

It took several years to evaluate this integral. From 1977–1984 we could derive some asymptotic results ($m \to \infty$, n fixed) under special distributions for the a_i.

THEOREM 5.6 (1977–1984). *There is a function $\varepsilon(m, n)$ with $\varepsilon(m, n) \to$ 0 asymptotically such that for*

Gaussian distribution on $\mathbf{R}^n$:

$$E_{m,n}(S) \leq \sqrt{\ln m}\, n^{3/2} 2^{3/2} \frac{20}{9}(1 + \varepsilon(m, n))$$

$$\geq \sqrt{\ln m}\, n^{3/2} 2^{3/2} \frac{3}{5}(1 - \varepsilon(m, n)).$$

Uniform distribution on unit ball:

$$E_{m,n}(S) \leq m^{1/(n+1)} n^2 2 \left(1 + \frac{1}{\sqrt{2}}\right)(1 + \varepsilon(m, n))$$

$$\geq m^{1/(n+1)} n^2 2 \frac{1}{3}(1 - \varepsilon(m, n)).$$

Uniform distribution on unit sphere:

$$E_{m,n}(S) \leq m^{1/(n-1)} n^2 2 \left(1 + \frac{1}{\sqrt{2}}\right)(1 + \varepsilon(m, n))$$

$$\geq m^{1/(n-1)} n^2 2 \frac{1}{7}(1 - \varepsilon(m, n)).$$

General distributions with bounded support:

$$E_{m,n}(S) \leq m^{1/(n-1)} n^2 \sqrt{2\pi}(1 + \varepsilon_F(m, n)).$$

THEOREM 5.7 (Borgwardt 1979–1984). *For $m \to \infty$, n fixed,*

(1) *there exist distributions according to (5.2) such that $E_{m,n}(S)$ converges to a constant $\mathscr{C}(n)$ in m.*

EXAMPLE. radial distribution $\widetilde{F}(r) = \begin{cases} 0 & r < 1 \\ 1 - \frac{1}{r} & r \geq 1. \end{cases}$

(2) *there is a distribution with the above property and $\mathscr{C}(n) \leq n^{5/2}$.*

EXAMPLE. $\widetilde{F}(r) = \begin{cases} 0 & r < 1 \\ 1 - \frac{1}{r^{n^2}} & r \geq 1. \end{cases}$

The dramatic difference in the order of growth for different distributions results from different redundancy rates. We can realize a redundancy rate of 0 by choosing the uniform distribution on the sphere, whereas the examples in Theorem 5.7 establish redundancy rates very close to 1. So it is not astonishing that these examples simulate the effect observed at the Sign-Invariance Model. Asymptotically, the influence of m disappears. And this holds here while every instance is feasible and almost all are bounded!

In 1981 I succeeded in proving polynomiality in both dimensions.

THEOREM 5.8 (1981). *For all distributions according to (5.2),*

$$E_{m,n}(S) \leq e\pi \left(\frac{\pi}{2} + \frac{1}{e}\right) m^{1/(n-1)} n^3.$$

We now come back to the inclusion of Phase I. We could introduce a (rather lengthy) method for doing the task of Phase I and Phase II by several applications of the shadow-vertex algorithm. This enables us to exploit our known results, since in every application the stochastic requirements (5.2) are still satisfied. By the way, all our trouble with the auxiliary objective $u^T x$ disappear. We do not need it any longer.

The complete method works as follows.

Initialization.

(1) Starting from the origin, find a vertex of the polyhedron $X^{(2)} = \{x | Ax \le b, x^3 = \cdots = x^n = 0\}$ (two-dimensional). This is easily done by running along a coordinate axis until a restricting hyperplane is hit. In case the axis does not hit, we invert the search direction. If we still do not hit, the original problem is unbounded. We can STOP. If a hitting point is found, it is located on an edge of $X^{(2)}$. Find a vertex of that edge. If the edge does not have vertices, the original problem is unbounded; we can STOP.

(2) Apply the shadow-vertex algorithm starting from the given vertex in order to maximize $v^T x$ on $X^{(2)}$. If $v^T x$ is unbounded, we can STOP (unbounded on X, too).

Typical step. Use the solution point $(x^1, \ldots, k^{k-1}, 0, \ldots, 0)^T$ of $X^{(k-1)}$. It is located on an edge of $X^{(k)}$.

(1) Find an adjacent vertex in $X^{(k)}$ to that edge.

(2) Apply the shadow-vertex algorithm by using $e_k^T x$ and $v^T x$ as objective and auxiliary objective, respectively. If $v^T x$ is unbounded on $X^{(k)}$ we can STOP.

(3) If $k < n$ we go to the next step and use the solution point with $k := k + 1$. If $k = n$ we STOP.

This method allowed evaluation, since it was in principle an $(n-1)$-fold application of the shadow-vertex algorithm. Our main result was

THEOREM 5.9 (1981). *For all distributions satisfying* (5.2) *our complete method solves the* LP *in not more than*

$$m^{1/(n-1)}(n+1)^2 n^2 \frac{e\pi}{4}\left(\frac{\pi}{2} + \frac{1}{e}\right)$$

steps on the average.

By this method we were also able to verify this result for problems including sign constraints $x \ge 0$.

Our result had a certain drawback. We had to know a feasible point in advance. But how should one handle problems of the form

$$\text{maximize} \quad v^T x$$

(5.10)   subject to $a_1^T x \le b^1, \ldots, a_m^T x \le b^m$

with arbitrary b^i (not necessarily positive)?

We call the polyhedron of (5.10) P_n. We reformulate our restrictions as follows:

$$
\begin{array}{ccccc}
a_1^T x \le b & & a_1^T x \le 1 - \tilde{b}^1 & & a_1^T x + \tilde{b}^1 \le 1 \\
\vdots & \overset{\text{set } b^i = 1 - \tilde{b}^i}{\longleftrightarrow} & \vdots & \longleftrightarrow & \vdots \\
a_m^T x \le b^m & \text{resp. } \tilde{b}^i = 1 - b^i & a_m^T x \le 1 - \tilde{b}^m & & a_m^T x + \tilde{b}^m \le 1
\end{array}
$$

Now we imbed our polyhedron P_n in $\mathbf{R}^{n+1}$ by writing

$$
(a_i^1, a_i^2, a_i^3, \ldots, a_i^n, \tilde{b}^i)
\begin{pmatrix}
x^1 \\
\vdots \\
x^n \\
x^{n+1}
\end{pmatrix}
\le 1 \iff a_i^T x + \tilde{b}^i x^{n+1} \le 1
$$

for $i = 1, \ldots, m$.

This inequality system defines a new polyhedron $P_{n+1} \subset \mathbf{R}^{n+1}$. Here we distinguish between different values of the additional variable x_{n+1}, the so-called levels.

In level $x_{n+1} = 1$ we find our original polyhedron P_n (augmented by 1's in the additional component), because $a_i^T x + \tilde{b}^i x_{n+1} \le 1$ and $x_{n+1} = 1$ means that $a_i^T x + \tilde{b}^i \le 1$.

In level $x_{n+1} = 0$ we find a polyhedron $\widetilde{P}_n$ (augmented by 0's in the additional component), because $a_i^T x + \tilde{b}^i x_{n+1} \le 1$ and $x_{n+1} = 0$ means that $a_i^T x \le 1$.

If all the vectors a_i satisfy the conditions of (5.2), we can immediately solve the $\widetilde{P}_n$ problem, and our known results hold. Now we solve what we can solve and proceed to level 1.

  (1) Solve the problem of $\widetilde{P}_n$. (By our old complete method, 0 is feasible.) If this problem has no solution, we can STOP. (Unboundedness of $v^T x$ leads to unboundedness of $v^T x$ also in P_n — if P_n is nonempty.)
  (2) Consider the solution of the $\widetilde{P}_n$ problem in level 0. It is located on an edge of P_{n+1}. Now apply the shadow-vertex algorithm to

$$\text{maximize} \quad x^{n+1}$$

(5.11)   subject to $a_i^T x + \tilde{b}^i x^{n+1} \le 1 \quad (1 = 1, \ldots, m)$.

Every point on the realized path is (v, e_{n+1})-cooptimal. Hence a path point of level $x^{n+1} = 1$ will be v-optimal in level $x_{n+1} = 1$. Then the truncated vector $(x^1, \ldots, x^n)$ is the optimal solution.

If we cannot reach level $x_{n+1} = 1$, then this proves that P_n is empty and that our original problem was infeasible.

Since this is only prolonging our known complete method, the results hold if the stochastic assumptions still hold.

The effort for Step 1 above is known from Theorem 5.9. We have to consider the additional effort for Step 2.

We have a first obvious result:

THEOREM 5.12. *If* $\binom{a_1}{\tilde{b}^1}, \ldots, \binom{a_m}{\tilde{b}^m}$ *are distributed on* $\mathbf{R}^{n+1}$ *according to* (5.2), *then*

$$E_{m,n}(s^t) \leq m^{1/n}(n+2)^4 \mathscr{C} \qquad \text{(for the total algorithm)}.$$

A special result.

THEOREM 5.13. *If* $a_1, \ldots, a_m, \tilde{b}^1, \ldots, \tilde{b}^m$ *are independent and all components of* A *and* $\tilde{b}$ *are Gaussian distributed, then*

$$E_{m,n}(s^t) \leq m^{1/n}(n+2)^4 \mathscr{C}.$$

For more general distributions of the right-hand sides $1 - \tilde{b}^i$, a lot of additional work has to be done. Up to now we know

THEOREM 5.14. *If* $a_1, \ldots, a_m, \tilde{b}^1, \ldots, \tilde{b}^m$ *are independent, the* a_i *are distributed according to* (5.2), *and if the* $\tilde{b}^i$ *are uniformly distributed over an interval* $[-q, q]$, *then Step 2 does not exceed the size of effort in Step 1, which is* $0(m^{1/(n-1)}n^4)$.

So, we have an algorithm that admits a probabilistic analysis solving all types of LP problems. Its average behavior is still polynomial, although we work under the "hard" rotational symmetry model.

REFERENCES

1. I. Adler, R. Karp, and R. Shamir, *A simplex variant solving an* $m \times d$ *linear program in* $0(\min(m^2, d^2))$ *expected number of pivot steps*, J. Complexity **3** (1987), 372–387.
2. I. Adler and N. Megiddo, *A simplex algorithm where the average number of steps is bounded between two quadratic functions of the smaller dimension*, J. ACM **32** (1985), 871–895.
3. K. H. Borgwardt, *Untersuchungen zur Asymptotik der mittleren Schrittzahl von Simplexverfahren in der Linearen Optimierung*, Dissertation Universität Kaiserslautern, 1977.
4. ____, *Some distribution-independent results about the asymptotic order of the average number of pivot steps of the Simplex Method*, Math. Oper. Res. **7** (1982), 441–462.
5. ____, *The average number of pivot steps required by the Simplex Method is polynomial*, Zeitschrift für Oper. Res. **26** (1982), 157–177.
6. ____, *The Simplex Method—a probabilistic analysis*, Springer-Verlag, Berlin, Heidelberg, New York, 1987.
7. M. Haimovich, *The simplex algorithm is very good!—on the expected number of pivot steps and related properties of random linear programs*, Columbia University, New York, April, 1983.
8. J. H. May and R. L. Smith, *Random polytopes: Their definition, generation and aggregate properties*, Math. Programming **24** (1982), 39–54.

9. R. Shamir, *The efficiency of the Simplex Method: a survey*, Management Sci. **33** (1987), 301–334.

10. S. Smale, *On the average speed of the Simplex Method*, Math. Programming **27** (1983), 241–262.

11. M. J. Todd, *Polynomial expected behavior of a pivoting algorithm for linear complementarity and linear programming problems*, Math. programming **35** (1986), 173–192.

12. I. Adler, *The expected number of pivots needed to solve parametric linear programs and the efficiency of the self-dual Simplex Method*, manuscript, Dept. of Industrial Engineering and Operations Research, University of California Berkeley, California, 1983.

13. N. Megiddo, *A note on the generality of the self-dual Algorithm with various starting points*, Methods Oper. Res. **49** (1985), 271–275.

14. I. Adler and S. E. Berenguer, *Generating random linear programs*, manuscript, Dept. of Industrial Engineering and Operations Research, University of California, Berkeley, California, 1983.

15. ___, *Random linear programs*, Operations Research Center Report No. 81–4, Dept. of Industrial Engineering and Operations Research, University of California, Berkeley, California, 1981.

DEPARTMENT OF MATHEMATICS, UNIVERSITY OF AUGSBURG, WEST GERMANY

Contemporary Mathematics
Volume **114**, 1990

On Solving the Linear Programming Problem Approximately

NIMROD MEGIDDO

ABSTRACT. This paper studies the complexity of some approximate solutions of linear programming problems with real coefficients.

1. Introduction

The general linear programming problem is to maximize a linear function over a set defined by linear inequalities and equations. There are many equivalent ways to represent instances of the linear programming problem. For example, consider the symmetric form

$$(\text{Sym}(A, b, c)) \qquad \begin{array}{ll} \text{Maximize} & c^T x \\ \text{subject to} & Ax \leq b \\ & x \geq 0. \end{array}$$

The dual is then

$$\begin{array}{ll} \text{Minimize} & b^T y \\ \text{subject to} & A^T y \geq c \\ & y \geq 0. \end{array}$$

Intuitively, two representations are equivalent if there is an easy way to transform solutions of one to solutions of the other and vice versa. We first mention some of the well-known equivalences. First, any set of linear inequalities and linear equations can be reduced to a set of linear equations with nonnegativity constraints or to a set of inequality and nonnegativity constraints. Also, any linear programming problem can be reduced to a linear programming problem with a nonempty set of solutions by using artificial variables. Moreover, any linear programming problem can be reduced to a problem of finding a solution to a system of linear inequalities (by combining the constraints of the primal and dual and adding the inequality $c^T x \geq b^T y$)

1980 *Mathematics Subject Classification* (1985 *Revision*). Primary 90C05.

or else concluding that the system has no solution. By the duality theorem, if we put the problem in the combined primal-dual form, every problem can be reduced to a problem that is either infeasible or feasible and bounded (but not unbounded). It is interesting to note that over any ordered field, every linear programming problem can be reduced to one that is both feasible and bounded. This is done as follows. Suppose the problem is given in the symmetric form $\text{Sym}(A, b, c)$. Consider the following:

$$
\begin{aligned}
\text{Minimize} \quad & c^T x - b^T y + t \\
\text{subject to} \quad & Ax - te \le b \\
& A^T y + te \ge c \\
& c^T x - b^T y \ge 0 \\
& x, y, t \ge 0,
\end{aligned}
$$

where e denotes a vector of 1's. It is easy to verify that the optimal value of the latter is zero if and only if the former has an optimal solution. In this case the latter provides optimal solutions for the former and its dual.

The equivalences mentioned above are valid as long as exact computation is feasible. In practice one usually works with finite precision and hence obtains results that are only "approximately true." However, the meaning of the last sentence depends on the particular representation of the practical problem. Indeed, a good approximate solution for one representation of the problem may transform into a very bad approximate solution for another "equivalent" representation of the same problem.

When two people talk about approximate solutions, they often think of different notions of approximation. It is quite likely though that they refer to one of the following:

(i) A feasible point (i.e., one that satisfies all of the constraints in the exact sense) and is close in a certain metric to an optimal point.

(ii) A feasible solution whose objective function value is close to the optimal value.

(iii) A point, not necessarily feasible, close to an optimal solution.

(iv) A point that approximately satisfies every constraint, and whose objective function value is close to the optimal value.

(v) A point close to the feasible domain, whose objective function value is close to the optimal value (called the "weak optimization problem" in [7]).

(vi) A basis where the simplex algorithm (using exact arithmetic) terminates, but the numerical values of variables are only approximate.

(vii) A basis where the simplex algorithm terminates due to a prescribed tolerance.

The choice of the right definition depends very much on the practical situation. In fact, practical considerations dictate which constraints must be satisfied and which may be approximately satisfied. In other words, the tolerance may be different for different constraints.

It is not known whether the $m \times n$ linear programming problem with real data can be solved in a polynomial number of arithmetic operations and comparisons in terms of m and n. (We will refer to this notion of complexity as strongly polynomial time, even though the usual definition of this concept also requires polynomial time in the usual sense.) Thus, another natural question is whether for any $\varepsilon > 0$, any of the ε-approximation problems can be solved in a polynomial number of operations in terms of m, n, and $-\log \varepsilon$. To deal with this question, we first have to define what we mean by an "ε-approximate" solution. In particular, such a definition should not make the second question trivially equivalent to the first one. Consider, for example, the concept suggested in (i) above. Thus, for the problem Maximize $c^T x$ subject to $Ax \leq b$ (assuming its maximum V^* exists) an ε-approximate solution is a point x such that $Ax \leq b$ and $c^T x \geq V^* - \varepsilon$. This definition is not satisfactory since it is not clear whether an ε-approximation algorithm is required to decide the existence of an x such that $Ax \leq b$, and the boundedness of the function $c^T x$ on the feasible domain. If indeed it is required to decide these questions, then in the worst-case sense this approximation problem is trivially equivalent to the exact problem.

The consequences of the ellipsoid algorithm with respect to approximation problems on convex sets are studied in [7]. It is not clear whether these results can be applied to achieve the type of results we seek here. The reason is that over the real numbers it seems difficult to obtain estimates of the radii of a circumscribing sphere and an inscribed sphere. The main complexity result on convex minimization in [7] (Theorem 2.2.15) assumes that the convex set is given with estimates of such radii. Our main interest here is the question of what is a reasonable sense of approximation when the algorithm fails to classify the instance correctly as feasible, unbounded, etc.

In Section 2 we give some preliminaries and discuss the difficulties involved in classifying the problem. In Section 3 we discuss approximate solutions based on satisfying a termination criterion within some tolerance. In Section 4 we discuss a notion of approximation that is based on solving a perturbed instance exactly. Section 5 gives an analysis of complexity for various notions of complexity.

2. Preliminaries

We pointed out in the introduction that the practical situation usually dictates the right notion of approximate solution. For a theoretical discussion it is often convenient to consider the problem in the symmetric form $\text{Sym}(A, b, c)$. (See Section 1.) Traditionally, an exact algorithm for this problem (for example, the simplex method) is supposed to provide the user with information as follows. It has to classify the problem into one of the following three categories:

(i) Infeasible. (The domain X defined by $Ax \leq b$ and $x \geq 0$ is empty.)

(ii) Feasible and bounded. (There exists a maximizer of $c^T x$ over X.)

(iii) Feasible and unbounded. (The function $c^T x$ is unbounded over X.)

In case (ii) the algorithm has to provide an optimal solution. The algorithm may also be required in case (iii) to provide a ray contained in X along which $c^T x$ tends to infinity. A nice property of the simplex method is that it also solves the dual problem

$$\text{Minimize} \quad b^T y$$
$$\text{subject to} \quad A^T y \geq c$$
$$y \geq 0.$$

Thus, besides providing such a ray in case (iii), the algorithm also provides in case (ii) an optimal solution to the dual, and in case (i) a "certificate" in the form of a ray of a related problem.

In fact, the (exact) simplex method always computes a *basis* that provides the required information. Specifically, it provides a representation of the problem (by a suitable linear transformation of the space) from which the classification and the numerical values of both the primal and the dual variables are transparent. Thus, the simplex method classifies problems into one of *four* categories (even though the commonly used variants do not distinguish between IF and II):

(i) FF: primal feasible, dual feasible.

(ii) FI: primal feasible, dual infeasible (that is, unbounded primal).

(iii) IF: primal infeasible, dual feasible (here the dual is unbounded).

(iv) II: primal infeasible, dual infeasible.

It is quite common to include this classification in the requirements from an exact algorithm for the general linear programming problem. We refer to it later as the classification problem of linear programs.

An approximation algorithm should be expected sometimes to fail in classifying the input into the categories FF, FI, IF, and II. Interestingly, the existence of a strongly polynomial algorithm for the classification problem implies the existence of one for the problem itself. (See page 445 in [1].)

So far we have discussed the subject of approximation under the assumption that the result should be "close" to the true one. However, a different approach can sometimes be useful. We may allow the algorithm to be totally wrong in a small number of cases. This approach is approximate when the output space of the algorithm is discrete and has no natural metric associated with it. For example, consider the following trivial problem: Given two numbers α, β, recognize whether $\alpha > \beta$ or $\alpha \leq \beta$. Suppose the comparison of α to β can be performed with arbitrary finite precision. Thus, for any given $\varepsilon > 0$, we can recognize either that

$$\alpha \leq \beta + \varepsilon$$

or that

$$\alpha \geq \beta - \varepsilon.$$

The algorithm reports $\alpha \leq \beta$ in the first case and $\alpha \geq \beta$ in the second one. Thus, the algorithm gives the correct answer if

$$|\beta - \alpha| > \varepsilon$$

but may fail otherwise. The grey area is the set of pairs (α, β) such that $|\alpha - \beta| \leq \varepsilon$. Of course, the smaller ε the smaller the grey area. Thus, the measure of the grey area reflects the quality of the approximation.

The problem of the preceding paragraph can be cast as a linear programming problem

$$\text{Maximize} \quad x_1$$
$$\text{subject to} \quad \alpha x_1 \leq x_2 \leq \beta x_1 .$$

Here the point $(0, 0)$ is feasible for any α and β. The problem is unbounded if and only if $\alpha \leq \beta$. This suggests that the grey area approach would be suitable for the classification problem of linear programs.

A general linear programming problem in standard form

$$\text{Maximize} \quad c^T x$$

$$(\text{SF}(A, b, c)) \qquad \text{subject to} \quad Ax = b$$

$$x \geq 0$$

is determined by $A \in R^{m \times n}$, $b \in R^m$, and $c \in R^n$. There is a one-to-one correspondence between problems of order $m \times n$ and points of $\Omega = R^{mn+m+n}$. The classification corresponds to a partition of Ω into four sets: FF, FI, IF, and II, as discussed above. For example, IF is the set of triples (A, b, c) that determine infeasible primal problems whose dual problems are feasible (and hence unbounded).

Let Ω' denote the union of the boundaries of these four sets. Obviously, an approximation algorithm (for the classification problem) may fail if the input (A, b, c) is close to Ω'. For example, if an instance is close to the common boundary of FF and FI, but far from the boundaries of IF and II, then an approximation algorithm is expected to recognize that the problem is feasible, but is expected to fail in deciding whether it is bounded. Interestingly, there are more "pathological" cases. In fact, the intersection of all four boundaries, which we denote by Ω_4, is not empty. Thus, given an instance close to the intersection of the four boundaries, an approximation algorithm may not be able to recognize anything in terms of the above classification. This observation is obvious in view of the invariance of the classification under multiplication of columns and rows by positive scalars. Thus, the neighborhood of the origin is obviously pathological in this sense. The difficulties with the origin can be avoided by scaling rows and columns. However, it is easy to construct other examples with similar characteristics.

PROPOSITION 2.1. *The instance*

$$\text{Maximize} \quad x_1 - x_2$$
$$\text{subject to} \quad x_1 - x_2 \leq -1$$
$$(*) \qquad\qquad\qquad\qquad -x_1 + x_2 \leq -1$$
$$x_1, x_2 \geq 0 .$$

belongs to Ω_4.

PROOF. It is easy to see that $(*)$ itself is in IF. Using small perturbations, one can move from $(*)$ to instances in any of the other three classes. More precisely, if only c_1 is slightly increased then we can get instances in II. If only A_{21} is slightly decreased then we get instances in FF. Finally, if only A_{11} is slightly increased then an instance in FI is obtained. □

We note that most of the numerical difficulties in solving linear programming problems are due to the fact that many such problems are ill posed. It is well known in numerical analysis (see, e.g., [3]) that near-singularities in the matrix A can cause problems. However, in this paper we also discuss intrinsic aspects of approximate solutions that arise even when exact arithmetic is used. For example, if one is interested only in an approximate solution, what should be a good termination criterion? Because of such questions we have to deal with perturbations of the vectors b and c and not only of the matrix A.

Due to the classification aspect of the problem, we clearly cannot always measure the quality of the approximation by the distance between the exact solution and the approximate one (neither in terms of the solution vector nor in terms of the objective function value). Thus a different approach to approximation may be proposed for a general situation, where there is some natural metric on the input space, but there does not seem to exist one for the output space. The following definition is similar to backward analysis of errors in numerical analysis [9].

DEFINITION 2.2. Let $M = (S, d)$ be a metric space and let f be a mapping from S into some set T that does not necessarily have any metric associated with it. A mapping $g: S \to T$ is called *an ε-approximation to* f if for every $x \in S$, there exists an $x' \in S$ such that $d(x, x') < \varepsilon$ and $g(x) = f(x')$.

In linear programming the output space (including the classification) does have a metric structure. Besides the classification information, there are also numerical values associated with the variables. One might propose for the linear programming problem the following approach to approximation by posing the following problem:

PROBLEM 2.3. Given the problem $SF(A, b, c)$ and $\varepsilon > 0$, name a class $S \in \{FF, FI, IF, II\}$ and assign numerical values to the variables so that the following condition is satisfied: There exists an instance (A', b', c') in S for which the numerical values are correct within an error of ε, such that $\|(A, b, c) - (A', b', c')\|_\infty < \varepsilon$.

The approach represented by Problem 2.3 takes care of pathological cases where some other approaches fail. Consider, for comparison, a different notion of approximate solution of systems of inequalities reflected in the following problem:

PROBLEM 2.4. Given A, b and $\varepsilon > 0$, either give an x such that $Ax \leq b + \varepsilon e$ or conclude that there is no x such that $Ax \leq b - \varepsilon e$.

The approach represented by Problem 2.4 seems to be a natural generalization of the obvious approximate comparison of two real numbers. Its weakness is apparent in the following example. Consider the problem

$$\alpha x_1 - x_2 \le -1,$$
$$-x_1 + x_2 \le -1.$$

If $\alpha = 1$ then, obviously, for every $\varepsilon < 1$ the system

$$x_1 - x_2 \le -1 + \varepsilon,$$
$$-x_1 + x_2 \le -1 + \varepsilon$$

is infeasible. However, for any $\alpha \ne 1$ the system is feasible for every ε. Thus, in order to solve Problem 2.4 we have to know whether $\alpha = 1$. This example can easily be generalized so that in order to solve Problem 2.4 one has to know whether a certain matrix is singular. The latter involves some numerical difficulties in practice.

The weakness of the concept of Problem 2.4 is that it considers perturbations of the given problem only in a limited and quite arbitrary way. In Problem 2.3 we allow perturbations in all directions. We note that for certain classes of linear programming problems (e.g., the min-cost flow problem) certain coefficients have the values 1 or 0 throughout the class. In such cases we would allow only perturbations within the subject class. Thus, if the concept represented by Problem 2.3 were to apply to a min-cost flow problem $SF(A, b, c)$ then we would require that $A' = A$.

3. Tolerance-based approximation

In this section we discuss the issue of approximation as it arises in the context of the simplex method. There are two ways to look at the question. First, imagine we run the simplex algorithm using exact arithmetic but have to compute only "ε-approximate" solutions. Thus, rather than running the algorithm to the end, we seek to apply some stopping rule that guarantees our output to be ε-approximate. The interesting problem is of course to devise such stopping rules for various concepts of approximation. Another way to look at the question is to realize that on a machine we usually have numerical errors, and thus we almost always have to specify some "tolerance" within which we accept our results. It is important to know the implications of using a certain tolerance with regard to the results.

Consider the problem $SF(A, b, c)$. In the Appendix we review some properties of basic solutions and how the simplex method uses them. (See the Appendix for the notation.)

In practice, one usually works with some "tolerance" $\delta > 0$, so that any number $a \le \delta$ is accepted as nonpositive and any $a \ge -\delta$ is accepted as nonnegative. This suggests another approach to approximation, which may be called the "tolerance" approach:

DEFINITION 3.1. A basic solution $x = x(B) = (x_B, x_N)$ (where $x_B = B^{-1}b$ and $x_N = 0$) is *optimal with tolerance* δ for SF(A, b, c) if for every j,

$$x_j \geq -\delta \quad \text{and} \quad c_j \leq c_B^T B^{-1} A_j + \delta.$$

Obviously, such a solution is not necessarily feasible. It is only "approximately feasible" in the sense that the equality constraints are satisfied while the nonnegativity constraints are approximately satisfied. Analogously, the dual vector $y(B)$ is approximately feasible in the dual problem. Moreover, the vectors $x(B)$ and $y(B)$ satisfy the complementary slackness condition

$$x_j(y^T A_j - c_j) = 0,$$

which is necessary for optimality. This implies that both vectors yield the same objective function value in their respective problem

$$c^T x = c^T B^{-1} b = y^T b.$$

Note that the problem may also be feasible and unbounded but that an optimal solution with tolerance δ may still exist. One can also define a notion of an unbounded ray with tolerance δ.

4. Perturbation-based approximation

It is interesting to observe that from a solution that is optimal with small tolerance we can easily obtain an exact solution to an instance that is close to the given one. More precisely, we have the following proposition:

PROPOSITION 4.1. *Suppose* $x = x(B)$ *is a basic solution that is optimal with tolerance* δ. *Let* x' *be defined by* $x'_j = x_j + \delta$ *for* j *associated with* B *and* $x_j = 0$ *otherwise. Also, let* c' *be defined by* $c'_j = c_j$ *for* j *associated with* B *and* $c'_j = c_j - \delta$ *otherwise, and let* $b' = b + \delta Be$. *Under these conditions,* x' *is an optimal solution for the problem* SF(A, b', c'), *and* $y = B^{-T}c_B$ *is optimal for its dual problem.*

PROOF. We have

$$x'_B = B^{-1}b' = x_B + \delta e \geq 0,$$
$$A^T y \geq c'$$

and x' and y satisfy the complementary slackness conditions required in SF(A, b', c'). □

In simpler words, we have

COROLLARY 4.2. *If* x *is optimal with tolerance* δ *for* SF(A, b, c) *then there exist* b' *and* c' *and an optimal solution* x' *for* SF(A, b', c') *such that*

$$\|x' - x\|_\infty, \|c' - c\|_\infty \leq \delta,$$

and

$$\|b' - b\|_\infty \leq \min\{\delta\|A\|_\infty, mT\delta\}$$

where $\|A\|_\infty$ *is the usual operator norm, corresponding to the vector supremum norm* $\|\cdot\|$, T *is the maximum absolute value of any entry in* A, *and* m *is the number of rows of* A.

A analogous proposition can be proven with respect to the unbounded case:

PROPOSITION 4.3. *Suppose* $x = x(B)$ *is a basic solution such that*

$$x_j \geq -\delta$$

and for some k *not associated with* B,

$$c_k > c_B^T B^{-1} A_k$$

and

$$B^{-1} A_k \geq -\delta e.$$

Let c' *be the same as* c *except that* $c'_k = c_k + \delta c_B^T e$, *and let* $b' = b + \delta Be$. *Also, let* A' *be the same as* A *except that* $A'_k = A_k + \delta Be$. *Under these conditions, the problem* $SF(A', b', c')$ *is unbounded.*

PROOF. In the new problem the basis B certifies unboundedness since

$$B^{-1} b' \geq 0,$$
$$c'_k > (c')_B^T B^{-1} A'_k,$$

and

$$B^{-1} A'_k \geq 0. \quad \square$$

COROLLARY 4.4. *If* $SF(A, b, c)$ *is concluded within tolerance* δ *to be unbounded, then there exist* A', b', *and* c' *such that* $SF(A', b', c')$ *is unbounded,*

$$\|A' - A\|_1 \leq \delta \|A\|_\infty$$

(where $\|A' - A\|_1$ *is the operator norm corresponding to the vector norm* $\|\cdot\|_1$*),*

$$\|b' - b\|_\infty \leq \delta \|A\|_\infty,$$

and

$$\|c' - c\|_\infty \leq \delta \|c\|_1.$$

5. Complexity questions

We start this section with yet another variant of an approximation problem. Again, we consider approximation concepts that avoid the difficulties involved in the classification problem. Suppose the exact problem is given in the dual form

$(DF(A, b, c))$ 　　　　Minimize $c^T x$
　　　　　　　　　　　subject to $Ax \geq b$

where the output has to be one of the following:

(i) a point x^* that minimizes $c^T x$ subject to $Ax \geq b$,

(ii) a point x^* and a scalar $t^* > 0$ that minimize the value of t subject to $Ax + te \geq b$ (in which case the problem is infeasible), or

(iii) vectors x and u such that $Ax \geq b$, $c^T u < 0$, and $Au \geq 0$ (in which case the problem is unbounded).

The above motivates the definition of the following approximation problem:

PROBLEM 5.1. Denote the optimal value of a given problem $DF(A, b, c)$ by V^* (allowing $V^* = \pm\infty$) and let t^* denote the minimum of t subject to $Ax + te \geq b$. Given a number $\varepsilon > 0$, output one of the following:

(i) a point x such that $c^T x \leq V^* + \varepsilon$ and $Ax \geq b - \varepsilon e$,

(ii) a point x and a scalar t, $-\varepsilon \leq t \leq t^* + \varepsilon$ such that $Ax + te \geq b - \varepsilon e$, or

(iii) vectors x and u such that $Ax \geq b - \varepsilon e$, $c^T u < \varepsilon$, and $Au \geq -\varepsilon e$.

It is interesting to look at the question of the existence of an algorithm for the approximation problem, where the number of operations is expressed in terms of ε as well.

PROPOSITION 5.2. *Over any ordered field, if Problem 5.1 can be solved in $f(m, n, \varepsilon)$ field operations (including comparisons) then it can be solved in $g(m, n) = O(f(m, n, 1))$ operations.*

PROOF. Suppose $\mathscr{A}$ is an algorithm for Problem 5.1 that runs in $f(m, n, \varepsilon)$ field operations. Given A, b, c and $\varepsilon > 0$, let

$$\overline{A} = \varepsilon^{-1} A, \qquad \overline{b} = \varepsilon^{-1} b, \qquad \overline{c} = \varepsilon^{-1} c.$$

The instance $DF(\overline{A}, \overline{b}, \overline{c})$ is equivalent to $DF(A, b, c)$. Moreover, a valid output for $DF(\overline{A}, \overline{b}, \overline{c})$ with precision $\varepsilon = 1$ is also a valid output for $DF(A, b, c)$ with the prescribed precision ε. □

COROLLARY 5.3. *If there exists a polynomial $f(m, n, \varepsilon)$ such that Problem 5.1 with rational data can be solved in $f(m, n, \varepsilon)$ arithmetic operations, then the exact problem with rational data can be solved in a polynomial number of operations.*

PROOF. For a problem with rational data it is easy to determine a value ε such that an exact solution can be computed from a solution of Problem 5.1 in a polynomial number of operations. Thus, the problem can be scaled so that $\varepsilon = 1$ suffices for determining an exact solution. □

In view of Proposition 5.2 it is reasonable to ask whether Problem 5.1 can be solved in a polynomial number of operations in terms of m, n, and $\log R/\varepsilon$, where $R = R(A, b, c) > 0$ is some quantity such that for any positive scalar λ

$$R(\lambda A, \lambda b, \lambda c) = \lambda R(A, b, c)$$

(e.g., R equals the maximum absolute value of any input coefficients).

Consider first the feasibility problem

$$\text{Minimize} \quad t$$

(FB(A, b))   subject to $Ax + te \geq b$

$$t \geq -1,$$

and the associated approximation problem:

PROBLEM 5.4. Given FB(A, b) and $\varepsilon > 0$, find x and t such that $Ax + te \geq b$ and $t < t^* + \varepsilon$ (where t^* is the minimum of FB(A, b)).

Before stating the next proposition, recall that the set of solutions of the system $Ax \geq b$ is bounded for every b if and only if the rows of A span the space R^n in nonnegative linear combinations.

PROPOSITION 5.5. *If Problem 5.4 is given with real data such that the rows of A span the space R^n in nonnegative linear combinations, then it can be solved in a polynomial number of arithmetic operations in terms of m, n, and*

$$\rho = \log \left(\frac{b_{max} + 1}{\varepsilon} \right)$$

(assuming b_{max}, the maximal b_i, is positive).

PROOF. The point $x^0 = 0$, $t^0 = b_{max} + 1$ is in the interior of the feasible domain of FB(A, b). By our assumption, the set of optimal solutions of FB(A, b) is bounded. Several interior point algorithms are now known (e.g., [2], [10]) that can start from any interior point and reduce the value of the objective function to a value not greater than $t^* + \varepsilon$ in a polynomial number of iterations in terms m, n, and ρ, where each iteration takes a polynomial number of operations in terms of m and n. $\square$

An obvious consequence of Proposition 5.5 is the following:

COROLLARY 5.6. *Suppose the rows of A spanned the whole R^n in non-negative linear combinations. It takes a polynomial number of operations in terms of m, n, and ρ to either compute a vector x such that $Ax \geq b$ or conclude that there is no x such that $Ax \geq b + \varepsilon e$.*

PROOF. Run any of the polynomial interior point algorithms for a number of iterations that guarantees that $t < t^* + \varepsilon$. As soon as t becomes nonpositive, stop. (The current x is feasible.) If at the end t is still positive then $t^* > -\varepsilon$, and hence there is no x such that $Ax - \varepsilon e \geq b$. $\square$

Another consequence with respect to optimal solutions with tolerance can be stated conveniently when the problem is in the symmetric form Sym(A, b, c):

$$\text{Maximize} \quad c^T x$$
$$\text{subject to} \quad Ax \leq b$$
$$x \geq 0,$$

whose dual is

$$\text{Minimize} \quad b^T y$$
$$\text{subject to} \quad A^T y \geq c$$
$$y \geq 0.$$

Recall that by the duality theorem, a feasible solution x is optimal if and only if there exists a dual feasible solution y such that $c^T x = b^T y$. On the other hand, $c^T x \leq b^T y$ for any pair of feasible x and y. This suggests the following approximation problem:

PROBLEM 5.7. Given A, b, c and $\varepsilon > 0$, either find a pair of vectors x, y such that

$$Ax \leq b + \varepsilon e,$$
$$A^T y \geq c - \varepsilon e,$$
$$b^T y - c^T x \leq \varepsilon,$$
$$x \geq -\varepsilon e, \qquad y \geq -\varepsilon e$$

or conclude that $\mathrm{Sym}(A, b, c)$ does not have an optimal solution.

The notion of approximation presented in Problem 5.7 is very close to the one used in practice, where optimality criteria are applied without knowledge of proximity to the value of an optimal solution.

Note that Problem 5.7 is trivial if $c \leq 0 \leq b$. Thus, assume

$$\gamma = \max\{-b_{\min}, c_{\max}\} > 0$$

and denote

$$\rho^* = \log\left(\frac{\gamma + 1}{\varepsilon}\right).$$

PROPOSITION 5.8. *Suppose the rows of the matrix*

$$\begin{bmatrix} 0 & A^T \\ -A & 0 \\ c^T & -b^T \\ I & 0 \\ 0 & I \end{bmatrix}$$

span the space R^{m+n} *in nonnegative linear combinations. Problem 5.7 can be solved in a polynomial number of operations in terms of* m, n, *and* ρ^*.

PROOF. Consider the problem

$$\begin{aligned} \text{Minimize} \quad & t \\ \text{subject to} \quad & Ax - te \leq b \\ & A^T y + te \geq c \\ & b^T y - c^T x - t \leq 0 \\ & x + te, \; y + te \geq 0. \end{aligned}$$

Starting at $t = \gamma + 1$, $x = 0$, and $y = 0$, run a number of iterations that guarantees $t \leq t^* + \varepsilon$. This number is polynomial in m, n, and ρ^*, and ρ^* is trivial to compute. If $t \leq \varepsilon$ then the current x and y solve Problem 5.7. Otherwise, $t^* > 0$, and $\mathrm{Sym}(A, b, c)$ does not have an optimal solution. $\square$

Note that a solution of Problem 5.7 does not guarantee that $c^T x$ is close to the optimal value V^* when the latter exists. It seems much more difficult to solve the approximation problem in the latter sense. This difficulty can be explained by considering the following practical question, which arises when one applies the simplex method to the problem in standard form $SF(A, b, c)$. Suppose x and y satisfy

$$Ax = b,$$
$$x \geq -\delta e,$$
$$A^T y \geq c - \delta e,$$
$$c^T x = b^T y.$$

Assuming the problem has an optimum whose value is V^*, we are interested in finding what should be the value of δ in order to ensure that

$$c^T x \geq V^* - \varepsilon.$$

To answer the question, suppose x^* is an optimal solution and we get

$$c^T x = b^T y = y^T A x^* \geq c^T x^* - \delta e^T x^*.$$

Thus, we have to estimate the quantity $e^T x^*$.

As is known from the analysis of the ellipsoid algorithm, the coordinates of a basic solution x^* can be bounded as follows: First, x^* satisfies an equation $Bx^* = b$, where B is a nonsingular square submatrix of A. Thus,

$$x_i^* = \frac{\det(B^i)}{\det(B)}$$

where B^i is the matrix obtained from B by substituting b for the ith column. The known analyses do not use this relationship between B^i and B. In the case of integer coefficients an obvious lower bound on the absolute value of a nonzero determinant is 1. Thus, a lower bound can also be obtained in the case of rational coefficients. It is not known whether over the reals the following problem can be solved in a polynomial number of arithmetic operations and comparisons:

PROBLEM 5.9. Given a real matrix $A \in R^{m \times n}$, compute a positive α such that for every nonsingular submatrix $B \in R^{m \times m}$ of A, $|\det(B)| \geq \alpha$.

A tight upper bound on the absolute value of a determinant in terms of the maximum value M of any entry can be obtained over the reals as follows: Suppose the columns of B are $v^1, \ldots, v^m$. Obviously,

$$|\det(B)| \leq \|v^1\| \cdots \|v^m\| \leq (\sqrt{m}M)^m.$$

At least for values of m for which there exist Hadamard matrices (i.e., matrices of orthogonal columns consisting of ± 1's), there exist matrices with the determinant $(\sqrt{m}M)^m$.

Without exploiting the relationship between B^i and B the only claim we can prove is that

$$x_i^* \leq \frac{(\sqrt{m}M)^m}{\eta(A)}$$

where $\eta(A)$ is any positive lower bound on the absolute value of the determinant of any nonsingular $m \times m$ submatrix of A. We note in passing that in the case of integral coefficients the bounds on x_i^* cannot be improved dramatically. For example, if

$$B = \begin{bmatrix} 1 & & & & \\ -M & 1 & & & \\ -M & -M & 1 & & \\ \vdots & \vdots & \ddots & \ddots & \\ -M & -M & \cdots & -M & 1 \end{bmatrix}$$

and $b = (M, \ldots, M)^T$ then $x_m^* = \sum_{i=1}^m M^i$. Also, if

$$B = \begin{bmatrix} M & M & M & \cdots & M \\ -1 & M & M & \cdots & M \\ & -1 & M & \cdots & M \\ & & \ddots & \ddots & \vdots \\ & & & -1 & M \end{bmatrix}$$

and $b = (1, 0, \ldots, 0)^T$ then $x_m^* = 1/(M(M+1)^{m-1})$.

Recall that to guarantee an ε-approximation in terms of the function value, we have to choose δ such that

$$\delta \le \frac{\varepsilon}{e^T x^*} .$$

Hence we get the estimate

$$\delta \le \frac{\varepsilon \eta(A)}{m(\sqrt{m}T)^m} .$$

Another estimate can be derived by

$$e^T x^* \le \sqrt{m}\|x^*\| = \sqrt{m}\|B^{-1}b\| \le \frac{\sqrt{m}\|b\|}{\sqrt{\lambda_{\min}(BB^T)}}$$

where $\lambda_{\min}$ denotes the least eigenvalue. Thus, what we would need is a lower bound on the least eigenvalue of any matrix of the form BB^T where B is a basis. Note that our estimates depend on properties of an optimal basis rather than one that supports the approximate solution. So, the optimal basis may be ill conditioned, this fact being unknown to the user, and the terminal basis well conditioned and satisfying the optimality conditions within tolerance.

Obviously, if any $\eta(A)$ is known then we can solve the ε-approximation problem in a polynomial number of operations in terms of m, n, and $-\log \delta$. However, in general η is not known. It is interesting to note that for a practical solution of problems with thousands of variables, even with a sparse and well-structured problem with small integral coefficients, the required δ may be too small to be practical. It is not clear that an approximate solution based on tolerance has a value close to the optimal. In fact, the value may be far from the optimum even though the duality gap is very small, since the vectors x and y are only approximately feasible.

The following example illustrates the difficulties described above. Suppose we solve a problem in standard form, and the representation using the current basis is

$$\text{Maximize} \quad \tilde{c}_N^T x_N$$
$$\text{subject to} \quad x_B + \widetilde{N} x_N = e$$
$$x_B, x_N \geq 0,$$

where $\widetilde{N} \in R^{m \times (n-m)}$ contains the following $m \times m$ submatrix:

$$B' = \begin{bmatrix} 1 & & & & \\ -2 & 1 & & & \\ & -2 & 1 & & \\ & & \ddots & \ddots & \\ & & & -2 & 1 \end{bmatrix}.$$

Suppose further that the coordinates of $\tilde{c}_N$ are negative except for those corresponding to the columns of B' that are all equal to zero except for the last one, which equals 10^{-20}. Suppose $m = 1000$, which is quite common in practice. Assuming 10^{-50} is considered nonpositive within tolerance, the solution $x = e$ is accepted as optimal, with an objective function value of 0. However, the basis B' determines a feasible solution x where $x_{B_i'} = 2^i - 1$ and the objective function value is greater than 10^{250}. Note that the current basis B is very well conditioned. Moreover, the underlying matrix is very sparse and well structured.

Appendix

We review some known characteristics of the simplex method for problems in standard form $SF(A, b, c)$. Suppose, for simplicity, we run only "phase II" of the algorithm; i.e., we start from a basic feasible solution and attempt to find an optimal one or conclude that the problem is unbounded. (It is well known that if a problem in standard form has a feasible solution then it has a basic feasible one; the "phase I" problem of finding a basic feasible solution, or concluding that none exists, can be formulated as a problem in standard form with a known basic feasible solution.) Assuming exact arithmetic, the algorithm then terminates with a feasible basis. The termination criterion is stated in terms of signs of certain entries of the "tableau." As is common, let $B \in R^{m \times m}$ denote the nonsingular submatrix of A whose columns constitute the current basis, and let N denote the matrix consisting of the other columns. Let x_B and c_B denote the restrictions of the vectors x and c, respectively, to the indices corresponding to the columns of B. Let x_N and c_N denote the complementary restrictions of the vectors. The "tableau" is essentially a representation of the problem in an equivalent form:

$$\text{Maximize} \quad (c_N^T - c_B^T B^{-1} N) x_N$$
$$\text{subject to} \quad x_B + B^{-1} N x_N = B^{-1} b$$
$$x_B \geq 0, \quad x_N \geq 0.$$

Given a basis B, the corresponding basic primal solution $x = x(B)$ is given by

$$x_B = B^{-1}b, \qquad x_N = 0.$$

The basic dual vector $y = y(B)$ associated with B is given by the equation

$$y^T B = c_B^T.$$

The dual problem is

$$\text{Minimize} \quad y^T b$$
$$\text{subject to} \quad y^T A \geq c^T.$$

Thus, $y(B)$ is feasible in the dual problem if and only if

$$c_N^T \leq c_B^T B^{-1} N,$$

in which case y is optimal for the dual.

Assuming that the algorithm (namely, the primal simplex method) works with exact arithmetic, for every B occurring in the process,

$$B^{-1}b \geq 0.$$

The algorithm terminates in one of the following cases:

(i) $c_N^T \leq c_B^T B^{-1} N$, in which case B is optimal, or

(ii) there exists a column N_j such that $c_j > c_B^T B^{-1} N_j$ and $B^{-1}N_j \leq 0$, in which case the problem is unbounded.

In either case the basis B is said to be *terminal*. The termination criterion applies to signs of certain entries in the tableau.

REFERENCES

1. V. Chvátal, *Linear programming*, W. H. Freeman, New York, 1983.
2. D. M. Gay, *A variant of Karmarkar's linear programming algorithm for problems in standard form*, Math. Programming **37** (1987), 81–90.
3. G. H. Golub and C. Van Loan, *Matrix computations*, Johns Hopkins University Press, 1983.
4. C. C. Gonzaga, *Polynomial affine algorithms for linear programming*, Report ES-139/88, Dept. of Systems Engineering and Computer Sciences, COPPE, Rio de Janeiro, 1988.
5. M. Grötschel, L. Lovász, and A. Schrijver, *The ellipsoid method and its consequences in combinatorial optimization*, Combinatorica **1** (1981), 169–197; Corrigendum: Combinatorica **4** (1984), 291–295.
6. L. G. Khachiyan, *A polynomial algorithm in linear programming*, Soviet. Math. Dokl. **20** (1979), 191–194.
7. L. Lovász, *An algorithmic theory of numbers, graphs and convexity*, SIAM, Philadelphia, 1986.
8. N. Megiddo, *Combinatorial optimization with rational objective functions*, Math. Oper. Res. **4** (1979), 414–424.
9. J. H. Wilkinson, *Error analysis of floating-point computation*, Numer. Math. **2** (1960), 219–340.
10. Y. Ye and M. Kojima, *Recovering optimal dual solutions in Karmarkar's polynomial algorithm for linear programming*, Math. Programming **39** (1987), 305–317.

IBM RESEARCH DIVISION, ALMADEN RESEARCH CENTER SAN JOSE, CALIFORNIA 95120–6099, AND SCHOOL OF MATHEMATICAL SCIENCES, TEL AVIV UNIVERSITY, TEL AVIV, ISRAEL
E-mail address: MEGIDDO@IBM.COM

Contemporary Mathematics
Volume **114**, 1990

Riemannian Geometry Underlying Interior-Point Methods for Linear Programming

NARENDRA KARMARKAR

ABSTRACT. This paper studies the continuous trajectories underlying the projective and affine algorithms for linear programming, particularly the projective trajectories, using the techniques of differential geometry. We formulate a projectively invariant Riemannian metric defined on the feasible region of the linear programming problem and study the scalar curvature of the trajectories with respect to this metric. We present reasons why the total scalar curvature of trajectories should approximate the number of iterations needed by an interior-point algorithm following these trajectories. We show that the total scalar curvature of the projective trajectories is small in a special case and present evidence that this may be true in the general case. A relation between the vector field underlying the projective algorithm and geodesics in this metric is proved.

1. Interior-point methods for linear programming

This paper is motivated by the problem of determining the theoretical limit of the computational complexity of interior-point methods for linear programming. Interior-point methods can be viewed as making discrete approximations following a set of continuous trajectories to the optimal solution of the linear program. There are two basic ingredients that matter in such algorithms: choosing the continuous trajectories and choosing the method of making discrete approximation to the continuous trajectories.

This paper concentrates on studying the continuous trajectories associated to the projective algorithm that I introduced in 1984 [6]. The worst-case bounds known for the number of iterations taken by the projective algorithm contrast with the few iterations observed empirically. In §1.2 we give a rule for selecting the step length that may have good worst-case properties. That step-length rule suggests that the running time of the algorithm should be proportional to the total scalar curvature of the underlying trajectories measured from the starting point to a point suitably near the optimum.

1980 *Mathematics Subject Classification* (1985 *Revision*). Primary 90C05; Secondary 53B21.

The main goal of the paper is to study the geometric behavior of these continuous trajectories using techniques of differential geometry. Rather than study them using the Euclidean metric, we sutdy them using a projectively invariant Riemannian metric that reflects the projective invariance properties of the algorithm. This paper shows that the scalar curvature of the trajectories in this metric is small. This information may shed some light on the remarkably few iterations that the projective algorithm takes in practice.

In this section, we briefly review the projective and affine forms of interior-point methods for linear programming. We give a description of the algorithm as a discrete iterative process, then define the associated vector fields and continuous trajectories and give differential equations governing the trajectories. We then propose a rule for choosing the length of the steps taken in a discrete algorithm that approximately follows trajectories. This motivates studying the scalar curvature of trajectories as an approximation to the number of iterations needed by discrete algorithms.

In the rest of the paper, §2 describes a good Riemannian metric to study projective algorithm trajectories, §3 describes geodesics of this metric and relation to the projective algorithm vector field, and §§4 and 5 estimate the scalar curvature of the projective trajectories in this metric. There remain a number of open problems that must be solved to obtain bounds for the number of iterations taken by discrete interior-point algorithms, and some of these are stated in §§4.2 and 5.

1.1. Canonical problem form and projective algorithm. The projective algorithm is conveniently defined for linear programs given in a particular canonical form.

CANONICAL PROBLEM FORM (PROJECTIVE CASE).

$$Minimize \quad \mathbf{c}^T \mathbf{x}$$
$$Subject\ to \quad A\mathbf{x} = 0,$$
$$(1.1.1) \qquad\qquad\qquad \mathbf{e}^T \mathbf{x} = 1,$$
$$\mathbf{x}^i \geq 0.$$

Here $\mathbf{x}, \mathbf{c}, \mathbf{e} \in \mathbf{R}^n$, $\mathbf{e}$ is the vector of all 1's, and A is an $m \times n$ real matrix. The center of the simplex $\mathbf{x}^i = 1/n$ is assumed to be feasible; i.e., $A\mathbf{e} = 0$. Furthermore, the optimal objective value is assumed to be zero.

(In the affine case, the constraint $\mathbf{e}^T \mathbf{x} = 1$ is not needed, and linear equations are allowed to be nonhomogeneous: $A\mathbf{x} = \mathbf{b}$ and the objective function can have any optimal value.) The constraints $\{\mathbf{e}^T \mathbf{x} = 1, \mathbf{x}^i \geq 0\}$ define a simplex S_n, and $\Omega = \{\mathbf{x} \mid A\mathbf{x} = 0\}$ defines a subspace.

We are interested in their intersection $P = \Omega \cap S_n$, which defines the feasible region for the linear programming problem. Figure 1 illustrates the region P in both the *unconstrained case* where $P = S_n$ and the *general case* where Ω is nontrivial.

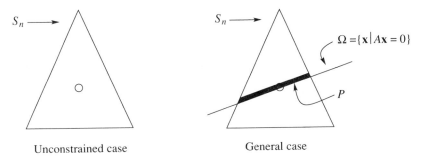

S_n $\longrightarrow$

S_n $\longrightarrow$

$\Omega = \{\mathbf{x} \mid A\mathbf{x} = 0\}$

P

Unconstrained case

General case

FIGURE 1. Canonical form problem.

The projective algorithm creates a sequence of feasible points $\mathbf{x}^{(0)}$, $\mathbf{x}^{(1)}$, ..., $\mathbf{x}^{(k)} \in S_n \cap \Omega$, as follows:

(1) Take the center as the starting point $\mathbf{x}^{(0)} = \mathbf{e}/n$
(2) At the kth step obtain $\mathbf{x}^{(k+1)}$ from $\mathbf{x}^{(k)}$ by these steps:
   (a) Define a diagonal matrix $D = \text{diag}\{\mathbf{x}_1^{(k)}, \mathbf{x}_2^{(k)}, \dots\}$.
   (b) Apply the projective transformation $\mathbf{y} = T(\mathbf{x})$ given by

(1.1.2)
$$\mathbf{y} = \frac{D^{-1}\mathbf{x}}{\mathbf{e}^T D^{-1}\mathbf{x}}.$$

This transformation maps the simplex S_n onto itself, each face of the simplex onto the same face, and the strict interior of the simplex onto itself. It maps the current point $\mathbf{x}^{(k)}$ onto the center of the simplex. The linear objective $\mathbf{c}^T\mathbf{x}$ and the null space $A\mathbf{x} = 0$ are mapped as follows:

$$\mathbf{c}^T\mathbf{x} \to \frac{(D\mathbf{c})^T\mathbf{y}}{(D\mathbf{e})^T\mathbf{y}},$$

$$\left\{ \begin{array}{c} \text{Null space} \\ A\mathbf{x} = 0 \end{array} \right\} \to \left\{ \begin{array}{c} \text{Null space} \\ AD\mathbf{y} = 0 \end{array} \right\}.$$

The inverse transformation is given by $\mathbf{x} = T(\mathbf{y})$ with

$$\mathbf{x} = \frac{D\mathbf{y}}{\mathbf{e}^T D\mathbf{y}}.$$

(c) Project the vector $D\mathbf{c}$ appearing in the numerator of transformed objective functions on the transformed null space $\left\{ \begin{smallmatrix} AD\mathbf{y}=0 \\ \mathbf{e}^T\mathbf{y}=0 \end{smallmatrix} \right\}$:

$$\mathbf{c}_p = \mathbf{P}_{\left[\begin{smallmatrix} AD \\ e \end{smallmatrix}\right]} \cdot D\mathbf{c} = \left[I - \frac{1}{n}\mathbf{e}\mathbf{e}^T \right][I - DA^T(AD^2A^T)^{-1}AD]D\mathbf{c}.$$

(d) Follow the projected direction $\hat{\mathbf{c}}_p$, up to a suitable distance α; from the center

$$\mathbf{x}' = \frac{\mathbf{e}}{n} - \alpha\hat{\mathbf{c}}_p.$$

(e) Apply the inverse transform to $\mathbf{x}'$ to get the next point in the sequence: $\mathbf{x}^{(k+1)}$.

$$\mathbf{x}^{(k+1)} = \frac{D\mathbf{x}'}{\mathbf{e}^T D\mathbf{x}'}$$

(In the *affine algorithm*, we use the linear tranformation

$$\mathbf{y} = D^{-1}\mathbf{x}$$

instead of the projective transformation (1.1.2).)

In the analysis I published in 1984 [6], the number of iterations of the projective algorithm was bounded by $O(nL)$ to obtain a reduction in the objective function by a factor 2^{-L}. Subsequently, many variants of the basic procedure have been proven to have an $O(\sqrt{n} \cdot L)$ bound [13]. However, the number of iterations of the projective algorithm in practice appears to be much fewer, perhaps even $O(\max(L, \log n))$. Efforts to understand the reasons for such a small number of iterations lead naturally into questions in differential geometry.

1.2. Continuous trajectories. In the discrete iterative process defined above the next point in the sequence is obtained by following a certain distance the direction computed at each step. Instead, if we were to follow the direction an infinitesimal distance and then make a new transformation to obtain a new direction, we would get the "continuous" or infinitesimal analogue of the discrete algorithm. If the discrete form of the method produces an iteration of the type

$$\mathbf{x}^{(k+1)} \leftarrow \mathbf{x}^{(k)} + \alpha \cdot \mathbf{v}^{(k)} + O(\alpha^2)$$

then by taking the limit as $\alpha \to 0$, we get the infinitesimal version, whose continuous trajectories are given by the differential equation

$$\frac{d\mathbf{x}}{d\alpha} = \mathbf{v}.$$

Here, $\mathbf{v}$ as a function of $\mathbf{x}$ defines a vector field.

For the projective method, the differential equation is

(1.2.1) $$\frac{d\mathbf{x}}{dt} = -[D - \mathbf{x}\mathbf{x}^t]P_{AD} \cdot D\mathbf{c},$$

where $D = \text{diag}\{\mathbf{x}_1, \mathbf{x}_2, \dots, \mathbf{x}_n\}$ and

$$P_{AD} = I - DA^T(AD^2A^T)^{-1}AD.$$

(For the affine algorithm,

(1.2.2) $$\frac{d\mathbf{x}}{dt} = -DP_{AD} \cdot D\mathbf{c}$$

is the associated differential equation; see [1].)

These trajectories have been studied by several authors from different perspectives; see [1], [2], [7], [9], and [11].

In order to understand the reasons behind the remarkably few iterations taken by the projective algorithm in practice, we are led to consider the following question: *Given a system of continuous trajectories how best should we make a discrete approximation?*

Intuitively, it seems that we can take "long" steps when the path is relatively straight, but we want to take short steps when the path is bending more. This leads us to consider the following simple step-length rule (see Figure 2).

Step-length rule: As long as the angle between the straight-line segment we are following and the direction given by the vector field does not exceed a fixed limit θ_{max}*, continue on the current segment.*

More generally, one can consider such a step-length rule in which the angle is measured with respect to a general Riemannian metric.

Returning to the step-length rule above, observe that in the limit as $\theta_{max} \to 0$ the step length is the distance moved on the current trajectory until its tangent turns by θ_{max}. Hence the scalar curvature $|d\theta/ds|$ is the "infinitesimal step cost" at a point, and the integral of the scalar curvature $\int |d\theta/ds|ds$ is a surrogate measure for the performance of a discrete algorithm. (More precisely, one expects it to provide a lower bound for such performance up to a multiplicative constant, and perhaps also on upper bounds under special circumstances.) This motivates the study of the following question:

Given a Riemannian metric defined on the interior of the feasible region, what is the integral of the scalar curvature of the continuous trajectories for projective and affine variants, until we reach a point that achieves a reduction of 2^{-L} *in the objective function?*

While the continuous trajectories of the projective algorithm are invariant with respect to projective transformations, the discrete approximation to these trajectories, make according to the step-length rule described above,

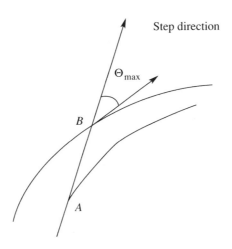

FIGURE 2. Step-length rule.

would remain projectively invariant if and only if the metric used for measuring angles is also projectively invariant.

In §2.2 we introduce such a projectively invariant Riemannian metric. In §4, we show that the integral of the scalar curvature with respect to this Riemannian metric is bounded by $O(L)$ for the unconstrained case where the problem has no equality constraints, and in §5 we present a result suggesting that a similar bound may hold for the general case.

It remains to relate such scalar curvature bounds to the performance of the discretized version of the algorithm. In §4.2 we indicate questions remaining to be answered regarding the relation between the discrete and continuous version of the algorithm.

The usefulness of the Riemannian metric that we introduce is also demonstrated by the fact that there is a natural characterization of the vector field underlying the projective algorithm in terms of the geodesics in this metric, given in §3.3.

2. Riemannian metrics and affine connections

In the projective algorithm, in order to decide how to proceed from a given point on the curve, we first make a projective transformation (at least conceptually) to bring that point to the center of the simplex. Then how long a step we can take from this point should depend on how much the path bends in this *transformed* coordinate system.

Instead of using an infinite number of different coordinate systems to analyze the performance of the infinitesimal version of the algorithm, an equivalent approach is to construct a *Riemannian metric* $g_{ij}(\mathbf{x})d\mathbf{x}^i d\mathbf{x}^j$ on the original space, which is invariant with respect to projective transformations of the simplex onto itself. There is an essentially unique choice of such a metric, and we study it here.

In the rest of the paper, in formulae for Riemannian metrics and affine connections, we generally use the *Einstein summation convention*, in which upper and lower indices that match in any expression are to be summed over; e.g., the expression $g_{ij}(\mathbf{x})d\mathbf{x}^i d\mathbf{x}^j$ for a Riemannian metric abbreviates $\sum_{i=1}^{n} \sum_{j=1}^{n} g_{ij}(\mathbf{x})d\mathbf{x}^i d\mathbf{x}^j$. However, this convention is *not* used in formulae containing explicit summation signs, e.g., (3.2.2).

2.1. Projectively invariant Riemannian metric. A projectively invariant choice of metric is based on the observation that the *strict interior* of the simplex is a *Lie group* with the following definition for multiplication:

$$\mathbf{x}, \mathbf{y} \in S_n, \qquad (\mathbf{x} \circ \mathbf{y})^i = \frac{\mathbf{x}^i \mathbf{y}^i}{\sum_i \mathbf{x}^i \mathbf{y}^i}.$$

The identity element is the center of the simplex:

$$\mathbf{x}^i = \frac{1}{\mathbf{n}}.$$

The inverse of an element $\mathbf{x}$ is

$$(\mathbf{x}^{-1})^i = \frac{\frac{1}{\mathbf{x}^i}}{\sum_i \frac{1}{\mathbf{x}^i}}.$$

For any element $\mathbf{x}$ of the group *translation* with respect to $\mathbf{x}$ is denoted by $L_{\mathbf{x}}: G \to G$, where

$$L_{\mathbf{x}}(g) = \mathbf{x} \circ g,$$

and its inverse is

$$L_{\mathbf{x}}^{-1}(g) = \mathbf{x}^{-1} \circ g.$$

Note that $L_x^{-1} = L_{x^{-1}}$. (The notation $L_{\mathbf{x}}$ represents *left* translation. But in our case, the group is abelian; hence, we need not distinguish between left and right translations.)

Each iteration of the projective algorithm is equivalent to applying the translation $L_{\mathbf{x}}^{-1}$, then optimizing over the ball of infinitesimal radius, then applying the translation $L_{\mathbf{x}}$.

Once we define the metric at the identity, the translation $L_{\mathbf{x}}$ gives us a unique way of extending it to all interior points of the simplex, since the derivative of $L_{\mathbf{x}}$ gives a natural way to map the tangent space at the identity to the tangent space at $\mathbf{x}$:

$$DL_{\mathbf{x}}: T_{(\mathbf{e})} \to T_{(\mathbf{x})}.$$

Thus we get the following Riemannian metric on $\operatorname{Int}(S_n) = \{\mathbf{x}: \sum \mathbf{x}^i = 1$, all $\mathbf{x}^i > 0\}$ for the projective case:

$$(2.1.1) \qquad g_{ij}(\mathbf{x}) = \frac{1}{\mathbf{x}^i \mathbf{x}^j}\left[\delta_{ij} - \frac{1}{n}\right].$$

In the canonical problem form with the extra constraints $A\mathbf{x} = 0$ added, the associated Riemannian metric is the restriction to this subspace. In the affine case, we get the following Riemannian metric defined on the positive orthant and further restricted to the subspace $A\mathbf{x} = \mathbf{b}$,

$$(2.1.2) \qquad g_{ij}(\mathbf{x}) = \frac{1}{\mathbf{x}^i \mathbf{x}^j}\delta_{ij}.$$

In these formulae $g_{ij}(\mathbf{x})$ is a covariant, symmetric tensor of degree 2. The corresponding contravariant tensor is denoted by $g^{ij}(\mathbf{x})$ and satisfies

$$g^{ij} g_{jk} = \delta_k^i.$$

Note that although the dimension of the manifold we are working with is $n-1$ due to the constraint $\sum \mathbf{x}^i = 1$, we are using n coordinates $i, j = 1, \ldots, n$ in g_{ij} for convenience. It is understood that in the quadratic form $g_{ij}\mathbf{u}^i\mathbf{u}^j$, $\mathbf{u}^i$ is a contravariant vector satisfying $\sum \mathbf{u}^i = 0$. Similarly, the composition $g^{ij} g_{jk}$ needs to behave like the identity only on an appropriate $n-1$ dimensional subspace.

2.2. Affine connection, parallel translation. Corresponding to a Riemannian metric, the coefficients of the associated affine connection are given by

$$(2.2.1) \qquad L^i_{jk} = -\frac{1}{2} g^{il} \left[\frac{\partial g_{jl}}{\partial \mathbf{x}^k} + \frac{\partial g_{kl}}{\partial \mathbf{x}^j} - \frac{\partial g_{jk}}{\partial \mathbf{x}^l} \right].$$

An affine connection gives us a definition of parallel translation of a vector along a curve. Given a pair of points A and B, and a curve connecting the two, we can move each vector in the tangent space at A to the tangent space at B along the curve by means of the affine connection. Thus it "connects" tangent spaces at two different point, and the map is linear, hence the name. In general the map depends on the choice of curve; however, in the unconstrained case analyzed in §4, it is path-independent.

If a contravariant vector $\mathbf{u}^i$ situated at point $\mathbf{x}$ is moved to a point $\mathbf{x} + \Delta\mathbf{x}$ while keeping parallel to itself, its components at $\mathbf{x} + \Delta\mathbf{x}$ are

$$\mathbf{u}'^i = \mathbf{u}^i + L^i_{jk} \Delta\mathbf{x}^j \mathbf{u}^k + O(\Delta\mathbf{x}^2).$$

For the projective algorithm, one has

$$(2.2.2) \qquad L^i_{jk} = \delta_{jk} \left[\frac{\delta_{ij}}{\mathbf{x}^i} - \frac{\mathbf{x}^i}{\mathbf{x}^k} \right].$$

3. Geodesics and vector field of the algorithm

In this section we derive properties of geodesics of the projectively invariant Riemannian geometry introduced in §2. In the unconstrained case in which the feasible region of the linear program is the simplex S_n we determine the geodesics explicitly in §3.2, and also determine the global distance function between points in S_n induced by this Riemannian geometry. In §3.3 we characterize the projective algorithm vector field, in terms of geodesics. The projective algorithm trajectories are not geodesics of this geometry in general.

3.1. Geodesics. Geodesics can be defined in two ways:

    (1) variational form,
    (2) in terms of parallel translation.

The variational form is obtained by observing that if we want to find the shortest curve joining a pair of points A, B, then the first variation of the length integral along such a curve must vanish. A curve satisfying this necessary condition is called a geodesic [8, p. 130]. (It need not be the shortest, and a shortest path need not be unique.) The length integral is given by

$$\int_0^1 L\left(\mathbf{u}, \frac{d\mathbf{u}}{dt}\right) dt, \quad \text{where } L(\mathbf{u}, \mathbf{v}) = \sqrt{g_{ij}(\mathbf{u}) \mathbf{v}^i \mathbf{v}^j}.$$

The variational conditions give Euler-Lagrange equations defining the differential equation for geodesic curves:

(3.1.1)
$$\frac{d}{dt}\left(\frac{\partial L}{\partial \mathbf{v}^i}\right) = \left(\frac{\partial L}{\partial \mathbf{u}^i}\right), \qquad 1 \leq i \leq n.$$

Geodesics can alternatively be defined in terms of parallel translation using an affine connection. Let $T = d\mathbf{x}/dt$ be the tangent to a curve $\mathbf{x}(t)$ where t is any smooth parametrization of the curve, not necessarily the arc-length parametrization. Then

$$T(\mathbf{x} + \Delta\mathbf{x}) = T(\mathbf{x}) + \frac{d^2\mathbf{x}}{dt^2}\Delta\mathbf{x} + O(\Delta\mathbf{x}^2).$$

If we translate T parallel to itself then

$$T'^i = T^i + L^i_{jk}T^j\Delta\mathbf{x}^k.$$

Equating the two,

(3.1.2)
$$\frac{d^2\mathbf{x}^i}{dt^2} = L^i_{jk}T^j\frac{d\mathbf{x}^k}{dt} = L^i_{jk}\frac{d\mathbf{x}^j}{dt}\frac{d\mathbf{x}^k}{dt}.$$

Thus it is possible to find the differential equation for a geodesic in terms of an affine connection alone. Hence if two different Riemannian metrics give the same affine connection then they have the same set of geodesics. (Given a Riemannian metric, there is only one affine connection compatible with it, but a given affine connection can be compatible with more than one Riemannian metric.)

3.2. Global distance function in the unconstrained case. In the special case of (1.1.1) when the polytope is just the simplex S_n for the projective algorithm, it is easy to compute the geodesics either by solving the differential equation or by exponentiating the elements of Lie algebra associated with the Lie group. We show that the distance between $\mathbf{x}$ and $\mathbf{y}$ in this metric is given by

(3.2.1)
$$d^2(\mathbf{x}, \mathbf{y}) = \sum_i \left[\ln\frac{\mathbf{x}^i}{\mathbf{y}^i} - \frac{1}{n}\sum_j \ln\frac{\mathbf{x}^j}{\mathbf{y}^j}\right]^2.$$

To derive this formula we first develop a relationship between the arc-length parameter s and an arbitrary parameter t, for any curve $\mathbf{x}(t)$, $t \in [t_1, t_2]$. We have

$$\left(\frac{ds}{dt}\right)^2 = g_{ij}(\mathbf{x})\frac{d\mathbf{x}^i}{dt}\frac{d\mathbf{x}^j}{dt}.$$

Substituting for the Riemannian metric $g_{ij}(\mathbf{x})$ from (2.1.1) we get

$$\left(\frac{ds}{dt}\right)^2 = \sum_{i,j=1}^{n} \frac{1}{\mathbf{x}^i\mathbf{x}^j}\left[\delta_{ij} - \frac{1}{n}\right]\frac{d\mathbf{x}^i}{dt}\frac{d\mathbf{x}^j}{dt}$$

(3.2.2)

$$= \sum_{i=1}^{n}\left[\frac{1}{\mathbf{x}^i}\frac{d\mathbf{x}^i}{dt}\right]^2 - \frac{1}{n}\left[\sum_{k=1}^{n}\frac{1}{\mathbf{x}^k}\frac{d\mathbf{x}^k}{dt}\right]^2$$

$$= \sum_{i=1}^{n}\left\{\frac{1}{\mathbf{x}^i}\frac{d\mathbf{x}^i}{dt} - \frac{1}{n}\sum_{k=1}^{n}\frac{1}{\mathbf{x}^k}\frac{d\mathbf{x}^k}{dt}\right\}^2.$$

While this relationship holds for any curve, we now specialize the result for a geodesic. The differential equation of a geodesic is given by

$$\frac{d^2\mathbf{x}^i}{dt^2} = L^i_{jk}\frac{d\mathbf{x}^j}{dt}\frac{d\mathbf{x}^k}{dt}.$$

Substituting for the coefficients of affine connection L^i_{jk} from (2.2.1) we get

$$\frac{d^2\mathbf{x}^i}{dt^2} = \sum_{j,k=1}^{n}\delta_{jk}\left[\frac{\delta_{ij}}{\mathbf{x}^i} - \frac{\mathbf{x}^i}{\mathbf{x}^k}\right]\frac{d\mathbf{x}^j}{dt}\frac{d\mathbf{x}^k}{dt}$$

$$= \sum_{j=1}^{n}\left[\frac{\delta_{ij}}{\mathbf{x}^i} - \frac{\mathbf{x}^i}{\mathbf{x}^j}\right]\left(\frac{d\mathbf{x}^j}{dt}\right)^2$$

$$= \frac{1}{\mathbf{x}^i}\left(\frac{d\mathbf{x}^i}{dt}\right)^2 - \mathbf{x}^i\sum_{j=1}^{n}\frac{1}{\mathbf{x}^j}\left(\frac{d\mathbf{x}^j}{dt}\right)^2.$$

Therefore

$$\frac{1}{\mathbf{x}^i}\frac{d^2\mathbf{x}^i}{dt^2} - \left(\frac{1}{\mathbf{x}^i}\frac{d\mathbf{x}^i}{dt}\right)^2 = -\sum_{j=1}^{n}\frac{1}{\mathbf{x}^j}\left(\frac{d\mathbf{x}^j}{dt}\right)^2,$$

which is

$$\frac{d}{dt}\left[\frac{1}{\mathbf{x}^i}\frac{d\mathbf{x}^i}{dt}\right] = -\sum_{j=1}^{n}\frac{1}{\mathbf{x}^j}\left(\frac{d\mathbf{x}^j}{dt}\right)^2.$$

Note that the right-hand side is independent of the index i. Summing over the index i on the left-hand side, we get

$$\frac{d}{dt}\left\{\frac{1}{n}\sum_{i=1}^{n}\frac{1}{\mathbf{x}^i}\frac{d\mathbf{x}^i}{dt}\right\} = -\sum_{j=1}^{n}\frac{1}{\mathbf{x}^j}\left(\frac{d\mathbf{x}^j}{dt}\right)^2.$$

Comparing the last two equations yields

$$\frac{d}{dt}\left\{\frac{1}{\mathbf{x}^i}\frac{d\mathbf{x}^i}{dt} - \frac{1}{n}\sum_{k=1}^{n}\frac{1}{\mathbf{x}^k}\frac{d\mathbf{x}^k}{dt}\right\} = 0.$$

Thus the quantity in parenthesis remains constant along the geodesic, and we set

$$\alpha_i := \frac{1}{\mathbf{x}^i}\frac{d\mathbf{x}^i}{dt} - \frac{1}{n}\sum_{k=1}^{n}\frac{1}{\mathbf{x}^k}\frac{d\mathbf{x}^k}{dt}.$$

which is the same as the vector field (1.2.1) used by the projective algorithm. Finally, observe that the vector field is independent of the source hypersurface $f(\mathbf{x}) = \beta$. □

An analogous theorem can be proved for the affine trajectories; we omit the details.

The projective (or affine) trajectories are *not* geodesics in the corresponding Riemannian metric, in general. This holds because a geodesic is obtained by minimizing the distance between two fixed points, whereas the curves used in constructing the vector field above are obtained by keeping one endpoint fixed and allowing the other endpoint to vary on the equipotential surface. Also, note the asymmetric roles played by source hypersurface and target hypersurface: If $\mathbf{y}$ is the closest point on the target hypersurface to the given point $\mathbf{x}$ on the source hypersurface, then $\mathbf{x}$ *need not* be the closest point on the source hypersurface to the point $\mathbf{y}$ on the target hypersurface, with respect to the Riemannian metric. On the other hand, if we were to allow *both* points $\mathbf{x}$ and $\mathbf{y}$ to vary on the corresponding hypersurfaces and asked for the closest *pair* of points, then the two hypersurfaces would play a symmetric role.

4. Analysis of scalar curvature for the unconstrained case of the problem

In the general problem (1.1.1), the feasible region is the intersection of the simplex $S_n : \{\mathbf{x} \mid \mathbf{x}^i \geq 0$ and $\sum_{i=1}^{n} \mathbf{x}^i = 1\}$ and the affine subspace $\Omega = \{\mathbf{x} \mid A\mathbf{x} = 0\}$. We define the *unconstrained case* of the problem by dropping the subspace Ω, recall Figure 1. In this case, we prove below an $O(L)$ bound on total scalar curvature with respect to the Riemannian metric (2.1.1). Even in this case, it seems that one can get a bound of only $O(\sqrt{n}L)$ on the number of iterations by other proof techniques (see §3.2).

4.1. Total integral curvature for the unconstrained case.
In this section we analyze the integral of the scalar curvature in the unconstrained case of the problem and prove the following result.

THEOREM 2. *Let* $\mathbf{u} = \mathbf{e}/n$ *be the starting point of the continuous version of the projective algorithm for the unconstrained case of the problem and let* $\mathbf{v}$ *be an ending point with the property* $\mathbf{v}_i \geq 2^{-L}$ *for all* i . *The integral of the scalar curvature of the continuous trajectory between* $\mathbf{u}$ *and* $\mathbf{v}$ *is* $O(L)$.

PROOF. For a curve parametrized by arc-length s , in the Riemannian metric, geodesic curvature [7, p. 136] is given by

$$(4.1.1) \qquad p^i = \frac{d^2 \mathbf{x}^i}{ds^2} - L^i_{jk} \frac{d\mathbf{x}^j}{ds} \frac{d\mathbf{x}^k}{ds}.$$

This is a contravariant vector; its length gives the scalar curvature

$$\frac{d\theta}{ds} = \sqrt{g_{ij} p^i p^j}$$

of the curve at $\mathbf{x}$. After substituting

$$g_{ij}(\mathbf{x}) = \frac{1}{\mathbf{x}^i \mathbf{x}^j} \left[\delta_{ij} - \frac{1}{n} \right],$$

we get

$$
\begin{aligned}
\left(\frac{d\theta}{ds} \right)^2 &= \frac{p^i p^j}{\mathbf{x}^i \mathbf{x}^j} \left[\delta_{ij} - \frac{1}{n} \right] \\
&= \sum_{i=1}^{n} \left(\frac{p^i}{\mathbf{x}^i} \right)^2 - \frac{1}{n} \left[\sum_{k=1}^{n} \frac{p^k}{\mathbf{x}^k} \right]^2 \\
&= \sum_{i=1}^{n} \left\{ \frac{p^i}{\mathbf{x}^i} - \frac{1}{n} \sum_{k=1}^{n} \frac{p^k}{\mathbf{x}^k} \right\}^2 .
\end{aligned}
$$

(4.1.2)

Substituting $L_{jk}^i = \delta_{jk} \left[\frac{\delta_{ij}}{\mathbf{x}^i} - \frac{\mathbf{x}^i}{\mathbf{x}^k} \right]$ in the expression for the geodesic curvature p^i we get

$$
\begin{aligned}
p^i &= \frac{d^2 \mathbf{x}^i}{ds^2} - \delta_{jk} \left[\frac{\delta_{ij}}{\mathbf{x}^i} - \frac{\mathbf{x}^i}{\mathbf{x}^k} \right] \frac{d\mathbf{x}^j}{ds} \frac{d\mathbf{x}^k}{ds} \\
&= \frac{d^2 \mathbf{x}^i}{ds^2} - \left[\frac{\delta_{ij}}{\mathbf{x}^i} - \frac{\mathbf{x}^i}{\mathbf{x}^j} \right] \left(\frac{d\mathbf{x}^j}{ds} \right)^2 \\
&= \frac{d^2 \mathbf{x}^i}{ds^2} - \frac{1}{\mathbf{x}^i} \left(\frac{d\mathbf{x}^i}{ds} \right)^2 + \mathbf{x}^i \sum_{j=1}^{n} \frac{1}{\mathbf{x}^j} \left(\frac{d\mathbf{x}^j}{ds} \right)^2 .
\end{aligned}
$$

(4.1.3)

Hence,

$$
\begin{aligned}
\frac{p^i}{\mathbf{x}^i} &= \frac{1}{\mathbf{x}^i} \frac{d^2 \mathbf{x}^i}{ds^2} - \frac{1}{\mathbf{x}^{i2}} \left(\frac{d\mathbf{x}^i}{ds} \right)^2 + \sum_{j=1}^{n} \frac{1}{\mathbf{x}^j} \left(\frac{d\mathbf{x}^j}{ds} \right)^2 \\
&= \frac{d}{ds} \left[\frac{1}{\mathbf{x}^i} \frac{d\mathbf{x}^i}{ds} \right] + \sum_{j=1}^{n} \frac{1}{\mathbf{x}^j} \left(\frac{d\mathbf{x}^j}{ds} \right)^2 .
\end{aligned}
$$

Therefore,

$$\frac{p^i}{\mathbf{x}^i} - \frac{1}{n} \sum_{k=1}^{n} \frac{p^k}{\mathbf{x}^k} = \frac{d}{ds} \left[\frac{1}{\mathbf{x}^i} \frac{d\mathbf{x}^i}{ds} - \frac{1}{n} \sum_{k=1}^{n} \frac{1}{\mathbf{x}^k} \frac{d\mathbf{x}^k}{ds} \right].$$

Substituting this in the expression (4.1.2) for scalar curvature,

(4.1.4) $$\left(\frac{d\theta}{ds} \right)^2 = \sum_{i=1}^{n} \left\{ \frac{d}{ds} \left[\frac{1}{\mathbf{x}^i} \frac{d\mathbf{x}^i}{ds} - \frac{1}{n} \sum_{k=1}^{n} \frac{1}{\mathbf{x}^k} \frac{d\mathbf{x}^k}{ds} \right] \right\}^2 .$$

Changing the parametrization of the curve from arc-length parameter s to an arbitrary monotonically increasing parameter t we get

(4.1.5) $$\left(\frac{d\theta}{dt} \right)^2 = \sum_{i=1}^{n} \left\{ \frac{d}{dt} \left[\frac{1}{\dot{s}} \left(\frac{1}{\mathbf{x}^i} \frac{d\mathbf{x}^i}{ds} - \frac{1}{n} \sum_{k=1}^{n} \frac{1}{\mathbf{x}^k} \frac{d\mathbf{x}^k}{ds} \right) \right] \right\}^2 ,$$

where $\dot{s}$, as shown in (3.2.2), is given by

$$(4.1.6) \qquad \dot{s}^2 = \left(\frac{ds}{dt}\right)^2 = \sum_{i=1}^{n}\left\{\frac{1}{\mathbf{x}^i}\frac{d\mathbf{x}^i}{ds} - \frac{1}{n}\sum_{k=1}^{n}\frac{1}{\mathbf{x}^k}\frac{d\mathbf{x}^k}{ds}\right\}^2.$$

Define

$$(4.1.7) \qquad \mathbf{y}^i = \frac{1}{\mathbf{x}^i}\frac{d\mathbf{x}^i}{dt} - \frac{1}{n}\sum_{k=1}^{n}\frac{1}{\mathbf{x}^k}\frac{d\mathbf{x}^k}{dt}.$$

Therefore,

$$\dot{s}^2 = \sum_{i=1}^{n}(\mathbf{y}^i)^2,$$

and

$$\left(\frac{d\theta}{dt}\right)^2 = \sum_{i=1}^{n}\left\{\frac{d}{dt}\left(\frac{\mathbf{y}^i}{\dot{s}}\right)\right\}^2$$

$$= \sum_{i=1}^{n}\left\{\frac{d}{dt}\left(\frac{\mathbf{y}^i}{\sqrt{\sum_{j=1}^{n}(\mathbf{y}^j)^2}}\right)\right\}^2.$$

Differentiating and rearranging terms, we get

$$(4.1.8) \qquad \left(\frac{d\theta}{dt}\right)^2 = \frac{\sum_{i=1}^{n}\left(\frac{d\mathbf{y}^i}{dt}\right)^2}{\sum_{i=1}^{n}(\mathbf{y}^i)^2} - \frac{\left(\sum_{i=1}^{n}\mathbf{y}^i\frac{d\mathbf{y}^i}{dt}\right)^2}{\left(\sum_{i=1}^{n}(\mathbf{y}^i)^2\right)^2}.$$

Now we can apply this formula to the continuous trajectory of the algorithm. In the general case of the problem, the differential equation governing the trajectory is

$$\frac{d\mathbf{x}}{dt} = -(D - \mathbf{x}\mathbf{x}^T)P_{AD} \cdot D\mathbf{c}.$$

From this we get in the *unconstrained case*

$$(4.1.9) \qquad \frac{d\mathbf{x}^i}{dt} = -\mathbf{c}_i(\mathbf{x}^i)^2 + \mathbf{x}^i\sum_{j=1}^{n}\mathbf{c}_j(\mathbf{x}^j)^2.$$

Hence, (4.1.7) becomes

$$(4.1.10) \qquad \mathbf{y}^i = -\mathbf{c}_i\mathbf{x}^i + \frac{1}{n}\sum_{k=1}^{n}\mathbf{c}_k\mathbf{x}^k.$$

Differentiating and using (4.1.9) gives

$$\frac{d\mathbf{y}^i}{dt} = \mathbf{c}_i^2(\mathbf{x}^i)^2 - \frac{1}{n}\sum_{k=1}^{n}\mathbf{c}_k^2(\mathbf{x}^k)^2 + \left(\sum_{j=1}^{n}\mathbf{c}_j(\mathbf{x}^j)^2\right)\left(\frac{1}{n}\sum_{k=1}^{n}\mathbf{c}_k\mathbf{x}^k - \mathbf{c}_i\mathbf{x}^i\right).$$

If we set

(4.1.11) $$\mathbf{z}^i = \mathbf{c}_i^2(\mathbf{x}^i)^2 - \frac{1}{n}\sum_{k=1}^n \mathbf{c}_k^2(\mathbf{x}^k)^2$$

and use (4.1.10), this equation becomes

$$\frac{d\mathbf{y}^i}{dt} = \mathbf{z}^i + \left(\sum_{j=1}^n \mathbf{c}_j(\mathbf{x}^j)^2\right)\mathbf{y}^i.$$

Substituting this in the expression (4.1.8) for $d\theta/dt$, we get (after a calculation)

(4.1.12) $$\left(\frac{d\theta}{dt}\right)^2 + \left(\frac{\mathbf{y}\cdot\mathbf{z}}{\mathbf{y}\cdot\mathbf{y}}\right)^2 = \frac{\mathbf{z}\cdot\mathbf{z}}{\mathbf{y}\cdot\mathbf{y}}.$$

Hence,

(4.1.13) $$\frac{d\theta}{dt} \le \sqrt{\frac{\mathbf{z}\cdot\mathbf{z}}{\mathbf{y}\cdot\mathbf{y}}}.$$

Now we invoke the following inequality:

PROPOSITION 4.1. *For any real* a_i, $i = 1, \dots, n$,

(4.1.14) $$\sum_{i=1}^n a_i^4 - \frac{1}{n}\left(\sum_{i=1}^n a_i^2\right)^2 \le 4\max_i\{a_i^2\}\cdot\left[\sum_{i=1}^n a_i^2 - \frac{1}{n}\left(\sum_{i=1}^n a_i\right)^2\right].$$

We apply this inequality in (4.1.13), taking $a_i = \mathbf{c}_i\mathbf{x}^i$ and using the definitions of $\mathbf{y}^i$ and $\mathbf{z}^i$ to obtain

(4.1.15)
$$\begin{aligned}
\frac{d\theta}{dt} &\le \left(\frac{\sum_{i=1}^n\left(a_i^2 - \frac{1}{n}\sum_{k=1}^n a_k^2\right)^2}{\sum_{i=1}^n a_i^2 - \frac{1}{n}\left(\sum_{k=1}^n a_k^2\right)^2}\right)^{1/2} \\
&\le \left(4\max_i\{(\mathbf{c}_i\mathbf{x}^i)^2\}\right)^{1/2} \\
&= 2\max_i\{\mathbf{c}_i\mathbf{x}^i\}.
\end{aligned}$$

The integral of the scalar curvature with respect to the metric (2.1.1) is then bounded by

$$\int_0^T \frac{d\theta}{dt}\,dt \le 2\int_0^T \max_i\{\mathbf{c}_i\mathbf{x}^i\}\,dt.$$

Now suppose that the objective function $\mathbf{c}$ is *normalized*, i.e., all $\mathbf{c}^i \ge 0$ and some $\mathbf{c}^i = 0$, and without loss of generality $0 = \mathbf{c}_1 \le \mathbf{c} \le \cdots \le \mathbf{c}_n$. One can directly verify that

(4.1.16) $$\mathbf{x}^i(u) = \frac{1}{1 + \mathbf{c}_i u}(\Delta(u))^{-1}$$

with

(4.1.17)
$$\Delta(u) = \sum_{i=1}^{n} \frac{1}{1 + \mathbf{c}_i u}$$

satisfies

$$\frac{1}{\mathbf{x}^i} \frac{d\mathbf{x}^i}{du} = \Delta(u) \left(-c_i \mathbf{x}^i + \sum_{j=1}^{n} \mathbf{c}_j (\mathbf{x}^j)^2 \right),$$

which is a scaled version of the differential equation (4.1.9). Hence, $\mathbf{x}^i(t)$ is a reparametrized version of (4.1.16), with t related to u by

$$\frac{du}{dt} = \Delta(u).$$

One immediately sees from (4.1.16) that

$$\max\{\mathbf{c}_i \mathbf{x}^i\} = \mathbf{c}_n \mathbf{x}^n.$$

Hence,

$$\int_0^T \max\{\mathbf{c}_i \mathbf{x}^i\}\, dt = \int_0^T \mathbf{c}_n \mathbf{x}^n(u) \frac{dt}{du}\, du$$

$$= \int_0^T \Delta(u)^{-2} \frac{\mathbf{c}_n}{1 + \mathbf{c}_n u}\, du.$$

Since $\mathbf{c}$ is *normalized* with $\mathbf{c}_1 = 0$, one has $\Delta(u) \geq 1$ for all u; hence,

$$\int_0^T \max\{\mathbf{c}_i \mathbf{x}^i\}\, dt \leq \int_0^T \frac{\mathbf{c}_n}{1 + \mathbf{c}_n u}\, du$$

$$= \log(1 + \mathbf{c}_n T)$$

$$\leq \max_i \left[\log \frac{1}{\mathbf{x}^i(T)} \right].$$

But when we terminate the algorithm.

$$\mathbf{x}^i(T) \geq 2^{-L}.$$

Substitution of this bound in the equation above completes the proof of Theorem 2. □

4.2. Comments on the relation between continuous and discrete versions of the algorithm. A bound on the integral of scalar curvature of the continuous trajectory does not by itself imply a bound on the number of iterations of the discrete algorithm, for several reasons:

(1) The kth step of the discrete algorithm makes a tangential approximation to the continuous trajectory going through $\mathbf{x}^{(k)}$ to arrive at the next point $\mathbf{x}^{(k+1)}$. At the next step, we do not go back to the trajectory going through $\mathbf{x}^{(k)}$, but instead a tangential approximation is made to the new trajectory going through $\mathbf{x}^{(k+1)}$. Thus it is necessary

to show that "nearby" trajectories have similar curvature in order to relate behavior of the discrete algorithm to the corresponding continuous trajectories.

(2) Each step of the algorithm proceeds by first making a transformation that maps the current point to the center. At the center, the Euclidean metric and the Riemannian metric are the same. As one moves away from the center, the difference between the two metrics grows, and the corresponding geodesics gradually go apart. One needs to make a quantitative estimate of this difference for the step length of the discrete algorithm, since the latter is following geodesic segments in the Euclidean metric.

Both of these issues deal with analysis of the relation between the discrete and continuous versions of the algorithm, which we do not address in this paper.

5. Local comparison theorem between projective trajectories of the special and general case

We have seen that in the special case the integral of scalar curvature with respect to the metric (2.1.1) of the projective trajectory is $O(L)$ (Theorem 2). It appears that the curvature in the general case is *no more* than in the special case. We prove a "local" comparison theorem giving some evidence for this in this section. Consequently, it seems possible that an $O(L)$ bound for the scalar curvature of projective trajectories may hold in the general case.

We are going to make a "local" comparison between the continuous trajectories of the general version and the unconstrained versions of the problem. The Riemannian metric (2.1.1) for the unconstrained problem induces the Riemannian metric for the constrained problem. However, the induced affine connection for the constrained problem is different from the expressions (2.2.1). While the Riemannian space for the unconstrained problem is "flat," this is not so, in general, for the constrained problem. Since we are interested in comparing the continuous trajectories for the two cases, we view both sets of trajectories as curves in a common Riemannian space, namely, the unconstrained space. Now extend this common framework further:

(1) We think of the collection of all objective functions and the corresponding vector fields as one entity called a "fiber bundle" F (see [12]).

Base space: $\text{int}(S_n) \cap \{\mathbf{x} \mid A\mathbf{x} = 0\}$;

Fiber: R^d where $d = \text{co-dim}(\Omega) - 1$ for the projective case.

Cross-section: Each objective function gives a vector field or cross-section of F.

(2) We also extend the domain of definition of the general case to all interior points in the simplex as follows. Take any point $\mathbf{y}$ interior to

the simplex. Construct an affine space $\Omega' = \{x: Ax = Ay\}$ containing y, "parallel" to $\Omega = \{x \mid Ax = 0\}$ of the original problem. Now we have a new problem. In order to define vector fields and the fiber bundle for this problem, which is not in the standard form, we first express the constraints in homogeneous form. Define a matrix $A' := A - Aye^T$, so that

$$\Omega' = \{x \mid Ax = Ay, \ e^Tx = 1\} = \{x \mid A'x = 0, \ e^Tx = 1\}.$$

We then take any interior point $x^{(0)} \in \Omega'$ and apply the projective transformation T that maps $x^{(0)}$ to the center $1e/n$. Vector fields for this transformed problem are defined as in §1.1. Applying the inverse transform T^{-1} to these vector fields, we obtain vectors fields defined on Ω' and, putting them together as in Step 1 above, we get the required fiber bundle, on $\text{int}(S_n) \cap \{x \mid Ax = Ay\}$.

Suppose we do this construction for *all* interior points y and put together all fiber bundles into one.

Now for any point x in the strict interior of the simplex, and any direction y in the null space of A, there is a unique integral curve $c_g(t)$ for the general case such that

$$(5.1.1) \qquad \begin{aligned} c_g(0) &= x, \\ \frac{dc_g}{dt}(0) &= y. \end{aligned}$$

Similarly, there is a unique integral curve $c_s(t)$ for the unconstrained problem with

$$(5.1.2) \qquad c_s(0) = x, \qquad \frac{dc_s}{dt}(0) = y.$$

We consider scalar curvature of the general and unconstrained curves at a point x measured with respect to the Riemannian metric (2.2.1).

THEOREM 3. *The scalar curvature τ_g of the general case curve going through point x and having the tangent y is equal to the scalar curvature τ_s of the special case curve going through the same point and having the same tangent:*

$$\tau_g = \tau_s.$$

PROOF. As before, let P_e and P_{AD} denote the projection operators

$$P_e = I - \frac{1}{n}ee^T$$

and

$$P_{AD} = I - DA^T(AD^2A^T)^{-1}AD.$$

We also introduce a "reflection" operator

$$(5.1.3) \qquad R_{AD} = I - 2DA^T(AD^2A^T)^{-1}AD = 2P_{AD} - I.$$

These operators satisfy the following identities:

(5.1.4) $$P_e P_{AD} = P_{AD} P_e ,$$

(5.1.5) $$P_e R_{AD} = R_{AD} P_e ,$$

(5.1.6) $$P_{AD} R_{AD} = R_{AD} P_{AD} = P_{AD} ,$$

(5.1.7) $$P_{AD}^2 = P_{AD} , \qquad P_e^2 = P_e , \qquad R_{AD}^2 = I.$$

Furthermore, when D is a function of time t, the derivative of the projection operator P_{AD} is given by

(5.1.8) $$\frac{dP_{AD}}{dt} = R_{AD}\left(D^{-1}\frac{dD}{dt}\right)P_{AD} - P_{AD}\left(D^{-1}\frac{dD}{dt}\right).$$

To express the scalar curvature formula more concisely, we define

(5.1.9) $$V = D^{-1}\frac{dD}{dt} = \text{diag}\left(\frac{1}{\mathbf{x}^1}\frac{d\mathbf{x}^1}{dt} , \dots , \frac{1}{\mathbf{x}^n}\frac{d\mathbf{x}^n}{dt}\right),$$

(5.1.10) $$\mathbf{v} = V\mathbf{e} \quad \text{and} \quad \mathbf{w} = V^2 \mathbf{e}.$$

Using the expression (4.17) for $\mathbf{y}$, we get

(5.1.11) $$\mathbf{y} = P_e \mathbf{v},$$

and

$$\frac{d\mathbf{y}}{dt} = \frac{d}{dt}(P_e \mathbf{v}) = P_e \frac{d\mathbf{v}}{dt}.$$

We can now write the equation (4.1.6) for ds/dt and the equation (4.1.8) for $d\theta/dt$ in terms of $\mathbf{v}$:

(5.1.12) $$\left(\frac{ds}{dt}\right)^2 = \mathbf{v}^T P_e^2 \mathbf{v} = \mathbf{v}^T P_e \mathbf{v}$$

and

(5.1.13) $$\left(\frac{d\theta}{dt}\right)^2 = \frac{\left(\frac{d\mathbf{v}^T}{dt} P_e \frac{d\mathbf{v}}{dt}\right)(\mathbf{v}^T P_e \mathbf{v}) - \left(\mathbf{v}^T P_e \frac{d\mathbf{v}}{dt}\right)^2}{(\mathbf{v}^T P_e \mathbf{v})^2}.$$

To evaluate these expressions for the general case, we use the differential equation of the curve:

(5.1.14) $$\frac{d\mathbf{x}}{dt} = -(D - \mathbf{x}\mathbf{x}^T)P_{AD}D\mathbf{c}.$$

Substituting this in expression (5.1.10) for $\mathbf{v}$,

(5.1.15) $$\mathbf{v} = -(I - \mathbf{e}\mathbf{x}^T)P_{AD}D\mathbf{c}.$$

The actions of the operators P_e, P_{AD}, and R_{AD} on $\mathbf{v}$ are given by

(5.1.16) $$P_e \mathbf{v} = -P_e P_{AD} D\mathbf{c}, \qquad P_{AD}\mathbf{v} = \mathbf{v}, \qquad R_{AD}\mathbf{v} = \mathbf{v}.$$

Differentiating $P_e\mathbf{v}$ and using (5.1.8) for dP_{AD}/dt, we get

$$
\begin{aligned}
P_e\frac{d\mathbf{v}}{dt} &= \frac{d}{dt}(P_e\mathbf{v}) \\
&= \frac{d}{dt}(-P_eP_{AD}D\mathbf{c}) \\
(5.1.17) \quad &= -P_eR_{AD}D^{-1}\frac{dD}{dt}P_{AD}D\mathbf{c} + P_eP_{AD}D^{-1}\frac{dD}{dt}D\mathbf{c} - P_eP_{AD}\frac{dD}{dt}\mathbf{c} \\
&= -P_eR_{AD}VP_{AD}D\mathbf{c} \\
&= -R_{AD}P_eVP_{AD}D\mathbf{c}.
\end{aligned}
$$

We can express $P_{AD}D\mathbf{c}$ in terms of $\mathbf{v}$ by using equation (5.1.15) and setting $\alpha := \mathbf{x}^TP_{AD}D\mathbf{c}$:

$$
\begin{aligned}
P_{AD}D\mathbf{c} &= -\mathbf{v} + \mathbf{e}\mathbf{x}^TP_{AD}D\mathbf{c} \\
&= -\mathbf{v} + \alpha\mathbf{e}.
\end{aligned}
$$

Substituting this in equation (5.1.17), we get

$$
\begin{aligned}
P_e\frac{d\mathbf{v}}{dt} &= -R_{AD}P_eV(-\mathbf{v} + \alpha\mathbf{e}) \\
(5.1.18) \quad &= R_{AD}P_e(V^2\mathbf{e} - \alpha\mathbf{v}) \\
&= R_{AD}P_e(\mathbf{w} - \alpha\mathbf{v}).
\end{aligned}
$$

Therefore,

$$
\begin{aligned}
\mathbf{v}^TP_e\frac{d\mathbf{v}}{dt} &= \mathbf{v}^TR_{AD}P_e(\mathbf{w} - \alpha\mathbf{v}) \\
(5.1.19) \quad &= (R_{AD}\mathbf{v})^TP_e(\mathbf{w} - \alpha\mathbf{v}) \\
&= \mathbf{v}^TP_e(\mathbf{w} - \alpha\mathbf{v}) \\
&= (\mathbf{v}^TP_e\mathbf{w}) - \alpha(\mathbf{v}^TP_e\mathbf{v}).
\end{aligned}
$$

Similarly, using (5.1.18), we have

$$
\begin{aligned}
\frac{d\mathbf{v}^T}{dt}P_e\frac{d\mathbf{v}}{dt} &= \left(P_e\frac{d\mathbf{v}}{dt}\right)^T\left(P_e\frac{d\mathbf{v}}{dt}\right) \\
(5.1.20) \quad &= (\mathbf{w} - \alpha\mathbf{v})^TP_eR_{AD}^2P_e(\mathbf{w} - \alpha\mathbf{v}) \\
&= (\mathbf{w} - \alpha\mathbf{v})^TP_e(\mathbf{w} - \alpha\mathbf{v}) \\
&= (\mathbf{w}^TP_e\mathbf{w}) + \alpha^2(\mathbf{v}^TP_e\mathbf{v}) - 2\alpha(\mathbf{v}^TP_e\mathbf{w}).
\end{aligned}
$$

Substituting equations (5.1.19) and (5.1.20) into equation (5.1.13) for $(d\theta/dt)^2$ and simplifying yields

$$
(5.1.21) \quad \left(\frac{d\theta}{dt}\right)^2 = \frac{(\mathbf{v}^TP_e\mathbf{v})(\mathbf{w}^TP_e\mathbf{w}) - (\mathbf{v}^TP_e\mathbf{w})^2}{(\mathbf{v}^TP_e\mathbf{v})^2}.
$$

Now the scalar curvature τ_g for the general case is given by

$$\tau_g = \frac{d\theta}{ds} = \frac{\frac{d\theta}{dt}}{\frac{ds}{dt}}.$$

Substituting for $d\theta/dt$ and ds/dt from equations (5.1.21) and (5.1.12) we get

(5.1.22)
$$\tau_g = \sqrt{\frac{(\mathbf{v}^T P_e \mathbf{v})(\mathbf{w}^T P_e \mathbf{w}) - (\mathbf{v}^T P_e \mathbf{w})^2}{(\mathbf{v}^T P_e \mathbf{v})^3}}.$$

Now we are ready to compare the scalar curvatures of the unconstrained and general cases of the problem. Since both curves pass through the same point $\mathbf{x}$ and have the same tangent $d\mathbf{x}/dt$, the vector $\mathbf{v}$ defined above is the same; i.e.,

$$\mathbf{v}_g = \mathbf{v}_s.$$

Hence,

$$V_g = V_s = V,$$

and

$$\mathbf{w}_g = V_g^2 \mathbf{e} = V_s^2 \mathbf{e} = \mathbf{w}_s.$$

Using the formula (5.1.22) for the scalar curvature in the two cases, we get $\tau_g = \tau_s$ as required. $\square$

We can bound the scalar curvature in the general case further, as in the following theorem.

THEOREM 4. *For the general case curve, $d\theta/dt$ at the point $\mathbf{x} = (\mathbf{x}^1, \ldots, \mathbf{x}^n)$ is bounded by*

(5.1.23)
$$\frac{d\theta}{dt} \le 2 \max_i \left| \frac{d}{dt} (\ln \mathbf{x}^i) \right|.$$

PROOF. We have

$$\frac{d\theta}{dt} = \sqrt{\frac{(\mathbf{w}^T P_e \mathbf{w})(\mathbf{v}^T P_e \mathbf{v}) - (\mathbf{w}^T P_e \mathbf{v})^2}{(\mathbf{v}^T P_e \mathbf{v})^2}}$$

$$\le \sqrt{\frac{\mathbf{w}^T P_e \mathbf{w}}{(\mathbf{v}^T P_e \mathbf{v})}} = \sqrt{\frac{\mathbf{e}^T V^2 P_e V^2 \mathbf{e}}{(\mathbf{e}^T V P_e V \mathbf{e})}}.$$

Hence, using Proposition 4.1,

$$\tau_g \le \sqrt{\frac{\sum_i \mathbf{v}_i^4 - \frac{1}{n}(\sum \mathbf{v}_i^2)^2}{\left[\sum \mathbf{v}_i^2 - \frac{1}{n}(\sum \mathbf{v}_i)^2\right]}} \le 2 \max_i |\mathbf{v}_i|.$$

Therefore,

$$\tau_g \le 2 \max_i \left| \frac{1}{\mathbf{x}^i} \frac{d\mathbf{x}^i}{dt} \right| = 2 \max_i \left| \frac{d}{dt} (\ln \mathbf{x}^i) \right|.   \square$$

While Theorems 3 and 4 provide local bounds on scalar curvature, proving a "global" bound on the integral of scalar curvature in the general case continues to be a challenging open problem in the analysis of interior-point methods for linear programming.

Acknowledgments

We are indebted to Jeff Lagarias and an anonymous referee for several suggestions that improved the exposition of this paper.

REFERENCES

1. D. A. Bayer and J. C. Lagarias, *The nonlinear geometry of linear programming.* I, *Affine and projective scaling trajectories*, Trans. Amer. Math. Soc. **314** (1989), 499–526.
2. ___, *The nonlinear geometry of linear programming.* II, *Legendre transform coordinates and central trajectories*, Trans. Amer. Math. Soc. **314** (1989), 527–581.
3. Boothby, *Introduction to differentiable manifolds and Riemannian geometry*, Academic Press, New York, 1975.
4. C. Chevalley, *Theory of Lie groups*, Princeton Univ. Press, Princeton, NJ, 1946.
5. L. Eisenhart, *Riemannian geometry*, Princeton Univ. Press, Princeton, NJ, 1925.
6. N. Karmarkar, *A new polynomial-time algorithm for linear programming*, Combinatorica **4** (1984), 373–395.
7. J. C. Lagarias, *The nonlinear geometry of linear programming.* III, *Projective Legendre transform coordinates and Hilbert geometry*, Trans. Amer. Math. Soc. **320** (1990), 193–225.
8. T. Levi-Civita, *The absolute differential calculus (calculus of tensors)*, Dover Publications, 1926.
9. N. Megiddo, *Pathways to the optimal set in linear programming*, Progress in Mathematical Programming, Interior-Point and Related Methods (N. Megiddo, ed.), Springer-Verlag, New York, 1989, pp. 131–158.
10. M. M. Postnikov, *The variational theory of geodesics*, Dover Publications, New York, 1967.
11. Gy. Sonnevand, *An "analytic center" for polyhedrons and new classes of global algorithms for linear (smooth, convex) programming*, Proc. 12th IFIP Conf. System Modelling, Budapest, 1985, Lectures Notes in Computer Science, 1986.
12. N. Steenrod, *The topology of fibre bundles*, Princeton Univ. Press, Princeton, NJ, 1950.
13. P. M. Vaidya, *An algorithm for linear programming which requires* $O(((m + n)n^2 + (n + n)^{1.5}n)L)$ *arithmetic operations*, Proceedings of the 19th Annual ACM Symposium on Theory of Computing, May 1987, pp. 29–38.

AT&T BELL LABORATORIES, MURRAY HILL, NEW JERSEY 07974

Contemporary Mathematics
Volume **114**, 1990

Steepest Descent, Linear Programming,
and Hamiltonian Flows

A. M. BLOCH

ABSTRACT. In this paper we analyze the structure of the gradient-like flows associated with several problems of numerical importance. In particular, we review the flows associated with the QR algorithm and the total least squares problem, the A-trajectories associated with the affine scaling algorithm that is a variant of Karmarkar's algorithm for linear programming, and certain flows on the orthogonal group associated with geometrical matching problems and linear programming. We show that all these flows are equivalent, under a nonlinear transformation, to flows of integrable Hamiltonian systems or are closely related to them.

1. Introduction

Recently there has been some interest in the continuous flows associated with numerical algorithms. (See, for example, [**17**].) A by now classic case is the relationship between the Toda lattice flow and the QR algorithm.

We have observed a common thread among the flows associated with this algorithm and with several other algorithms that are either of steepest descent type or closely related—namely that these flows are Hamiltonian. Further, these flows are integrable. Moreover, several of the flows are associated with linear programming. In this paper we review a number of flows of this type and we also introduce some new results.

We begin by discussing the QR algorithm for diagonalizing symmetric matrices, which can be shown to be equivalent to the time-1 mapping of the Toda lattice flow. (See [**17**], [**19**], [**27**], [**28**].) The Toda flow is the flow of an integrable Hamiltonian system, but as can be seen in the original work of Moser [**25**], the Toda flow (in the nonperiodic case) is, in fact, essentially a gradient flow.

1980 *Mathematics Subject Classification* (1985 *Revision*). Primary 34A05, 49D35, 70H05.
This paper was submitted to the editors for publication in October 1988.
Supported in part by grants from the National Science Foundation and the Air Force Office of Scientific Research, by the U. S. Army Research Office through the Mathematical Sciences Institute of Cornell University, and by a Seed grant from Ohio State University.

In Section 4 we point out that the gradient flow associated with another important numerical problem, the total least squares (TLS) problem of fitting linear to data points in n-space, is just the Toda lattice flow. (In the complex case the TLS problem has another interesting Hamiltonian flow of spinning top type associated with it. (See [7], [8].) We also note that the solution of the TLS problem can be shown to be equivalent to a linear programming problem.

In Section 5 we discuss recent work of D. Bayer and J. Lagarias. In [5] and [6] they showed that the so-called A-trajectories associated with the affine scaling algorithm, a variant of Karmarkar's algorithm for solving linear programming problems, are the integral curves of integrable Hamiltonian systems. These flows are also closely related to gradient flows.

Finally we discuss some gradient flows recently analyzed by R. Brockett. (See [12], [13].) These flows also arise from problems of least squares type—namely "matching" in a least squares sense two n-tuples of points in space by rotating one of the n-tuples. In the simplest case the gradient flow gives rise to a Riccati equation that may be integrated by a "Hamiltonian pair." A more complex example gives rise to an equation in the "Lax pair" form common to most integrable systems. Recently we have shown that this equation gives the Toda lattice flow on tridiagonal matrices as a special case. Brockett has shown that, like the Toda lattice flow, this flow can diagonalize matrices and, moreover, is capable of "solving" linear programming problems if the vertices are given. We also show how a related equation solves the total least squares problem.

Recent joint work of the author with R. Brockett, H. Flaschka, and T. Ratiu on the Toda lattice, which utilizes some of the ideas discussed here, may be found in [9], [10], and [11].

In the following pages we begin by reviewing the theory of Hamiltonian systems and complete integrability, and then we examine the various flows discussed above and some of the mathematical features they have in common.

2. Hamiltonian systems

We begin with a brief review of the theory of Hamiltonian systems. (For more details see, for example, [1] or [3].)

A Hamiltonian system on a $2n$-dimensional manifold M equipped with a nondegenerate bilinear 2-form ω is defined to be the triple (H, M, ω). Here H is a function on M, and M is said to be a symplectic manifold with symplectic form ω. The Hamiltonian vector field X_H corresponding to H is defined by $DH \cdot Y = \omega(X_H, Y)$.

Locally, by the Darboux theorem ([1]), canonical coordinates

$$(q_1, \ldots, q_n, p_1, \ldots, p_n)$$

can be found such that $\omega = \sum_{i=1}^{n} dq_i \wedge dp_q$, and the Hamiltonian equations

$\dot{x} = x_H(x)$ take their usual classical form

$$\dot{q}_i = \frac{\partial H}{\partial p_i},$$

$$\dot{p}_i = -\frac{\partial H}{\partial q_i} \qquad i = 1, \ldots, n.$$

The Poisson bracket of two functions on M is defined by $\{F, G\} = \omega(X_F, X_G)$. Again, in local coordinates the bracket reduces to the classical

$$\{F, G\} = \sum_{i=1}^{n} \frac{\partial F}{\partial q_i} \frac{\partial G}{\partial p_i} - \frac{\partial G}{\partial q_i} \frac{\partial F}{\partial p_i}.$$

On $\mathbb{R}^{2n}$ these local coordinate expressions are valid globally.

In this paper we are interested in systems of Hamiltonian equations that can be solved explicitly. We call such systems completely integrable.

Liouville's theorem [3] tells us that a Hamiltonian system with Hamiltonian H_1 is completely integrable if there exist n independent (almost everywhere) functions $H_1, \ldots, H_n$ that Poisson commute with H_1. (H_i and H_j are said to Poisson commute with each other or to be in involution if $\{H_i, H_j\} = 0$.) Such systems can be solved explicitly by means of elementary integrations and inversion of functions. Further, consider a level set of the functions H_i, $M_h = \{x : H_i(x) = h_i, \; i = 1, \ldots, n\}$, on which the H_i are independent. Then M_h is a smooth manifold invariant under the flow of H_1, and if it is connected it is diffeomorphic to a cylinder $\mathbb{R}^k \times T^{n-k}$. In the compact case it is diffeomorphic to a torus, but we shall be concerned here mainly with the noncompact case. There is a beautiful theory of these systems that has undergone major developments in recent years. One characteristic of such systems is that they can very often be described by a so-called Lax pair, i.e., written

$$\dot{L}(t) = [B(t), L(t)] = BL - LB$$

where L and B are $n \times n$ matrices. (See, for example, [26] and [27] and the references in [10].) Given such a pair, there is usually a canonical way of arriving at the integrals in involution as well as carrying out the explicit integration. In particular, the Toda lattice has such a pair associated with it, as we discuss in the following pages.

3. The QR algorithm and the Toda lattice flow

Let us begin by considering the Toda lattice flow.

The original Toda lattice Hamiltonian on $\mathbb{R}^{2n}$ with coordinates

$$(x_1, \ldots, x_n, y_1, \ldots, y_n)$$

is given by (see [25])

(3.1)
$$H = \frac{1}{2} \sum_{k=1}^{n} y_k^2 + \sum_{k=1}^{n-1} e^{(x_k - x_{k+1})}.$$

The resulting system of equations with nonperiodic boundary conditions may be written

(3.2)
$$\dot{x}_k = H_{y_k} = y_k, \qquad k = 1, \ldots, n$$
$$\dot{y}_k = e^{x_{k-1} - x_k} - e^{x_k - x_{k+1}}, \qquad k = 1, \ldots, n$$

if we set $e^{x_0 - x_1} = 0$ and $e^{x_n - x_{n+1}} = 0$.

The energy surface for this system is noncompact, and the "particles" described by this system behave asymptotically like free particles.

Moser in [25] showed that this Hamiltonian system can be mapped by a one-to-one mapping to the following gradient system:

(3.3)
$$\frac{d\lambda_k}{dt} = 0,$$
$$\frac{dr_k}{dt} = -\frac{\partial V}{\partial r_k}, \qquad k = 1, \ldots, n$$

where

(3.4)
$$V = \sum_{k=1}^{n} \frac{\lambda_k r_k^2}{2 \sum_{I=1}^{n} r_k^2}$$

and $\lambda_1 < \lambda_2 \cdots < \lambda_n$, $\sum_{k=1}^{n} r_k^2 = 1$, $r_k > 0$.

These equations are explicitly solvable.

Flaschka [19] showed that the equations could also be recast in the Lax pair form $\dot{A} = [B, A]$. One makes the transformation

$$a_k = \frac{1}{2} e^{(x_k - x_{k+1})/2} \qquad b_k = -\frac{1}{2} y_k.$$

Then the resulting equations can be expressed as $\dot{A} = [B, A]$ where

(3.5)
$$A = \begin{bmatrix} b_1 a_1 & & & \\ a_1 b_2 & \ddots & & 0 \\ & & b_{n-1} a_{n-1} & \\ 0 & & a_{n-1} b_n & \end{bmatrix} \qquad B = \begin{bmatrix} 0 & a_1 & & \\ -a_1 & \ddots & & \\ & & & a_{n-1} \\ & & -a_{n-1} & 0 \end{bmatrix}.$$

One can check that this yields an isospectral flow; i.e., the (distinct) eigenvalues λ_k of A are conserved. (Set $dU/dt = BU$, $B(0) = I$. Then $L(t) = U^{-1}L(0)U$.) These λ_k are the λ_k of the gradient flow described above and, in fact, give us the n integrals in involution that we need to show that the system is integrable. Another basis of integrals is $\operatorname{Tr} A^k$, $k = 1, \ldots, n$, where Tr denotes the trace.

The Lax pair equation can also be written $\dot{A} = [\pi_s A, A]$ where $\pi_s A$ is the skew-symmetric part of A in the unique decomposition of A, as $A = \pi_s A + \pi_L A$ where $\pi_L A$ is lower triangular. One can show in fact that the Toda flow is Hamiltonian on the dual of an adjoint orbit of lower triangular matrices that may be identified with an orbit of symmetric matrices. The corresponding Hamiltonian is just $\frac{1}{2} \operatorname{Tr} A^2$.

How does this relate to the QR algorithm for diagonalizing tridiagonal or symmetric matrices? A nice account of this may be found in Symes [**28**]. (See also [**16**], [**18**], and [**24**].)

First we recall that for any real matrix A, A invertible, there exists a unique decomposition $A = QL$ where Q is orthogonal and L is lower triangular with nonnegative diagonal entries.

Let A be our Toda matrix defined above and let $Q(t)L(t) = \exp(tA(0))$. Then (see [**28**])

$$(3.6) \qquad\qquad A(t) = Q(t)^T A(0) Q(t).$$

The QR algorithm works as follows. Let $Y = Y^{(0)}$ be an invertible $n \times n$ matrix. Let $Y^{(0)} = Q^{(0)} L^{(0)}$ as defined above and let $Y^{(1)} = L^{(0)} Q^{(0)}$. Then $Y^{(1)} = Q^{(0)T} Y^{(0)} Q(0)$ and thus $Y^{(1)}$ has the same spectrum as $Y^{(0)}$. Similarly, define $Y^{(k+1)}$ from $Y^{(k)}$ by setting $Y^{(k)} = Q^{(k)} L^{(k)}$ and $Y^{(k+1)} = L^{(k)} Q^{(k)}$. Then one can show

THEOREM. *If Y is positive-definite symmetric with simple spectrum $\{\mu_1 > \mu_2, \ldots, \mu_n > 0\}$, the QL sequence above tends, as $k \to \infty$, to $Y^{(\infty)} = \operatorname{diag}(\mu_1 \cdots \mu_n)$.*

Let $A(t)$ be our Jacobi matrix under the Toda flow. Using (3.6) we can see that if $\exp A(k) = Q^{(k)} L^{(k)}$, $k = 0, 1, \ldots$, then $\exp A(k+1) = L^{(k)} Q^{(k)}$— i.e., the QR algorithm is the time-1 map of the Toda lattice! In fact, this works for any symmetric matrix A.

4. The total least squares problem, Toda flows, and linear programming

The total least squares (TLS) problem is the problem of fitting d-planes to a set of p points in real or complex n-space. The idea is to minimize the total perpendicular distance of the p points onto the d-plane. From the point of view of error analysis, this corresponds to the maximum likelihood estimate (MLE) when all the components of the points are measured subject to identically distributed normal errors. (This is in contrast to ordinary least squares where all components except one are supposed to be known exactly.)

We can write down the TLS function as follows. Let e_j, $j = 1, \ldots, n$ be an orthonormal basis for F^n ($\mathbb{R}^n$ or $\mathbb{C}^n$) and let $x_i = \sum_j \lambda_{ij} e_j$, $i = 1, \ldots, p$, be p points. Each λ_{ij} is measured with additive observation error ϵ_{ij}, $\epsilon_{ij} \sim N(0, 1)$ independent. Let $P = I - Q$ where Q is the orthogonal projection onto a d-plane.

Then we can show (see, e.g., [**7**] or [**15**])

LEMMA. *The total perpendicular distance of the p points onto the d-plane is $H(P) = \operatorname{Tr} CP = \operatorname{Tr} C(I - Q)$, where $C = \Lambda^* \Lambda$ and Λ is the data matrix (λ_{ij}).*

$H(P)$ is thus the restriction of a linear functional to the Grassmann manifold of d-planes in n-space, $Q = I - P$ representing a point in the Grassmannian.

The TLS problem can be seen to be equivalent to a linear programming problem as follows.

Choose a basis in which C is diagonal. The $H(P) = \sum_{i=1}^{n} c_i p_i$ where $C = \text{diag}(c_1, \ldots, c_n)$ and p_i are the diagonal entries of P. Then (see Byrnes and Willems [15]) the set of all p_i is precisely equal to the convex set $C(d, n) = \{(p_1, \ldots, p_n) \in \mathbb{R}^n : 0 \le p_i \le 1, \sum_{i=1}^{n} p_i = n - d\}$. (This follows from the Schur-Horn theorem [2], [21]. In the complex case this also follows from the convexity of the moment map associated with the natural torus action on the Grassmannian. (See [2], [20], [7], or [15] for Atiyah's and Guillemin and Sternberg's work on the moment map.) Hence we have a linear programming problem.

(The Schur-Horn theorem is the following: Let H be a Hermitian matrix and let $\boldsymbol{\lambda} = (\lambda_1, \ldots, \lambda_n)$, $\lambda_1 \ge \lambda_2 \ge \cdots \ge \lambda_n$, and $\mathbf{a} = (a_1, \ldots, a_n)$ denote the eigenvalues and diagonal entries of H, viewed as vectors in $\mathbb{R}^n$. Let Σ_n denote the permutation group acting on $\mathbb{R}^n$. Then $\mathbf{a} \subset \hat{\Sigma}_n \boldsymbol{\lambda}$, where $\hat{}$ denotes convex hull and $\hat{\Sigma}_n \boldsymbol{\lambda}$ is thus the convex hull of all permutations of $\boldsymbol{\lambda}$. Further, given any $\boldsymbol{\lambda}$ and $\mathbf{a} \subset \hat{\Sigma}_n \boldsymbol{\lambda}$, there exists a Hermitian matrix H with eigenvalues $\boldsymbol{\lambda}$ and diagonal entries $\mathbf{a}$. We recall this result again in Section 6.)

Let us restrict ourselves now (although the results generalize) to the real case and the fitting of lines.

We can show

THEOREM. *Consider the TLS function* $\text{Tr}\,CP$ *in the real, line-fitting case and suppose that* C *has distinct eigenvalues. Then its gradient flow converges to one of the vertices of* $C(d, n)$, *and the flow is equivalent to the Toda lattice flow.*

PROOF. Consider the function $\text{Tr}\,CQ$. Here Q is a rank 1 projection matrix that one may write as $r \otimes r = rr^T$ where $r = [r_1, \ldots, r_n]$ is a vector and $\Sigma r_i^2 = 1$. Then one easily checks that minimizing $\text{Tr}\,CQ$ is equivalent to minimizing $V = \sum_{I=1}^{n} c_i r_i^2 / 2\Sigma r_i^2$. Thus we may consider the gradient flow of this system, which will converge generically in Q-space to one of the vertices of the simplex $C(1, n)$. But V is just equal to the V in (3.4), which gives rise to the gradient flow formulation of the Toda lattice equations described in Section 3.

There is an alternative method of obtaining this result. (See Section 6.)

REMARKS. (1) In the case where the eigenvalue corresponding to the minimal critical value of $\text{Tr}\,CP$ is a multiple eigenvalue, the gradient flow will converge to the edges spanned by the corresponding critical vertices. The gradient flow in this case is still equivalent to a Toda flow (see Chu [16]), where in this case some of the b_i in (3.5) are zero.

(2) In the complex case $\operatorname{Tr} CP$ may be regarded as a Hamiltonian on the Grassmannian viewed as an adjoint orbit of the unitary group $U(n)$. This yields the (integrable) Hamiltonian system $\dot{P} = [P, C] = PC - CP$, which has interesting geometric and statistical properties. (See [7] or [8], for example.) This also gives an approach to solving the TLS problem, which we discuss briefly in Section 6.

5. Karmarkar's algorithm, A-trajectories, and Hamiltonian systems

Karmarkar's algorithm, an interior point method for linear programming, has received much attention in the literature over the past few years and has spawned many variants. Recently Bayer and Lagarias [5], [6] have shown that the A-trajectories of the affine scaling field associated with the affine scaling algorithm (see [4] or [31]) can be viewed as trajectories of integrable Hamiltonian systems. We briefly review these results here.

In [6], a linear program in $\mathbb{R}^n$ in inequality form is considered:

$$\min \langle c, x \rangle - c_0$$
$$\langle a_j, x \rangle \geq b_j, \quad 1 \leq j \leq m$$

where the given polytope P_H is full dimensional and the vectors $[a_1, \ldots, a_m]$ span $\mathbb{R}^n$. One also supposes that the polytope of feasible solutions is bounded with nonempty interior. (The assumption of boundedness may be relaxed, however.)

There are several ways of describing the A-trajectories (see [6]), but one may show that they are given by

$$\{x \in \operatorname{int}(P_H) : \phi_H(x) = \phi_H(x_0) + tc, \text{ some } t \in \mathbb{R}\}$$

where x_0 is the starting point and $\phi_H(x)$ are the "Legendre transform" coordinates given by

$$\phi_H(x) = v_{f_H}(x)$$

where $f_H(x)$ is the logarithmic barrier function

$$f_H(x) = -\sum_{j=1}^{m} \log(\langle a_j, x \rangle - b_j).$$

Intuitively, the A trajectories correspond to affine transformations of the steepest descent direction of the objective function subject to the constraints. Now one can find two Hamiltonian dynamical systems, the first of which yields all A-trajectories for a fixed objective function and the second of which yields A-trajectories for all objective functions.

In the first instance, Bayer and Lagarias prove

THEOREM. *The Hamiltonian*

$$H_1(p, q) = \langle p, \phi_H^{-1}(p) \rangle - \langle c, q \rangle - \sum_{j=1}^{m} \log(\langle a_j, \phi_H^{-1}(p) \rangle - b_j)$$

yields all the A-trajectories for the given objective function, and the resulting Hamiltonian system is completely integrable on the phase space $\mathbb{R}^{2n}$.

Direct calculation shows $\dot{q} = \phi_H^{-1}(c + c_0)$ as required. A complete set of commuting integrals is given by

$$F_k(p, q) = \langle c^{(k)}, p \rangle, \qquad 1 \le k \le n - 1$$
$$F_n(p, q) = H(p, q)$$

where the $c^{(k)}$ are a basis for the orthogonal complement of c in $\mathbb{R}^n$.

The second Hamiltonian system is defined on $\mathbb{R}^{4n}$. One has

THEOREM. *The Hamiltonian*

$$H_2 = \frac{1}{2}\langle \overline{p}, \overline{p} - 2q \rangle + \langle p, \phi_H^{-1}(p) \rangle + \sum_{j=1}^{m} \log(\langle a_j, \phi_H^{-1}(p) \rangle - b_j)$$

has $\dot{q}$-trajectories that realize A-trajectories for all possible objective functions $\langle c, x \rangle$ with the given constraints. Further, the system is completely integrable on the invariant open set $\mathbb{R}^{4n} - \{(p, \overline{p}, q, \overline{q}) : \overline{p}_1 = 0\}$.

In this case the complete set of integrals in involution is given by

$$F_k = \overline{p}_k, \qquad 1 \le k \le n$$
$$F_{n+k} = p_1\overline{p}_{k+1} - p_{k+1}\overline{p}_1, \qquad 1 \le k \le n - 1$$
$$F_{2n} = H_2.$$

Thus the A-trajectories of the affine scaling algorithm can be exhibited as the flows of a Hamiltonian system. Further, this system, like the nonperiodic Toda system, has noncompact level sets. As we saw in the previous section, there is a connection between Toda flows themselves and linear programming, and we shall see a further aspect of this in Section 6.

6. Gradient flows on the orthogonal group, Toda flows, and linear programming

Gradient flows on the orthogonal group arise in the least squares matching problem discussed in the introduction. Recent analysis of this problem by R. Brockett ([12] and [13]) has yielded some very interesting results.

The most general situation requires examining the critical points of the function $\text{Tr}(Q\theta N\theta^T - 2M\theta^T)$ where we shall take $\theta \in SO(n)$, the special orthogonal group, and Q, N, and M to be given $n \times n$ symmetric (real) matrices. (One can also look at arbitrary compact Lie groups.) To obtain the solution to the problem, i.e., to find the minimum of the given function, we examine the gradient with respect to the natural Riemannian metric on $SO(n)$ given by the trace inner product $\langle \Omega_1, \Omega_2 \rangle = \text{tr}\,\Omega_1\Omega_2$. Brockett examines in detail the two special cases $N = 0$ and $M = 0$.

First consider the case $N = 0$. We want the gradient flow of $\operatorname{Tr} M\theta^T$. In [12] it is shown that the gradient flow satisfies the equation

$$(6.1) \qquad \dot{\theta} = (\theta M \theta - M).$$

This quadratic equation may be solved explicitly by setting $\theta = XY^{-1}$ where X and Y satisfy the linear matrix equation

$$(6.2) \qquad \begin{bmatrix} \dot{X} \\ \dot{Y} \end{bmatrix} = A \begin{bmatrix} X \\ Y \end{bmatrix}$$

where A is the Hamiltonian matrix

$$\begin{bmatrix} 0 & M \\ M & 0 \end{bmatrix}.$$

Now let us consider the gradient flow corresponding to the function $\operatorname{Tr}(Q\theta N\theta^T)$. In [12] this is shown to satisfy the equation

$$(6.3) \qquad \dot{\theta} = \theta N \theta^T Q \theta - Q\theta N,$$

which is cubic in θ.

In [13] Brockett observes that by making the change of variables $H = \theta^T Q \theta$ one can rewrite the equation as an elegant equation on the space of symmetric matrices:

$$(6.4) \qquad \dot{H} = [H, [N, H]].$$

This equation is quadratic and in Lax pair form. Further, it turns out to have some remarkable properties.

In [9] Brockett proves

THEOREM. *Let N be a real diagonal matrix with distinct eigenvalues. Then if $H(0)$ is symmetric and*

$$(6.5) \qquad \dot{H} = [H, [H, N]],$$

$\lim_{t \to \infty} H(t)$ *exists and is a diagonal matrix.*

Since the flow is isospectral (since $H = \theta^T Q\theta$), this gives a method of diagonalizing H. The proof of the theorem rests on the observation that $\operatorname{Tr}(HN)$ is increasing along the flow.

The above property is reminiscent of the properties of Toda flows and, in fact, recently we showed

THEOREM. *Let N be the matrix $N = \operatorname{diag}(1, 2, 3, \ldots, n)$ and let H be tridiagonal. Then the equation*

$$\dot{H} = [H, [H, N]]$$

gives the Toda flow on tridiagonal matrices.

PROOF. Let

$$H = \begin{bmatrix} b_1 a_1 & & & 0 \\ a_1 b_2 & \ddots & & \\ 0 & & b_{n-1} a_{n-1} \\ & & a_{n-1} b_n \end{bmatrix}.$$

Then if N is as in the theorem, $[H, N]$ is precisely the matrix B of (3.5).

Thus we produce Flaschka's form of the Toda lattice equations, which are Hamiltonian with Hamiltonian $\mathrm{Tr}H^2/2$ on a (nongeneric) orbit of symmetric matrices viewed as the dual of an orbit of lower triangular matrices.

Brockett has also observed that the equation (6.5) can solve a linear programming problem, given a knowledge of the vertices.

Let $X = \{x | Ax \leq b\}$ be a convex bounded polytope in $\mathbb{R}^n$ with nonempty interior, and suppose we wish to maximize $\langle c, x \rangle$ over X. Then Brockett's result is as follows:

THEOREM. *Let T be the matrix $[a_1 | a_2 \cdot | a_k]$, where the a_i are the vertices of X in $\mathbb{R}^n$. Let $Q = \mathrm{diag}(1, 0, 0, \ldots 0)$ and $N = \mathrm{diag}(\mu_1, \mu_2, \ldots, \mu_k)$ with $\mu = cT$. Then for almost all $\theta \in SO(n)$*

$$\dot{H} = [H, [H, N]], \quad H(0) = \theta^T Q \theta$$

converges to $H = \mathrm{diag}(d_1, d_2, \ldots, d_k)$ with the optimal x being given by Td.

The proof rests on the following observations. Firstly, T maps the "standard simplex" in $\mathbb{R}^K$ (i.e., $S = \{x | x \in \mathbb{R}^k; x \geq 0; \Sigma x_i = 1\}$) to the given polytope. Secondly, since Q has eigenvalues $(1, 0, 0, \ldots)$, as t evolves, $H = \theta^T Q \theta$ fills out the standard simplex. Thirdly, as observed earlier, $\mathrm{Tr}\,HN$ is monotone increasing. Of course, this is not an efficient algorithm for linear programming, but it is very suggestive.

Note that we can also deduce (from the Schur-Horn discussed in Section 4) that the Toda flow may be mapped into a convex polytope, with the equilibria of the flow being mapped to the vertices of the polytope. More on this aspect of the Toda flows may be found in [18], [30], [32], [10], and [11].

Finally, we point out a remarkable connection of the above analysis with the Hamiltonian approach to the complex total least squares problem discussed in Section 4.

We can show in fact

THEOREM. *The gradient flow of the TLS function $\mathrm{Tr}\,CP$ with respect to the Kaehler metric on the complex Grassmannian viewed as an adjoint orbit of the unitary group is given by*

$$(6.6) \qquad\qquad \dot{P} = [P, [P, C]].$$

Note here that P and C should be taken to be i (Hermitian matrix), so that they lie in the Lie algebra $u(n)$ of $U(n)$. $\mathrm{Tr}\,CP$ is still real however! We remark that the Kaehler structure of the manifold can be used to derive this gradient flow from the Hamiltonian flow $\dot{P} = [P, C]$ that we mentioned in Remark 2 at the end of Section 4. On a Kaehler manifold the gradient flow and Hamiltonian corresponding to a given function are related by letting J, the complex structure, act on the Hamiltonian vector field. We can show that J here is just $[P, \cdot]$. Thus J acting on $[P, C]$ gives $[P, [P, C]]$! For further details see [9] and [11].

Thus the equations $\dot{P} = [P, [P, C]]$ solve the linear programming problem associated with the TLS problem. We recall from Section 4 that the fact that solving the TLS problem reduces to linear programming follows also from a result in Hamiltonian systems, the convexity of the moment map associated with a torus action.

Note further that if we take p to be rank 1, $p = iz \otimes \bar{z}$, $z = (z_1, \ldots, z_n)$, the equations for $|z_i|$ are precisely Moser's equations (3.3)! (For further details see [9] and [11].)

In joint work of the author with R. Brockett, H. Flaschka, and T. Ratiu, some of the ideas above have been used to give new results on the gradient-flow and convexity properties of the generalized (Lie-algebraic) Toda lattice equations. (See [10] and [11].) There is also an infinite-dimensional (partial differential equations) version of some of these results. (See [14].)

In conclusion, we remark that this paper has explored a number of different connections between gradient flows, integrable Hamiltonian systems, and linear programming. There appears to be a rich interplay between these subjects, which we intend to explore further. As far as the Hamiltonian structure is concerned, a key feature is the noncompactness of level sets and the resultant "scattering" behavior of the flows. Numerically, a common feature is that the algorithms explore the interior of a relevant polytope in moving towards a solution.

Acknowledgments

I would like to thank R. Brockett, C. Byrnes, H. Flaschka, J. Lagarias, J. Marsden, R. Montgomery, T. Ratiu, and J. Renegar for helpful conversations.

I would also like to thank J. Lagarias for inviting me to the meeting where this paper was presented and for useful discussions there on gradient flows and integrable systems.

REFERENCES

1. R. A. Abraham and J. E. Marsden, *Foundations of mechanics*, Benjamin Cummings, Reading, Massachusetts, 1978.
2. M. F. Atiyah, *Convexity and commuting Hamiltonians*, Bull. London Math Soc. **14** (1982), 1–15.
3. V. I. Arnold, *Mathematical methods of classical mechanics*, Springer-Verlag, Berlin and New York, 1978.
4. E. R. Barnes, *A variation on Karmarkar's algorithm for solving linear programming problems*, Math. Programming **36** (1986), 174–182.
5. D. A. Bayer and J. C. Lagarias, *The nonlinear geometry of linear programming*, I. *Affine and projective scaling trajectories*, Trans. Amer. Math. Soc. **314** (1989), 499–526.
6. D. A. Bayer and J. C. Lagarias, *The nonlinear geometry of linear programming*, II. *Legendre transform coordinates and central trajectories*, Trans. Amer. Math. Soc. **314** (1989), 527–581.
7. A. M. Bloch, *Estimation, principal components and Hamiltonian systems*, Systems Control Lett. **6** (1985), 103–108.

8. A. M. Bloch, *An infinite-dimensional Hamiltonian system on projective Hilbert space*, Trans. Amer. Math. Soc. **302** (1987), 787–796.

9. A. M. Bloch, *The Kahler structure of the total least squares problem, Brockett's steepest descent equations and constrained flows*, Proc. of the 9th MTNS Symposium (to appear).

10. A. M. Bloch, R. W. Brockett, and T. Ratiu, *A new formulation of the generalized Toda lattice equations and their fixed point analysis via the moment map*, Bull. Amer. Math. Soc. (N.S.) **23** (2) (1990), 477–485.

11. A. M. Bloch, H. Flaschka, and T. Ratiu, *A convexity theorem for isospectal manifolds of Jacobi matrices in a compact Lie algebra*, Duke Math J. (to appear).

12. R. W. Brockett, *Least squares matching problems*, Linear Algebra and Its Applications **122/123/124** (1989), 761–777.

13. R. W. Brockett, *Dynamical systems that sort lists and solve linear programming problems*, Proc. of the 27th IEEE Conf. on Decision and Control, IEEE, New Jersey, 1988, 799–803.

14. R. W. Brockett and A. M. Bloch, *Sorting with the dispersionless limit of the Toda lattice*, Proc. of the CRM Workshop on Hamiltonian Systems, Transformation Groups, and Spectral Transform Methods (J. Harnad and J. E. Marsden, eds.), Publications CRM, Montreal, 103–112.

15. C. I. Byrnes and J. C. Willems, *Least squares estimation, linear programming, and momentum*, unpublished manuscript, 1984.

16. M. T. Chu, *On the global convergence of the Toda lattice for real normal matrices and its applications to the eigenvalue problem*, SIAM J. Math. Anal. **15** (1984).

17. M. T. Chu, *On the continuous realization of iterative processes*, SIAM Review, 30, No. **3** (1988), 375–389.

18. P. Deift, T. Nanda, and C. Tomei, *Differential equations for the symmetric eigenvalue problem*, SIAM J. Numer. Anal. **20** (1983), 1–22.

19. H. Flaschka, *The Toda lattice*, Phys. Rev. B(3) **9** (1976), 1924–1925.

20. V. Guillemin and S. Sternberg, *Convexity properties of the moment mapping* I, Invent. Math **67** (1982), 491–513.

21. A. Horn, *Doubly stochastic matrices and the diagonal of a rotation matrix*, Amer. J. Math **76** (1956), 620–630.

22. N. Karmarkar, *A new polynomial time algorithm for linear programming*, Combinatorica **4** (1984), 373–395.

23. B. Kostant, *The solution to a generalized Toda lattice and representation theory*, Adv. Math. **34** (1979), 195–338.

24. J. Lagarias, *Monotonicity properties of the Toda flow, QR flow and subspace iteration*, SIAM J. Matrix Anal. Appl. (to appear).

25. J. Moser, *Finitely many mass points on the line under the influence of an exponential potential. An integrable system*, Dynamic Systems Theory and Applications (J. Moser, ed.), Springer-Verlag, Berlin, New York, 1975, pp. 467–497.

26. T. Ratiu, *Involution theorems*, Lecture Notes in Math., vol. 775, 1980, pp. 221–257.

27. W. W. Symes, *Hamiltonian group actions and integrable sytems*, Physica **1D** (1980), 339–374.

28. W. W. Symes, *The QR algorithm and scattering for the finite nonperiodic Toda lattice*, Physica **2D** (1982), 275–280.

29. W. W. Symes, *Systems of Toda type, inverse spectral problems and representation theory*, Invent. Math. **59** (1982), 13–51.

30. C. Tomei, *The topology of isospectral manifolds of tridiagonal matrices*, Duke Math. J. **51** (1984), 981–996.

31. R. J. Vanderbei, M. J. Meketon, and B. A. Freedman, *A modification of Karmarkar's linear programming alogrithm*, Algorithmica **1** (1986), 395–407.

32. P. Van Moerbeke, *The spectrum of Jacobi matrices*, Invent. Math. **37** (1976), 45–81.

DEPARTMENT OF MATHEMATICS, THE OHIO STATE UNIVERSITY, COLUMBUS, OHIO 43210
E-mail address: bloch@function.mps.ohio-state.edu

- 2 -
Interior-Point Methods
for Linear Programming

Contemporary Mathematics
Volume **114**, 1990

An $O(n^3L)$ Potential Reduction Algorithm for Linear Programming

YINYU YE

ABSTRACT. We describe a primal-dual potential function for linear programming

$$\phi(x, s) = \rho \ln(x^T s) - \sum_{j=1}^{n} \ln(x_j s_j)$$

where $\rho \geq n$, x is the primal variable and s is the dual-slack variable in the standard linear programming form. As a result, we develop an interior algorithm seeking reductions in the potential function with $\rho = n + \sqrt{n}$. The algorithm neither traces the central path nor uses projective transformations. It converges to the optimal solution set in $O(\sqrt{n}L)$ iterations and uses $O(n^3L)$ arithmetic operations.

1. Introduction

Since Karmarkar [**17**] proposed the polynomial interior algorithm for linear programming (LP), many developments have been made in the growing literature on interior algorithms: the projective algorithm, the affine scaling algorithm, and the path-following algorithm. All of these interior algorithms use the scaling technique and solve a least-squares problem at each iteration, and they are related to the classical barrier function method of Frisch [**9**], and Fiacco and McCormick [**8**]. (See, for examples, Gill, Murray, Saunders, Tomlin, and Wright [**12**], and Iri and Imai [**16**].)

Karmarkar first introduced the potential function to linear programming in his projective algorithm [**17**]. Then, Anstreicher [**2**], Gay [**10**], de Ghellinck and Vial [**11**], Todd and Burrell [**30**], and Ye and Kojima [**35**] proposed a primal projective algorithm using dual variables. The projective algorithm, including Karmarkar's original algorithm, uses potential functions to measure its iterative progress and converges in $O(nL)$ iterations and $O(n^{3.5}L)$

1980 *Mathematics Subject Classification* (1985 *Revision*). Primary 90C05, 90C06.

Key words and phrases. Linear programming, primal and dual, interior algorithms, potential functions.

arithmetic operations, where L is the data length and n is the number of variables in LP . In practice, far fewer iterations are required when a large-sized step is taken along the descent direction of the potential function.

Barnes [3], Kortanek and Shi [19], and Vanderbei, Meketon, and Freedman [33] updated the primal affine scaling algorithm that was originally proposed by Dikin [7]. Adler, Karmarkar, Resende, and Veiga [1] and Monma and Morton [24] then developed and implemented the dual affine scaling algorithm. The polynomial status of the affine scaling algorithm is still unknown, but it works well in practice by taking a large-sized step along the descent direction of the objective function.

Another polynomial interior algorithm, the (dual) path-following algorithm, was introduced by Renegar [27], who established the first $O(\sqrt{n}L)$-iteration interior algorithm for LP. Using Karmarkar's rank-one technique, Gonzaga [15] and Vaidya [32] further upgraded the algorithm's complexity to $O(n^3L)$. Renegar's algorithm is related to the "analytic center" of Sonnevend [29] and the central trajectories or pathways analyzed by Bayer and Lagarias [4], Megiddo [21], and Megiddo and Shub [22]. Finally, Kojima, Mizuno, and Yoshise [18] and Monteiro and Adler [26] developed the primal-dual path-following algorithm, Goldfarb and Liu [13] and Ye [34] developed the primal path-following algorithm, and Ben Daya and Shetty [5] and Mehrotra and Sun [23] developed the dual path-following algorithm for linear and/or convex quadratic programming. While remaining "centered," this algorithm seeks reductions in the objective function and converges in $O(\sqrt{n}L)$ iterations and $O(n^3L)$ arithmetic operations. Unfortunately, the need to trace closely the central path severely limits the permissible stepsize at any iteration. (A large-step variant of the primal-dual path-following algorithm has been implemented by McShane, Monma, and Shanno [20] with encouraging practical results, but the theoretical guarantee has gone.)

Recently, several efforts were made to improve the interior algorithms. Todd and Ye [31], [36] introduced a class of potential functions for linear programming and proposed a primal-dual projective algorithm, the centered projective algorithm, using a primal-dual potential function. They have shown that the moving direction of this algorithm is the gradient-projection of the potential function in the projective scaling frame. If the centering condition is satisfied, then the direction is also the direction of the path-following. The algorithm is motivated by seeking reductions in the potential function as the projective algorithms. It converges in $O(\sqrt{n}L)$ iterations but still has to follow the central path. Nevertheless, the approximate centering is an automatic by-product of the choice of the potential function. Monteiro, Adler, and Resende [25] simultaneously used the primal and dual affine scaling algorithm, resulting in an $O(nL^2)$-iteration algorithm. Gonzaga [14] used the steepest descent method for a potential function in the primal affine scaling frame, leading to an $O(n^2L)$-or-$O(nL)$-iteration algorithm. His potential function is a special subset of Todd and Ye's class of potential functions and uses the assumption of the known minimal objective value.

Therefore, the question remains open: Do we have to follow the central path to achieve $O(\sqrt{n}L)$-iteration convergence for linear programming, or can we obtain an $O(\sqrt{n}L)$-iteration algorithm based on potential reduction?

In this paper, we further study the primal-dual potential function described by Todd and Ye [31], [36]. As a result, we develop an interior algorithm directly minimizing the potential function in the LP standard form via the scaled-gradient projection method. (See Ye [34].) The algorithm seeks reductions in a suitable potential function like the projective algorithm, but without using the projective transformation; it converges in $O(\sqrt{n}L)$ iterations and $O(n^3L)$ arithmetic operations like the path-following algorithm, but without tracing the central path. We present the algorithm in two forms, the primal form and the dual form. We also show how our algorithm is related to the other interior algorithms mentioned above.

2. Potential function and linear programming

Linear programming is usually identified in the following standard form:

LP   minimize $c^T x$

          subject to $Ax = b$, $x \geq 0$,

where $c \in \mathbf{R}^n$, $A \in \mathbf{R}^{m \times n}$, and $b \in \mathbf{R}^m$ are given, $x \in \mathbf{R}^n$, and T denotes the transpose. The dual to LP can be written as

LD   maximize $b^T y$

          subject to $s = c - A^T y \geq 0$,

where vector $y \in \mathbf{R}^m$ and $s \in \mathbf{R}^n$. The components of s are called dual slacks. For all x and y that are feasible for LP and LD,

$$(1) \qquad b^T y \leq z^* \leq c^T x,$$

where z^* denotes the minimal (maximal) objective value of LP(LD) (Dantzig [6]).

In this paper, the upper-case letter X designates the diagonal matrix of the vector x in lower-case. We also assume that

A1   the relative interior of the feasible regions of LP and LD is nonempty.

A2 A has full rank.

The second assumption is merely added for simplicity.

Given an interior primal solution x^0 and dual solution y^0 such that

$$Ax^0 = b \quad \text{and} \quad x^0 > 0$$

and

$$s^0 = c - A^T y^0 > 0,$$

the primal affine scaling algorithm moves from x^0 along the descent direction

$$(2) \qquad -DP_{AD}Dc,$$

where

$$P_{AD} = I - DA^T (AD^2 A^T)^{-1} AD,$$

and the dual affine scaling algorithm moves from y^0 along the ascent direction

(3) $$(AD^2 A^T)^{-1} b,$$

where D is a diagonal scaling matrix. (See, for examples, [1], [7], [33].)

In the primal affine scaling algorithm $D = X^0$; in the dual affine scaling algorithm $D = (S^0)^{-1}$. Note that if x^0 and y^0 satisfy

$$X^0 s^0 = e$$

then $(A(S^0)^{-2} A^T)^{-1} b$ becomes $(A(X^0)^2 A^T)^{-1} b$. In general, we call (2) the primal affine direction and (3) the dual affine direction.

For the primal we consider the potential function

$$(P) \qquad \overline{\phi}(x, \underline{z}) = \rho \ln(c^T x - \underline{z}) - \sum_{j=1}^{n} \ln(x_j)$$

and for the dual we consider the potential function

$$(D) \qquad \underline{\phi}(y, \overline{z}) = \rho \ln(\overline{z} - b^T y) - \sum_{j=1}^{n} \ln(s_j),$$

while for the primal-dual we employ the joint potential function

$$(P - D) \qquad \phi(x, s) = \rho \ln(x^T s) - \sum_{j=1}^{n} \ln(x_j s_j),$$

where $\underline{z} \leq z^* \leq \overline{z}$ and $n \leq \rho < \infty$. The primal potential function was used by Karmarkar for $\rho = n + 1$ (see Ye and Kojima [35]) and by Gonzaga for $\rho \geq n$ and $\underline{z} = z^*$ [14]. The primal-dual potential function was introduced by Todd and Ye for $\rho = n + \sqrt{n}$ [31] and $\rho = 2n$ [36].

Since $x^T s = x^T (c - A^T y) = c^T x - b^T y$, both of the primal and dual potential functions are related to the primal-dual in the following way:

(4.1) $$\overline{\phi}(x, \underline{z}) = \phi(x, s) + \sum_{j=1}^{n} \ln(s_j)$$

and

(4.2) $$\underline{\phi}(y, \overline{z}) = \phi(x, s) + \sum_{j=1}^{n} \ln(x_j),$$

and the gradient vectors of (P) and (D) are

(5.1) $$\nabla \overline{\phi}(x, \underline{z}) = \nabla_x \phi(x, s) = \frac{\rho}{c^T x - \underline{z}} c - X^{-1} e$$

and

(5.2) $$\nabla \underline{\phi}(y, \overline{z}) = \nabla_y \phi(x, s) = -\frac{\rho}{\overline{z} - b^T y} b + AS^{-1}e,$$

where

$$\underline{z} = b^T y \quad \text{and} \quad \overline{z} = c^T x.$$

Furthermore, the primal-dual potential function $(P - D)$ can be written under the equivalent form (Todd and Ye [31])

$$\phi(x, s) = (\rho - n)\ln(x^T s) - \sum_{j=1}^{n} \ln\left(\frac{x_j s_j}{x^T s}\right).$$

From the inequality of the geometric mean and the arithmetic mean, we have

$$-\sum_{j=1}^{n} \ln\left(\frac{x_j s_j}{x^T s}\right) \geq n \ln n.$$

Hence,

(6) $\quad (\rho - n)\ln(c^T x - b^T y) = (\rho - n)\ln(x^T s) \leq \phi(x, s) - n\ln n \leq \phi(x, s).$

This tells the exact amount, $-(\rho - n)L$, by which ϕ should be reduced to reach

$$c^T x - b^T y \leq 2^{-L}.$$

Before going further, we state the following lemma, which essentially is due to Karmarkar [17].

LEMMA 1. *Let* $x \in \mathbf{R}^n$ *and* $\|x - e\|_\infty < 1$. *Then*

$$\sum_{j=1}^{n} \ln x_j \geq (e^T x - n) - \frac{\|x - e\|^2}{2(1 - \|x - e\|_\infty)},$$

where e *is the vector of all ones, and* $\| \cdot \|$ *(without subscript) denotes the* L_2 *norm.*

PROOF. For $1 \leq j \leq n$,

$$\ln x_j = \ln(1 + x_j - 1)$$

$$= (x_j - 1) - \frac{(x_j - 1)^2}{2} + \frac{(x_j - 1)^3}{3} - \frac{(x_j - 1)^4}{4} + \cdots$$

$$\geq (x_j - 1) - \frac{(x_j - 1)^2}{2}(1 + |x_j - 1| + |x_j - 1|^2 + \cdots)$$

$$= (x_j - 1) - \frac{(x_j - 1)^2}{2(1 - |x_j - 1|)} \geq (x_j - 1) - \frac{(x_j - 1)^2}{2(1 - \|x - e\|_\infty)}.$$

Summing up the inequality over j, we have

$$\sum_{j=1}^{n} \ln x_j \geq (e^T x - n) - \frac{\|x - e\|^2}{2(1 - \|x - e\|_\infty)}. \quad \text{Q.E.D.}$$

Due to the concavity of the first term of the potential functions, $\ln(c^T x - \underline{z})$ or $\ln(\overline{z} - b^T y)$, and Lemma 1, for any two points

$$x^0 > 0, \quad x^1 > 0 \quad \text{and} \quad \|(X^0)^{-1}(x^1 - x^0)\|_\infty < 1$$

we have
(7.1)

$$\overline{\phi}(x^1, \underline{z}) - \overline{\phi}(x^0, \underline{z}) \leq \nabla \overline{\phi}^T(x^0, \underline{z})(x^1 - x^0) + \frac{\|(X^0)^{-1}(x^1 - x^0)\|^2}{2(1 - \|(X^0)^{-1}(x^1 - x^0)\|_\infty)};$$

for any two points

$$s^0 = c - A^T y^0 > 0, \quad s^1 = c - A^T y^1 > 0, \quad \text{and} \quad \|(S^0)^{-1}(s^1 - s^0)\|_\infty < 1$$

we have
(7.2)

$$\underline{\phi}(y^1, \overline{z}) - \underline{\phi}(y^0, \overline{z}) \leq \nabla \underline{\phi}^T(y^0, \overline{z})(y^1 - y^0) + \frac{\|(S^0)^{-1}(s^1 - s^0)\|^2}{2(1 - \|(S^0)^{-1}(s^1 - s^0)\|_\infty)}.$$

The right-hand side of (7.1) ((7.2)) provides a quadratic overestimator for the reduction of the primal (dual) potential function.

3. The primal form

Let $\underline{z}^0 = b^T y^0$ for some $s^0 = c - A^T y^0 > 0$. Then, we minimize the linearized *primal* potential function subject to the ellipsoid constraint corresponding to the second-order term in (7.1):

$$\text{PP} \quad \text{minimize} \quad \nabla \overline{\phi}^T(x^0, \underline{z}^0)(x - x^0)$$
$$\text{subject to} \quad A(x - x^0) = 0$$
$$\|(X^0)^{-1}(x - x^0)\| \leq \beta < 1,$$

and denote by x^1 the minimal solution for PP. Thus, we have

$$(8) \qquad x^1 - x^0 = -\beta \frac{X^0 P_{AX^0} X^0 \nabla \overline{\phi}(x^0, \underline{z}^0)}{\|P_{AX^0} X^0 \nabla \overline{\phi}(x^0, \underline{z}^0)\|}.$$

Let

$$p(\underline{z}^0) = P_{AX^0} X^0 \nabla \overline{\phi}(x^0, \underline{z}^0).$$

Then

$$\nabla \overline{\phi}^T(x^0, \underline{z}^0)(x^1 - x^0) = -\beta \|p(\underline{z}^0)\|.$$

Hence, due to (7.1) the reduction of the primal potential function is

$$(9) \qquad \overline{\phi}(x^1, \underline{z}^0) - \overline{\phi}(x^0, \underline{z}^0) \leq -\beta \|p(\underline{z}^0)\| + \frac{\beta^2}{2(1 - \beta)}.$$

Now, we focus on the expression of $p(\underline{z}^0)$, which from (5.1) can be rewritten as

$$p(\underline{z}^0) = P_{AX^0} X^0 \left(\frac{\rho}{c^T x^0 - \underline{z}^0} c - (X^0)^{-1} e \right)$$

(10)
$$= (I - X^0 A^T (A(X^0)^2 A^T)^{-1} A X^0) \left(\frac{\rho}{c^T x^0 - \underline{z}^0} X^0 c - e \right)$$

$$= \frac{\rho}{c^T x^0 - \underline{z}^0} X^0 s(\underline{z}^0) - e$$

with

(11)
$$s(\underline{z}^0) = c - A^T y(\underline{z}^0)$$

and

$$y(\underline{z}^0) = (A(X^0)^2 A^T)^{-1} A X^0 \left(X^0 c - \frac{c^T x^0 - \underline{z}^0}{\rho} e \right)$$

$$= y_1 - \frac{c^T x^0 - \underline{z}^0}{\rho} y_2$$

where

$$y_1 = (A(X^0)^2 A^T)^{-1} A(X^0)^2 c$$

and

$$y_2 = (A(X^0)^2 A^T)^{-1} A X^0 e = (A(X^0)^2 A^T)^{-1} b.$$

It is clear that y_1 is related to the primal affine direction of (2), which corresponds to the gradient-projection of the linear objective function; y_2 is related to the dual affine direction of (3), which corresponds to the gradient-projection of the barrier function. Both directions are closely linked to the projective algorithm for the standard LP, too. (See, for example, Ye and Kojima [35].)

Regarding $\|p(\underline{z}^0)\|$, we have the following lemma.

LEMMA 2. *Let*

$$\Delta^0 = \frac{c^T x^0 - \underline{z}^0}{n} = \frac{(x^0)^T s^0}{n}, \qquad \Delta = \frac{(x^0)^T s(\underline{z}^0)}{n},$$

$\rho = n + \sqrt{n}$, *and* $\alpha < 1$. *If*

(12)
$$\|p(\underline{z}^0)\| < \min \left(\alpha \sqrt{\frac{n}{n + \alpha^2}}, 1 - \alpha \right),$$

then the following three inequalities hold:

(13.1)
$$s(\underline{z}^0) > 0,$$

(13.2)
$$\|X^0 s(\underline{z}^0) - \Delta e\| < \alpha \Delta,$$

and

(13.3)
$$\Delta < (1 - 0.5\alpha/\sqrt{n}) \Delta^0.$$

PROOF. The proof is by contradiction.

(i) If the inequality of (13.1) is not true, then $\exists\ j$ such that $s_j(\underline{z}^0) \le 0$ and

$$\|p(\underline{z}^0)\| \ge 1 - \frac{\rho}{n\Delta^0}x_j s_j(\underline{z}^0) \ge 1;$$

(ii) if the inequality of (13.2) does not hold, then

$$
\begin{aligned}
\|p(\underline{z}^0)\|^2 &= \left\| \frac{\rho}{n\Delta^0}X^0 s(\underline{z}^0) - \frac{\rho\Delta}{n\Delta^0}e + \frac{\rho\Delta}{n\Delta^0}e - e \right\|^2 \\
&= \left(\frac{\rho}{n\Delta^0}\right)^2 \|X^0 s(\underline{z}^0) - \Delta e\|^2 + \left\| \frac{\rho\Delta}{n\Delta^0}e - e \right\|^2 \\
&\ge \left(\frac{\rho\Delta}{n\Delta^0}\right)^2 \alpha^2 + \left(\frac{\rho\Delta}{n\Delta^0} - 1\right)^2 n \\
&\ge \alpha^2 \frac{n}{n + \alpha^2},
\end{aligned}
$$

(14)

where the last relation prevails since the quadratic term yields minimum at

$$\frac{\rho\Delta}{n\Delta^0} = \frac{n}{n + \alpha^2};$$

(iii) if the inequality of (13.3) is violated, then

$$\frac{\rho\Delta}{n\Delta^0} \ge \left(1 + \frac{1}{\sqrt{n}}\right)\left(1 - \frac{0.5\alpha}{\sqrt{n}}\right) \ge 1,$$

which in view of (14) leads to

$$
\begin{aligned}
\|p(\underline{z}^0)\|^2 &\ge \left(\frac{\rho\Delta}{n\Delta^0} - 1\right)^2 n \\
&\ge \left(\left(1 + \frac{1}{\sqrt{n}}\right)\left(1 - \frac{0.5\alpha}{\sqrt{n}}\right) - 1\right)^2 n \\
&= \left(1 - \frac{\alpha}{2} - \frac{\alpha}{2\sqrt{n}}\right)^2 \\
&\ge (1 - \alpha)^2. \quad \text{Q.E.D.}
\end{aligned}
$$

Based on the above lemmas, we have the following potential reduction theorem.

THEOREM 1. *Let x^0 and y^0 be any interior feasible solutions for LP and LD, and let $\rho = n + \sqrt{n}$, $\underline{z}^0 = b^T y^0$, x^1 be given by (8), and $y^1 = y(\underline{z}^0)$ and $s^1 = s(\underline{z}^0)$ of (11). Then, either*

$$\phi(x^1, s^0) \le \phi(x^0, s^0) - \delta \quad \text{or} \quad \phi(x^0, s^1) \le \phi(x^0, s^0) - \delta$$

where $\delta > 0.05$.

PROOF. If (12) does not hold, i.e.,

$$\|p(\underline{z}^0)\| \ge \min\left(\alpha\sqrt{\frac{n}{n + \alpha^2}}, 1 - \alpha\right),$$

then from (9)

$$\overline{\phi}(x^1, \underline{z}^0) \leq \overline{\phi}(x^0, \underline{z}^0) - \beta \min\left(\alpha\sqrt{\frac{n}{n+\alpha^2}}, 1-\alpha\right) + \frac{\beta^2}{2(1-\beta)}.$$

Hence, from (4.1),

$$\phi(x^1, s^0) \leq \phi(x^0, s^0) - \beta \min\left(\alpha\sqrt{\frac{n}{n+\alpha^2}}, 1-\alpha\right) + \frac{\beta^2}{2(1-\beta)}.$$

Otherwise, from Lemma 2 the inequalities of (13) hold:

   (i) (13.1) indicates that y^1 and s^1 are *interior* dual feasible solutions;
   (ii) using (13.2) and applying Lemma 1 to vector $X^0 s^1/\Delta$, we have

$$n\ln(x^0)^T s^1 - \sum_{j=1}^n \ln(x_j^0 s_j^1) = n\ln((x^0)^T s^1/\Delta) - \sum_{j=1}^n \ln(x_j^0 s_j^1/\Delta)$$

$$= n\ln n - \sum_{j=1}^n \ln(x_j^0 s_j^1/\Delta)$$

$$\leq n\ln n + \frac{\|X^0 s^1/\Delta - e\|^2}{2(1 - \|X^0 s^1/\Delta - e\|_\infty)}$$

$$\leq n\ln n + \frac{\alpha^2}{2(1-\alpha)}$$

$$\leq n\ln(x^0)^T s^0 - \sum_{j=1}^n \ln(x_j^0 s_j^0) + \frac{\alpha^2}{2(1-\alpha)};$$

(iii) according to (13.3), we have

$$\sqrt{n}(\ln(x^0)^T s^1 - \ln(x^0)^T s^0) = \sqrt{n}\ln\frac{\Delta}{\Delta^0} \leq -\frac{\alpha}{2}.$$

Adding the two inequalities in (ii) and (iii), we have

$$\phi(x^0, s^1) \leq \phi(x^0, s^0) - \frac{\alpha}{2} + \frac{\alpha^2}{2(1-\alpha)}.$$

Thus, by choosing $\alpha = 0.43$ and $\beta = 0.3$ we have the desired result.   Q.E.D.

Theorem 1 establishes an important fact: The *primal-dual* potential function can be reduced by a constant via solving PP on the interior of LP and LD, no matter where x^0 and y^0 are. In practice, one can perform the line search to minimize the primal-dual potential function. This results in the following primal algorithm.

PRIMAL ALGORITHM. *Given* $Ax^0 = b$, $x^0 > 0$, *and* $s^0 = c - A^T y^0 > 0$; *let* $\underline{z}^0 = b^T y^0$ *and set* $k = 0$;
   while $c^T x^k - b^T y^k \geq 2^{-L}$ **do**
      begin
      compute $s(\underline{z}^k)$ *of* (11) *and formulate* $p(\underline{z}^k)$ *of* (10);

if *the inequality of* (12) *does not hold* **then**

$$x^{k+1} = x^k - \beta^* X^k p(\underline{z}^k) \text{ with } \beta^* = \arg\min_{\beta \geq 0} \phi(x^k - \beta X^k p(\underline{z}^k), s^k);$$

$$s^{k+1} = s^k \text{ and } \underline{z}^{k+1} = \underline{z}^k;$$

else

$$s^{k+1} = s(\underline{z}^*) \text{ with } \underline{z}^* = \arg\min_{\underline{z} \geq \underline{z}^k} \phi(x^k, s(\underline{z}));$$

$$x^{k+1} = x^k \text{ and } \underline{z}^{k+1} = b^T y(\underline{z}^*);$$

end;

$$k = k + 1;$$

end.

The performance of the primal algorithm results from the following theorem.

THEOREM 2. *Let* $\rho = n + \sqrt{n}$ *and* $\phi(x^0, s^0) \leq O(\sqrt{n}L)$. *Then, the primal algorithm terminates in* $O(\sqrt{n}L)$ *iterations, and each iteration uses* $O(n^3)$ *arithmetic operations.*

PROOF. In $O(\sqrt{n}L)$ iterations

$$\phi(x^k, s^k) \leq -L\sqrt{n}.$$

Then, from (6)

$$\sqrt{n}\ln(c^T x^k - b^T y^k) < -L\sqrt{n};$$

i.e.,

$$c^T x^k - b^T y^k = (x^k)^T s^k < 2^{-L}. \quad \text{Q.E.D.}$$

The condition on the initial potential value in Theorem 2 is not critical. In fact, along the central path

$$\phi(x^0, s^0) = \sqrt{n}\ln(c^T x^0 - b^T y^0) + n\ln n.$$

Hence, $\phi(x^0, s^0) = O(\sqrt{n}L)$ while $c^T x^0 - b^T y^0 \leq 2^L$. Several papers on the path-following algorithm have shown how to transform an LP problem to an equalized LP problem with known centers x^0 and s^0. (See, for examples, Kojima et al. [18], Renegar [27], and Ye [34].) Also note that if $x^{k+1} = x^k$ in the algorithm, the projection matrix in (8) is unchanged and should be reused for the next iterate. In practice, a *strict* lower bound $\underline{z}^0 < z^*$ suffices to start the algorithm; i.e., the known y^0 and s^0 are not necessary. Moreover, a bidirectional search over β and $\underline{z}$ can be employed to update x and s simultaneously in minimizing $\phi(x, s)$.

The condition of (13.2) is the centering condition discussed in the path-following algorithm. While the condition is strictly enforced at any iteration of the path-following, our algorithm uses it as a signal to coordinate the movements of the primal and dual. If the algorithm is implemented the way it is, the iterative solutions will visit the central path many times; however, they are not required to stay at the central path—the next iterate may not even be close to the central path. If the bidirectional search over β and $\underline{z}$ is employed, the condition (13.2) may never be true, and the solution sequence may never visit the central path. By any means, the progress of the algorithm is uniquely measured by the potential function. It ignores following any particular path and concentrates on shrinking the level-set of the potential function, which is contained in the feasible set and contains the optimal solution set of the LP. We believe that the actual solution path generated by the algorithm depends on various implementation and line-search strategies, which is a subject for further research.

4. The dual form

Now we describe the algorithm in the dual form. Let $\bar{z}^0 = c^T x^0$ for some $x^0 > 0$ and $Ax^0 = b$. Then, we minimize the linearized *dual* potential function subject to the ellipsoid constraint corresponding to the second-order term in (7.2):

$$\text{PD} \qquad \text{minimize} \quad \nabla \underline{\phi}^T (y^0, \bar{z}^0)(y - y^0)$$

$$\text{subject to} \quad \|(S^0)^{-1} A^T (y - y^0)\| \le \beta.$$

Then, denoting by y^1 the minimal solution for PD, we have

$$(15) \qquad y^1 - y^0 = -\beta \frac{(A(S^0)^{-2} A^T)^{-1} \nabla \underline{\phi}(y^0, \bar{z}^0)}{\sqrt{\nabla \underline{\phi}^T (y^0, \bar{z}^0)(A(S^0)^{-2} A^T)^{-1} \nabla \underline{\phi}(y^0, \bar{z}^0)}}.$$

Let

$$p(\bar{z}^0) = (S^0)^{-1} A^T (A(S^0)^{-2} A^T)^{-1} \nabla \underline{\phi}(y^0, \bar{z}^0).$$

Then

$$\nabla \underline{\phi}^T (y^0, \bar{z}^0)(y^1 - y^0) = -\beta \|p(\bar{z}^0)\|.$$

Hence, due to (7.2) the reduction of the potential function is

$$\underline{\phi}(y^1, \bar{z}^0) - \underline{\phi}(y^0, \bar{z}^0) \le -\beta \|p(\bar{z}^0)\| + \frac{\beta^2}{2(1-\beta)}.$$

Now, we focus on the expression of $p(\bar{z}^0)$, which from (5.2) can be rewritten as

$$(16) \qquad p(\bar{z}^0) = -\frac{\rho}{\bar{z}^0 - b^T y^0} S^0 x(\bar{z}^0) + e$$

with

$$(17) \qquad x(\bar{z}^0) = x_1 + \frac{\bar{z}^0 - b^T y^0}{\rho} x_2$$

where

$$x_1 = (S^0)^{-2} A^T (A(S^0)^{-2} A)^{-1} b$$

and

$$x_2 = (S^0)^{-1} P_{A(S^0)^{-1}} (S^0)^{-1} c.$$

It is clear that x_1 is related to the dual affine scaling direction of (3), which corresponds to the gradient-projection of the linear objective function; x_2 is related to the primal affine scaling direction of (2), which corresponds to the gradient-projection of the barrier function. We emphasize that

$$Ax(\overline{z}^0) = b;$$

i.e., $x(\overline{z})$ satisfies the equality constraints of LP.

Parallel to Lemma 2, we have the following lemma, whose proof is omitted.

LEMMA 3. *Let*

$$\Delta^0 = \frac{\overline{z}^0 - b^T y^0}{n} = \frac{(s^0)^T x^0}{n}, \qquad \Delta = \frac{(s^0)^T x(\overline{z}^0)}{n},$$

$\rho = n + \sqrt{n}$, *and* $\alpha < 1$. *If*

$$(18) \qquad \qquad \|p(\overline{z}^0)\| < \min\left(\alpha\sqrt{\frac{n}{n+\alpha^2}}, 1 - \alpha\right),$$

then the following three inequalities hold:

$$x(\overline{z}^0) > 0,$$

$$\|S^0 x(\overline{z}^0) - \Delta e\| < \alpha\Delta,$$

and

$$\Delta < \left(1 - 0.5\alpha/\sqrt{n}\right)\Delta^0.$$

Similar to Theorem 1, we have

COROLLARY 1. *Let* x^0 *and* y^0 *be any interior feasible solutions for* LP *and* LD, *and let* $\rho = n + \sqrt{n}$, $\overline{z}^0 = c^T x^0$, $s^1 = c - A^T y^1$ *of* (15), *and* $x^1 = x(\overline{z}^0)$ *of* (17). *Then, either*

$$\phi(x^0, s^1) \leq \phi(x^0, s^0) - \delta$$

or

$$\phi(x^1, s^0) \leq \phi(x^0, s^0) - \delta$$

where $\delta > 0.05$.

Therefore, the dual algorithm can be described as follows.

DUAL ALGORITHM. *Given* $Ax^0 = b$, $x^0 > 0$, *and* $s^0 = c - A^T y^0 > 0$, *let* $\overline{z}^0 = c^T x^0$ *and set* $k = 0$;
   while $c^T x^k - b^T y^k \geq 2^{-L}$ **do**
      begin
         compute $x(\overline{z}^k)$ *of* (17) *and formulate* $p(\overline{z}^k)$ *of* (16);
         if *the inequality of* (18) *does not hold* **then**
            $s^{k+1} = s^k + \beta^* S^k p(\overline{z}^k)$ *with* $\beta^* = \arg\min_{\beta \geq 0} \phi(x^k, s^k + \beta S^k p(\overline{z}^k))$;
            $x^{k+1} = x^k$ *and* $\overline{z}^{k+1} = \overline{z}^k$;
         else

$$x^{k+1} = x(\bar{z}^*) \quad with \quad \bar{z}^* = \arg\min_{\underline{z} \le \bar{z}^k} \phi(x(\bar{z}), s^k);$$
$$s^{k+1} = s^k \quad and \quad \bar{z}^{k+1} = c^T x(\bar{z}^*);$$

      end;
      $k = k + 1;$
end.

The performance of the dual algorithm is identical to that of the primal algorithm.

COROLLARY 2. *Let $\rho = n + \sqrt{n}$ and $\phi(x^0, s^0) \le O(\sqrt{n}L)$. Then, the dual algorithm terminates in $O(\sqrt{n}L)$ iterations, and each iteration uses $O(n^3)$ arithmetic operations.*

Again, note that if $s^{k+1} = s^k$ in the dual algorithm, the projection matrix in (15) is unchanged and should be reused for the next iterate. In practice, a *strict* upper bound $\bar{z}^0 > \underline{z}^*$ suffices to start the algorithm; i.e., the known x^0 is not necessary. Moreover, a bidirectional search over β and $\bar{z}$ can be employed to update x and s simultaneously in minimizing $\phi(x, s)$.

5. Further discussion and concluding remark

Complexity analysis. Theorem 2 indicates that the potential reduction algorithm uses $O(n^{3.5}L)$ arithmetic operations. Applying Karmarkar's lower-rank scheme, we can employ a rank-one updating technique to update the projection matrix in (8). (See, for example, Shanno [28].) This can be implemented as follows:

Replacing X^0 in PP by a positive diagonal matrix D such that

$$\frac{1}{1.1} \le \frac{d_j}{x_j^0} \le 1.1 \quad for \ j = 1, \dots, n,$$

we have

$$x^1 - x^0 = -\beta \frac{D\hat{p}(\underline{z}^0)}{\|\hat{p}(\underline{z}^0)\|},$$

where

$$\hat{p}(\underline{z}^0) = P_{AD} D \nabla \bar{\phi}(x^0, \underline{z}^0).$$

Then

$$\nabla \bar{\phi}^T(x^0, \underline{z}^0)(x^1 - x^0) = -\beta \|\hat{p}(\underline{z}^0)\|.$$

Hence, from (7.1) the reduction of the potential function is

$$\bar{\phi}(x^1, \underline{z}^0) - \bar{\phi}(x^0, \underline{z}^0) \le -\beta \|\hat{p}(\underline{z}^0)\| + \frac{(1.1\beta)^2}{2(1 - 1.1\beta)},$$

since

$$\|(X^0)^{-1}(x^1 - x^0)\| = \|(X^0)^{-1}DD^{-1}(x^1 - x^0)\|$$
$$\le \|(X^0)^{-1}D\| \, \|D^{-1}(x^1 - x^0)\|$$
$$\le 1.1\|D^{-1}(x^1 - x^0)\| = 1.1\beta.$$

Now, $\hat{p}(\underline{z}^0)$ can be written as

$$(19) \qquad \hat{p}(\underline{z}^0) = \frac{\rho}{c^T x^0 - \underline{z}^0} Ds(\underline{z}^0) - D(X^0)^{-1}e = D(X^0)^{-1}p(\underline{z}^0),$$

where the expressions of $p(\underline{z}^0)$ and $s(\underline{z}^0)$ are again given by (10) and (11) with

$$(20) \qquad y(\underline{z}^0) = (AD^2A^T)^{-1}AD\left(Dc - \frac{c^T x^0 - \underline{z}^0}{\rho}D(X^0)^{-1}e\right).$$

Thus, we have

$$\|\hat{p}(\underline{z}^0)\| = \|D(X^0)^{-1}p(\underline{z}^0)\| \geq \|p(\underline{z}^0)\|/\|D^{-1}X^0\| \geq \|p(\underline{z}^0)\|/1.1.$$

Noting that Lemma 2 still holds for $p(\underline{z}^0)$, we only need to modify the first inequality in the proof of Theorem 1 by

$$\overline{\phi}(x^1, \underline{z}^0) \leq \overline{\phi}(x^0, \underline{z}^0) - \frac{\beta}{1.1}\min\left(\alpha\sqrt{\frac{n}{n+\alpha^2}}, 1-\alpha\right) + \frac{(1.1\beta)^2}{2(1-1.1\beta)}.$$

Therefore, upon choosing $\alpha = 0.43$ and $\beta = 0.25$, we see that Theorem 1 is still valid for $\delta > 0.04$. As a result, the following modified primal algorithm can be developed:

MODIFIED PRIMAL ALGORITHM. *Given* $Ax^0 = b$, $x^0 > 0$, *and* $s^0 = c - A^Ty^0 > 0$, *let* $\underline{z}^0 = b^Ty^0$ *and* $D = X^0$, *and set* $\alpha = 0.43$, $\beta = 0.25$, *and* $k = 0$;
    while $c^T x^k - b^T y^k \geq 2^{-L}$ **do**
        begin
        for $j = 1, \ldots, n$, *if* $d_j/x_j^k \notin [1/1.1, 1.1]$ *then* $d_j = x_j^k$;
        using $y(\underline{z}^k)$ *of* (20), *formulate* $s(\underline{z}^k)$ *of* (11), $p(\underline{z}^k)$ *of* (10), *and* $\hat{p}(\underline{z}^k)$ *of* (19);
        if *the inequality of* (12) *does not hold* **then**
            $x^{k+1} = x^k - \beta D\hat{p}(\underline{z}^k)/\|\hat{p}(\underline{z}^k)\|$;
            $s^{k+1} = s^k$ *and* $\underline{z}^{k+1} = \underline{z}^k$
        else
            $s^{k+1} = s(\underline{z}^*)$ *with* $\underline{z}^* = \arg\min_{\underline{z} \geq \underline{z}^k}\phi(x^k, s(\underline{z}))$;
            $x^{k+1} = x^k$ *and* $\underline{z}^{k+1} = b^Ty(\underline{z}^*)$;
        end;
        $k = k + 1$;
    end.

The projection matrix in (20) can be calculated using a rank-one updating technique whenever d_j is changed, and each update uses $O(n^2)$ arithmetic operations. Due to Karmarkar [17], Gonzaga [15], Vaidya [32], and many others, the total number of updates in $O(\sqrt{n}L)$ iterations is $O(nL)$. Therefore, we have

THEOREM 3. *Let* $\rho = n + \sqrt{n}$ *and* $\phi(x^0, s^0) \leq O(\sqrt{n}L)$. *Then, the modified primal algorithm terminates in* $O(\sqrt{n}L)$ *iterations and uses* $O(n^3 L)$ *total arithmetic operations.*

One can also develop a modified dual algorithm with the same performance. However, the modified algorithm may have only theoretical value, since a much larger step has been usually taken in practice.

Concluding remark. Finally, we have developed an $O(\sqrt{n}L)$-iteration and $O(n^3 L)$-operation potential reduction algorithm, comparing to the $O(nL)$-iteration and $O(n^{3.5}L)$-operation Karmarkar-type projective algorithm. Our algorithm is naturally equipped with the primal-dual potential function, which is uniquely used to measure the solution's progress. It does not need to trace any particular path as the path-following algorithm does, or to use the projective transformation as the projective algorithm does. There is *no* step-size-restriction during its iterative process: The greater the reduction of the potential function, the faster the convergence of the algorithm.

The algorithm itself works like a dynamic game, in which one player plays the leader and the other plays the follower. In the primal form the primal player, the leader, using only information about the current dual objective value from the dual player, reduces his potential function by a constant at each step until he is "stuck." Once the leader is "stuck," the follower can then make a move to reduce his potential function by a constant. No matter who moves, the joint primal-dual potential function is reduced by a constant. In this game, there is no winner or loser; the players arrive at the optimal solution set (Cournot-Nash equilibrium set to their potential functions) together. One can also develop a symmetric, cooperative strategy such that both players, the primal (x) and the dual (s), move simultaneously at each step to reduce their joint primal-dual potential function.

Acknowledgment

The author gratefully acknowledges Professors Michael Todd and Kurt Anstreicher, the editors, and the referees for their very helpful comments and remarks.

REFERENCES

1. I. Adler, N. Karmarkar, M. G. C. Resende, and G. Veiga, *An implementation of Karmarkar's algorithm for linear programming*, Math. Programming **44** (1989), 297–335.
2. K. M. Anstreicher, *A monotonic projective algorithm for fractional linear programming*, Algorithmica **1** (1986), 483–498.
3. E. R. Barnes, *A variation on Karmarkar's algorithm for solving linear programming problems*, Math. Programming **36** (1986), 174–182.
4. D. Bayer and J. C. Lagarias, *The nonlinear geometry of linear programming*, I. *Affine and projective scaling trajectories*, II. *Legendre transform coordinates and central trajectories*, Trans. Amer. Math. Soc. **314** (1989), 499–581.
5. M. Ben Daya and C. M. Shetty, *Polynomial barrier function algorithms for convex quadratic programming*, Report J 88-5, School of ISE, Georgia Institute of Technology, Atlanta, GA, 1988.

6. G. B. Dantzig, *Linear programming and extensions*, Princeton University Press, Princeton, NJ, 1963.

7. I. I. Dikin, *Iterative solution of problems of linear and quadratic programming*, Soviet Math. Dokl. **8** (1967), 674–675.

8. A. V. Fiacco and G. P. McCormick, *Nonlinear programming: Sequential unconstrained minimization techniques*, John Wiley and Sons, New York, NY, 1968.

9. K. R. Frisch, *The logarithmic potential method of convex programming*, Technical Report, University Institute of Economics, Oslo, Norway, 1955.

10. D. M. Gay, *A variant of Karmarkar's linear programming algorithm for problems in standard form*, Math. Programming **37** (1987), 81–90.

11. G. de Ghellinck and J.-P. Vial, *A polynomial Newton method for linear programming*, Algorithmica **1** (1986), 425–454.

12. P. E. Gill, W. Murray, M. A. Saunders, J. A. Tomlin, and M. H. Wright, *On projected Newton barrier methods for linear programming and an equivalence to Karmarkar's projective method*, Math. Programming **36** (1986), 183–209.

13. D. Goldfarb and S. Liu, *An $O(n^3 L)$ primal interior point algorithm for convex quadratic programming*, Technical Report, Department of IEOR, Columbia University, New York, NY, 1988.

14. C. C. Gonzaga, *Polynomial affine algorithms for linear programming*, Technical Report ES-139/88, Department of Systems Engineering and Computer Sciences, COPPE-Federal University of Rio de Janeiro, Rio de Janeiro, Brasil, 1988.

15. ____, *An algorithm for solving linear programming problems in $O(n^3 L)$ operations*, Progress in Mathematical Programming, Interior Point and Related Methods (N. Megiddo, ed.), Springer-Verlag, New York 1988, pp. 1–28.

16. M. Iri and H. Imai, *A multiplicative barrier function method for linear programming*, Algorithmica **1** (1986), 455–482.

17. N. Karmarkar, *A new polynomial-time algorithm for linear programming*, Combinatorica **4** (1984), 373–395.

18. M. Kojima, S. Mizuno, and A. Yoshise, *A polynomial-time algorithm for a class of linear complementarity problems*, Math. Programming **44** (1989), 1–26.

19. K. O. Kortanek and M. Shi, *Convergence results and numerical experiments on a linear programming hybrid algorithm*, European J. Oper. Res. **32** (1987), 47–61.

20. K. A. McShane, C. L. Monma, and D. F. Shanno, *An implementation of a primal-dual interior point method for linear programming*, ORSA Journal on Computing **1** (1989), 70–83.

21. N. Megiddo, *Pathways to the optimal set in linear programming*, Progress in Mathematical Programming, Interior Point and Related Methods (N. Megiddo, ed.), Springer-Verlag, New York, 1988, pp. 131–158.

22. N. Megiddo and M. Shub, *Boundary behavior of interior point algorithms in linear programming*, Math. Oper. Res. **14** (1989), 97–146.

23. S. Mehrotra and J. Sun, *An algorithm for convex quadratic programming that requires $O(n^{3.5} L)$ arithmetic operations*, Math. Oper. Res. **15** (1990), 342–363.

24. C. L. Monma and A. J. Morton, *Computational experimental with a dual affine variant of Karmarkar's method for linear programming*, Technical Report, Bell Communications Research, Morristown, NJ, 1987.

25. R. C. Monteiro, I. Adler, and M. C. Resende, *A polynomial-time primal-dual affine scaling algorithm for linear and convex quadratic programming and its power series extension*, Math. Oper. Res. **15** (1990), 191–214.

26. R. C. Monteiro and I. Adler, *Interior path following primal-dual algorithms. Part I: Linear programming*, Math. Programming **44** (1989), 27–42.

27. J. Renegar, *A polynomial-time algorithm, based on Newton's method, for linear programming*, Math. Programming **40** (1988), 59–93.

28. D. F. Shanno, *Computing Karmarkar's projection quickly*, Math. Programming **41** (1988), 61–71.

29. G. Sonnevend, *An 'analytic center' for polyhedrons and new classes of global algorithms for linear (smooth, convex) programming*, Lecture Notes in Control and Information Sciences, vol. 84, Springer-Verlag, New York, 1985, pp. 866–876.

30. M. J. Todd and B. P. Burrell, *An extension of Karmarkar's algorithm for linear programming using dual variables*, Algorithmica **1** (1986), 409–424.

31. M. J. Todd and Y. Ye, *A centered projective algorithm for linear programming*, Technical Report 763, School of ORIE, Cornell University, Ithaca, NY, 1987, Math. Oper. Res. (to appear).

32. P. M. Vaidya, *An algorithm for linear programming which requires* $O((m + n)n^2 + (m + n)^{1.5}n)L)$ *arithmetic operations*, manuscript, Bell Labs, Murray Hill, NJ, 1987.

33. R. J. Vanderbei, M. S. Meketon, and B. A. Freedman, *On a modification of Karmarkar's linear programming algorithm*, Algorithmica **1** (1986), 395–407.

34. Y. Ye, *Interior algorithms for linear, quadratic, and linearly constrained convex programming*, Ph. D. thesis, Department of Engineering-Economic Systems, Stanford University, Stanford, CA, 1987.

35. Y. Ye and M. Kojima, *Recovering optimal dual solutions in Karmarkar's polynomial algorithm for linear programming*, Math. Programming **39** (1987), 305–317.

36. Y. Ye and M. J. Todd, *Containing and shrinking ellipsoids in the path-following algorithm*, Math. Programming (to appear).

DEPARTMENT OF MANAGEMENT SCIENCES, THE UNIVERSITY OF IOWA, IOWA CITY, IOWA 52242

E-mail address: BBUYINVA@UIAMVS.BITNET

Contemporary Mathematics
Volume **114**, 1990

I. I. Dikin's Convergence Result
for the Affine-Scaling Algorithm

R. J. VANDERBEI AND J. C. LAGARIAS

ABSTRACT. The affine-scaling algorithm is an analogue of Karmarkar's linear programming algorithm that uses affine transformations instead of projective transformations. Although this variant lacks some of the nice properties of Karmarkar's algorithm (for example, it is probably not a polynomial-time algorithm), it nevertheless performs well in computer implementations. It has recently come to the attention of the western mathematical programming community that a Soviet mathematician, I. I. Dikin, proposed the basic affine-scaling algorithm in 1967 and published a proof of convergence in 1974. Dikin's convergence proof assumes only primal nondegeneracy, while all other known proofs require both primal and dual nondegeneracy. Our aim in this paper is to give a clear presentation of Dikin's ideas.

1. Introduction

In 1984, N. K. Karmarkar [7] discovered a polynomial-time algorithm for linear programming, which, unlike earlier such algorithms, was said to perform better than the traditional simplex method. This pioneering work inspired a huge amount of research on interior methods for linear programming. Today, these methods are divided into two basic classes. Algorithms in the first class are called projective-scaling algorithms (see [10] for a list of references). This class includes the original algorithm studied by Karmarkar. Roughly speaking, projective-scaling algorithms are hard to describe but easy to analyze. That is, the description of these algorithms involves technicalities such as logarithmic barrier functions, assumptions that the optimal objective function value is zero, etc., which although they originally seemed rather obscure are now seen as forcing the sequence of points generated by the algorithm to home in on something called the 'central trajectory.' Once these technicalities are understood, the analysis proceeds to a proof of polynomial-time convergence.

Algorithms in the second class are called affine-scaling algorithms (see, e.g.,

1980 *Mathematics Subject Classification* (1985 *Revision*). Primary 90C05.

[4], [13]). In comparison, these algorithms are easy to describe but hard to analyze. That is, affine-scaling algorithms have very simple, geometrically intuitive descriptions, but proving convergence is a difficult and interesting mathematical problem. Until recently, the only known proofs of convergence (in the west) involved assuming that the linear program was primal and dual nondegenerate. (Empirical evidence suggests that neither of these assumptions is necessary.) Furthermore, there is strong evidence [8] to support the belief that these algorithms are in fact exponential in the worst case.

This segregation into two classes is not intended to imply that projective-scaling algorithms are 'first-class' algorithms whereas the affine-scaling algorithms are 'second class.' In fact, all serious large scale implementations currently being pursued use affine-scaling methods ([1], [2], [9], [11], [12]). The reason is that the affine-scaling methods have certain advantages over the projective-scaling algorithms. For example, they apply directly to problems presented in standard form, and on the average they are computationally more efficient. Ironically, this brings us back to the old situation where what is theoretically best (worst case) is not practically best (average case).

A further point worth mentioning is that the distinction between these two classes of algorithms is being blurred as people discover polynomial-time algorithms that look more and more like affine-scaling algorithms. The most notable example of this is the recent paper by Monteiro et al. [10], where it is shown that the affine-scaling algorithm applied to the primal-dual problem is in fact polynomial if the initial point is close to the center of the polytope and the step size is not too big.

It has recently come to the attention of the western scientific community that a Soviet mathematician, I. I. Dikin, proposed the basic affine-scaling algorithm in the Soviet Mathematics Doklady in 1967 [5]. He published a proof of convergence in 1974 [6]. It turns out that, not only did Dikin predate the west by almost 20 years, but also his proof of convergence does not require the dual nondegeneracy assumption. The purpose of this paper is to give a clear presentation of Dikin's methods.

Perhaps the most interesting open problem in this area is to prove that the affine-scaling algorithm converges even if the problem is primal degenerate. Essentially, all real-world problems are both primal and dual degenerate, and yet practical experience shows that this does not present any difficulty (except that the code has to be able to solve a consistent system of equations even when there are dependent or almost dependent rows). In a recent paper [3] by Adler and Monteiro, it was shown that the continuous trajectories associated with the affine-scaling algorithm do indeed converge even when the problem is primal and/or dual degenerate. Hence the problem is the discreteness of the affine-scaling algorithm. Dikin chose a step size that is smaller than the one in [13] and was able to remove the dual nondegeneracy assumption. Perhaps by taking an even smaller (but noninfinitesimal) step size, it might be possible to remove the primal nondegeneracy assumption as well.

Acknowledgment. We would like to thank Mike Todd for several stimulating discussions. In particular, he pointed out the appropriate argument to show that $(AD_x^2A^T)^{-1}$ is bounded.

2. The main convergence result

In this section, we describe the basic affine-scaling algorithm studied by Dikin. The *primal linear program* is

(P) $$\min c \cdot x$$

$$Ax = b$$
$$x \geq 0.$$

The associated *dual linear program* is

(D) $$\max b \cdot w$$

$$A^T w \leq c.$$

We begin by introducing some notation. Let $\Omega = \{x \in R^n : Ax = b, x \geq 0\}$ denote the polytope for the primal problem; let $\Omega^0 = \{x \in \Omega : x > 0\}$ denote the relative interior of Ω; and let $\partial\Omega = \Omega - \Omega^0$ denote the boundary of Ω. Finally, given a vector x, let D_x denote the diagonal matrix having the components of x along its diagonal.

The motivation behind the affine-scaling algorithm can be found in many papers (see, e.g., [13]). Therefore, in this paper we assume that the reader has seen the motivation and we go straight to the definition of the algorithm. For this, we need to introduce three important functions. The first function $w : \Omega^0 \to R^m$ associates with each $x \in \Omega^0$ a vector of *dual variables*:

$$w(x) = (AD_x^2A^T)^{-1}AD_x^2c.$$

The second function $r : \Omega^0 \to R^n$ measures the slackness in the inequality constraints of the dual problem:

$$r(x) = c - A^T w(x).$$

Note that $w(x)$ is dual feasible if and only if $r(x) \geq 0$. The vector $r(x)$ is called the vector of *reduced costs*. The third function $y : \Omega^0 \to \Omega$ is given by

(1) $$y(x) = x - \frac{D_x^2 r(x)}{|D_x r(x)|}.$$

The affine-scaling algorithm is defined in terms of the function y:

$$x^{k+1} = \begin{cases} y(x^k), & x^k \in \Omega^0 \\ x^k, & x^k \in \partial\Omega. \end{cases}$$

The algorithm also generates a sequence of dual variables

$$w^k = \begin{cases} w(x^k), & x^k \in \Omega^0 \\ w^{k-1}, & x^k \in \partial\Omega \end{cases}$$

and a sequence of reduced costs

$$r^k = c - A^T w^k.$$

For notational convenience, put $D_k = D_{x^k}$. Note that if the sequence x^k hits the boundary of the polytope, then it becomes fixed at the point where it first hits. In contrast, the dual variables become fixed at the values associated with the last interior point before the boundary was hit.

This affine-scaling algorithm differs slightly from the ones studied in [4] and [13]. The difference is in the step size. In words, the algorithm studied here steps 100 percent of the way to the surface of the inscribed ellipsoid (see [4] for the definition of this ellipsoid), whereas the algorithm in [4] steps only a certain fraction α of the way to the surface. Hence, while the algorithm presented here can stop in a finite number of steps, the one in [4] always involves an asymptotic approach to the optimal solution. In contrast, the algorithm presented in [13] steps a certain fraction of the way to the nearest face. This means that the mapping $y(x)$ defined by (1) has to be changed to

$$y^{[13]}(x) = \frac{x - \alpha D_x^2 r(s)}{\gamma(x)},$$

where

$$\gamma(x) = \max_j x_j r_j(x).$$

It is easy to see that

$$\gamma(x) \leq |D_x r(x)|_\infty \leq |D_x r(x)|$$

(assumption (1) below implies that $\gamma(x) > 0$). Hence, the step length chosen in [13] is the longest, followed by the L^∞ norm, followed by the conservative L^2 step length studied here. It should be noted that all implementations use the step length described in [13], and so, in a sense, convergence proofs for that case are the most interesting. We do not study that case here, but it is easy to peruse the proof given here for the L^2 case and see that it can be easily modified to cover the L^∞ case as long as we also introduce a contraction factor $\alpha < 1$.

The following assumptions are made:

1. $-\infty < \min_\Omega c \cdot x < \max_\Omega c \cdot x$,
2. Ω^0 is nonempty; $x^0 \in \Omega^0$ is given.
3. A has full-row rank.
4. $AD_x^2 A^T$ is invertible for all $x \in \Omega$.

Assumption (4) is called *primal nondegeneracy*. Note that assumption (3) implies that $AD_x^2 A^T$ is invertible for all $x \in \Omega^0$. Hence, the primal nondegeneracy assumption is really an assumption about the boundary of Ω. Also note that the primal nondegeneracy assumption implies that the domain of the function w can be extended to all of Ω. Also, assumption (1) implies that $|D_x r(x)| \neq 0$ for all $x \in \Omega^0$, which in turn implies that $y(x)$ is well defined.

The main result is

THEOREM. *The sequence x^k converges to a primal feasible point $\bar{x}$. The sequence w^k converges to a dual feasible point $\bar{w}$. The pair consisting of the limiting primal variables $\bar{x}$ and the limiting reduced costs $\bar{r} = c - A^T \bar{w}$ satisfy strong complementarity:*

(2) $$\bar{x}_j = 0 \quad \text{if and only if} \quad \bar{r}_j > 0.$$

In the next section we prove this theorem assuming the polytope is bounded. Then, in Section 4, we remove the boundedness assumption.

3. Proof assuming compactness

In this section, we assume that Ω is bounded. For this case, we break the proof up into a series of steps.

Step 1. Primal feasibility is preserved. It is easy to check that $A D_x^2 r(x) = 0$, and therefore, since $Ax = b$, we see that

$$Ay(x) = Ax - \frac{A D_x^2 r(x)}{|D_x r(x)|} = b.$$

From the definition of $y(x)$, we see that the jth component is given by

(3) $$y_j(x) = x_j \left(1 - \frac{x_j r_j(x)}{|D_x r(x)|}\right).$$

The subtracted term above must lie between -1 and 1, and hence

(4) $$0 \leq y_j(x) \leq 2x_j.$$

Step 2. Hitting a face implies optimality. Suppose that $x^k \in \Omega^0$ and that $x_j^{k+1} = 0$. Then (3) implies that $x_j^k r_j^k = |D_k r^k|$. Hence, since $x_i^k > 0$ for all i, we see that $r_j^k > 0$ and $r_i^k = 0$ for all $i \neq j$. Since $\bar{x} = x^{k+1}$ and $\bar{w} = w^k$, we see now that (2) holds. Henceforth, we assume that $x^k \in \Omega^0$ for all k.

Step 3. The objective function decreases. We begin with the obvious:

$$c \cdot x - c \cdot y(x) = \frac{c \cdot D_x^2 r(x)}{|D_x r(x)|}.$$

Next, note that $D_x r(x) = P_x D_x c$, where $P_x = I - D_x A^T (A D_x^2 A^T)^{-1} A D_x$ denotes the projection onto the null space of $A D_x$. Hence,

$$c \cdot x - c \cdot y(x) = \frac{c \cdot D_x P_x D_x c}{|D_x r(x)|}$$

and, since P_x is a projection matrix, it is idempotent and symmetric, and so

$$c \cdot x - c \cdot y(x) = |P_x D_x c| = |D_x r(x)|.$$

Therefore,

$$c \cdot x^k - c \cdot x^{k+1} = |D_k r^k|.$$

Step 4. Complementary slackness holds in the limit. Since $c \cdot x^k$ converges (it is decreasing and bounded below), it follows from the previous step that

$$|D_k r^k| \to 0.$$

Step 5. The dual variables are bounded. Assumption (4) implies that $w(x)$ is a continuous function on the compact set Ω; hence $w(x)$ is bounded.

We now introduce some auxiliary notation. Let $\overline{w}$ be a limit point of w^k. Let k_p, $p = 1, 2, \ldots$, denote a subsequence along which w^k converges to $\overline{w}$. Put $\overline{r} = c - A^t \overline{w}$, and let

$$B = \{j : \overline{r}_j = 0\}, \qquad N = \{j : \overline{r}_j \neq 0\}.$$

The indices in B are called the *basic* indices and those in N are called *nonbasic*. Abusing notation, we use the same letters to denote the partition of A into its basic and nonbasic parts:

$$A = [B \mid N].$$

We denote the corresponding partitioning of n-vectors using subscripts:

$$x = \begin{bmatrix} x_B \\ x_N \end{bmatrix}.$$

Step 6. $\lim_{x_N \to 0} r(x) = \overline{r}$. First note that

$$\begin{aligned} r(x) &= [I - A^T(AD_x^2 A^T)^{-1} AD_x^2]c \\ &= [I - A^T(AD_x^2 A^T)^{-1} AD_x^2]\overline{r}, \end{aligned}$$

where the second equality follows from the definition of $\overline{r}$ and the fact that the bracketed matrix annihilates A^T. Rearranging the last equation and using the fact that the basic components of $\overline{r}$ vanish, we get

$$\overline{r} - r(x) = A^T(AD_x^2 A^T)^{-1} N D_{x_N}^2 \overline{r}_N.$$

Now, since $(AD_x^2 A^T)^{-1}$ exists and is continuous throughout the compact set Ω, it follows that it is bounded. Hence as x_N tends to zero the $D_{x_N}^2$ factor dominates and drives the difference to zero.

Step 7. The nonbasic components of x convergence to zero. Fix $j \in N$. By Step 4, we know that $x_j^k r_j^k$ tends to zero. We also know that r_j^k tends along the subsequence k_p to $\overline{r}_j$, which is a nonzero number. Hence, x_j^k must tend to zero along the subsequence k_p. Since this is true for any nonbasic index j, we see that

$$x_N^{k_p} \to 0.$$

Now suppose that x_N^k does not tend to zero along the entire sequence; i.e., there exists a $\delta > 0$ such that $x_N^k \in C_\delta^c$ infinitely often, where $C_\delta = \{x_N : 0 \leq x_i < \delta \text{ for all } i \in N\}$. (Given such a δ, any smaller value will also work.) Fix δ small enough that C_δ^c is visited infinitely often and

(5) $$|r_j^{k_p'}| \geq \tfrac{1}{2}|\bar{r}_j|$$

for all j and for all x such that $x_N \in C_\delta$. This, of course, is possible because of the previous step. Let $K = \{k : x_N^k \in C_\delta \text{ and } x_N^{k+1} \leq C_\delta^c\}$. Then clearly K is an infinite set. Now let $K_j = \{k \in K : x_j^{k+1} \geq \delta\}$. Then $K = \bigcup_{j \in N} K_j$. Since N is finite, there exists a j such that K_j is infinite. Let k_p' be the subsequence that enumerates K_j. Then

$$x_N^{k_p'} < \delta \quad \text{and} \quad x_j^{k_p'+1} \geq \delta.$$

Using formula (4), the second statement above implies that $x_j^{k_p'} \geq \delta/2$. From (5) we see that

$$|x_j^{k_p'} r_j^{k_p'}| \geq \tfrac{1}{4}\delta|\bar{r}_j| > 0$$

for all p. This contradicts Step 4 and so we are led to conclude that x_N^k actually tends to zero.

Step 8. The dual variables converge to $\overline{w}$. As in Step 6, we start by noting that

$$
\begin{aligned}
w^k - \overline{w} &= (AD_k^2 A^T)^{-1} AD_k^2(c - A^T\overline{w}) \\
&= (AD_k^2 A^T)^{-1} AD_k^2 \bar{r} \\
&= (AD_k^2 A^T)^{-1} N D_{x_N^k}^2 \bar{r}_N.
\end{aligned}
$$

Again, we use the fact that $(AD_x^2 A^T)^{-1}$ is bounded to conclude that the difference converges to zero, since x_N^k is now known to converge to zero.

Step 9. The limiting nonbasic reduced costs are positive. Suppose there exists a $j \in N$ such that $\bar{r}_j < 0$. Since $w^k \to \overline{w}$, there exists a K such that $r_j^k < 0$ for all $k \geq K$. Hence, we see that

$$x_j^{k+1} = x_j^k \left(1 - \frac{x_j^k r_j^k}{|D_k r^k|}\right) > x_j^k \quad \text{for all } k \geq K.$$

This contradicts Step 7.

Step 10. The basic components of x^k converge (say to $\overline{x}_B$). From the definition of the algorithm, we have

$$
(6) \qquad x_j^{k+1} - x_j^k = -\frac{(x_j^k)^2 r_j^k}{|D_k r^k|}.
$$

Therefore, x_j^k converges if

$$
\sum_{k=0}^{\infty} \frac{(x_j^k)^2 |r_j^k|}{|D_k r^k|} < \infty.
$$

For $j \in B$,

$$
r_j(x) = r_j(x) - \overline{r}_j = e_j^T A^T (AD_x^2 A^T)^{-1} N D_{x_N}^2 \overline{r}_N.
$$

Hence,

$$
x_j^2 r_j = \sum_{l \in N} \sigma_{jl}(x) x_l^2 \overline{r}_l,
$$

where

$$
\sigma_{jl}(x) = e_j^T D_x^2 A^T (AD_x^2 A^T)^{-1} N e_l.
$$

Continuity and compactness now imply that there are bounding constants $\overline{\sigma}_{jl} : |\sigma_{jl}(x)| \leq \overline{\sigma}_{jl}$ for all $x \in \Omega$. Therefore,

$$
(7) \qquad \sum_{k=0}^{\infty} \frac{(x_j^k)^2 |r_j^k|}{|D_k r^k|} \leq \sum_{l \in N} \overline{\sigma}_{jl} \sum_{k=0}^{\infty} \frac{(x_l^k)^2 \overline{r}_l}{|D_k r^k|}.
$$

Step 7 and equation (6) imply that, for any $l \in N$,

$$
\sum_{k=0}^{\infty} \frac{(x_l^k)^2 r_l^k}{|D_k r^k|}
$$

converges. Since $r_l^k \to \overline{r}_l$, we see that the right-hand side in (7) is finite.

Step 11. Strong complementarity holds. We only need to show that $\overline{x}_B > 0$. Fix $j \in B$. To show that $\overline{x}_j > 0$, it suffices to show that

$$
\sum_{k=0}^{\infty} \frac{\log x_j^{k+1}}{x_j^k}
$$

converges absolutely. Clearly, this sum converges absolutely if and only if

$$
\sum_{k=0}^{\infty} \frac{x_j^{k+1} - x_j^k}{x_j^k}
$$

converges absolutely. Using (3), we see that

$$
\sum_{k=0}^{\infty} \frac{x_j^{k+1} - x_j^k}{x_j^k} = \sum_{k=0}^{\infty} \frac{x_j^k r_j^k}{|D_k r^k|}.
$$

To see that this last sum converges absolutely, we use exactly the same argument as in the previous step, except that instead of σ_{jl} we get something similar.

$$\rho_{jl}(x) = e_j^T D_x A^T (AD_x^2 A^T)^{-1} Ne_l.$$

In the previous section, we found a bound for $\sigma_{jl}(x)$ that was valid throughout Ω. Here we settle for slightly less: we only need to show that $\rho_{jl}(x)$ is bounded along the sequence x^k. This follows from the fact that x_j^k converges and $(AD_x^2 A^T)^{-1}$ is bounded. This completes the proof for the case where Ω is assumed to be bounded.

4. Proof without compactness

The boundedness assumption (in conjunction with primal nondegeneracy) was used in three places:

1. In Step 5, showing that the dual variables function $w(x)$ is bounded.
2. In Steps 6, 8, and 11, showing that $(AD_x^2 A^T)^{-1}$ is bounded.
3. In Step 10, showing that the function $\sigma_{jl}(x)$ is bounded.

We will now show that the boundedness assumption is not necessary for any of these three statements to hold and that primal nondegeneracy is not necessary for the first and third.

We begin with the claim that $(AD_x^2 A^T)^{-1}$ is bounded. Let $\widehat{\Omega}$ denote the compact set consisting of the convex hull of all of the extreme points of Ω. Then any point x in Ω can be written as $x = \widehat{x} + \widehat{t}$, where $\widehat{x}$ belongs to $\widehat{\Omega}$ and $\widehat{t} \geq 0$, $A\widehat{t} = 0$. Hence, $AD_x^2 A^T \geq AD_{\widehat{x}}^2 A^T$ in the sense that the difference is positive semidefinite. Therefore, $\lambda_1(x) \geq \lambda_1(\widehat{x})$, where $\lambda_1(x)$ denotes the smallest eigenvalue of $AD_x^2 A^T$. Primal nondegeneracy and compactness of $\widehat{\Omega}$ imply that $\lambda_1(\widehat{x}) \geq \overline{\lambda}_1 > 0$ for all $\widehat{x} \in \widehat{\Omega}$. Since the L^2 norm of $(AD_x^2 A^T)^{-1}$ is exactly $1/\lambda_1(x)$, we see that

$$|(AD_x^2 A^T)^{-1}|_2 = \frac{1}{\lambda_1(x)} \leq \frac{1}{\lambda_1(\widehat{x})} \leq \frac{1}{\overline{\lambda}_1} < \infty.$$

Note that even though we have removed the boundedness assumption on Ω, we still have used primal nondegeneracy. This is essential, as the following example shows:

$$\min x_1$$
$$x_1 - x_2 = 0$$
$$x_1 \geq 0, \qquad x_2 \geq 0.$$

Clearly, $(x_1, x_2) = (0, 0)$ is a point in the polytope at which $(AD_x^2 A^T)^{-1} = 1/(x_1^2 + x_2^2)$ blows up.

The other two claims follow immediately from the following statement:

LEMMA. *For any n-vector* γ, *the function* $w^\gamma(x) = (AD_x^2 A^T)^{-1} AD_x^2 \gamma$ *is bounded.*

To prove this lemma, we use Cramer's rule and the Cauchy-Binet theorem to write the ith component as follows:
(8)

$$
\underset{i\text{th position}}{\downarrow}
$$

$$
w_i^\gamma = \frac{\sum_{1 \le j_1 < \cdots < j_m \le n} (x_{j_1} \cdots x_{j_m})^2 \det_{j_1, \ldots, j_m}(a_1, \ldots, a_m) \det_{j_1, \ldots, j_m}(a_1, \ldots, \gamma, \ldots, a_m)}{\sum_{1 \le j_1 < \cdots < j_m \le n} (x_{j_1} \cdots x_{j_m})^2 [\det_{j_1, \ldots, j_m}(a_1, \ldots, a_m)]^2},
$$

where $\det_{j_1, \ldots, j_m}(\alpha_1, \ldots, \alpha_m)$ denotes the determinant of the $m \times m$ matrix obtained by selecting columns $j_1, \ldots, j_m$ from the $m \times n$ matrix whose rows are the n-vectors $\alpha_1, \ldots, \alpha_m$ and where a_i denotes the ith row of A. To bound this expression, we use the following simple inequality:

$$
\frac{\left| \sum_{i=1}^{N} r_i \right|}{\sum_{i=1}^{N} s_i} \le \max_i \frac{|r_i|}{s_i},
$$

which is valid for any pair of sequences with $s_i > 0$. Since terms in the numerator in (8) vanish whenever terms in the denominator vanish, we can use this simple inequality to obtain the following bound:

$$
|w_i^\gamma(x)| \le \max_{\substack{1 \le j_1 < \cdots < j_m \le n \\ \text{denom} \ne 0}} \left| \frac{\det_{j_1, \ldots, j_m}(a_1, \ldots, \gamma, \ldots, a_m)}{\det_{j_1, \ldots, j_m}(a_1, \ldots, a_m)} \right|.
$$

The right-hand side is independent of x and hence furnishes a bound.

5. Dikin's proof

The ideas, methods, and results presented in this paper stem directly from Dikin's work. However, there are a few minor gaps in Dikin's paper that have been rectified here. The first is that he claimed that the limiting nonbasic reduced costs are positive (Step 9) before he showed that the dual variables converge (Step 8). His argument is extremely terse and combined with his proof that the nonbasic components of x converge to zero (Step 7), so it seems that at the very least he was leaving a lot to the reader. In any case, the proof of Step 9 here certainly depends on the result of Step 8. Mike Todd has a proof for Step 9 that does not depend on Step 8, but the proof given here is easier.

The other difficulty is that in Steps 6, 8, and 11, where the boundedness of $(AD_x^2 A^T)^{-1}$ is needed, Dikin argues that the desired result follows from Cramer's rule and the Cauchy-Binet theorem. We do not see how it is possible to establish the boundedness of this matrix using that approach.

REFERENCES

1. I. Adler, N. K. Karmarkar, and M. G. C. Resende, *An implementation of Karmarkar's algorithm for linear programming*, Math. Programming **44**, (1989), 297–335..
2. I. Adler, N. K. Karmarkar, M. G. C. Resende, and G. Veiga, *Data structures and programming techniques for the implementation of Karmarkar's algorithm*, ORSA J. Comp. **1** (1989), 84–106.
3. I. Adler and R. D. C. Monteiro, *Limiting behavior of the affine scaling continuous trajectories for linear programming problems*, this volume, pp. 189–211.
4. E. R. Barnes, *A variation on Karmarkar's algorithm for solving linear programming problems*, Math. Programming **36** (1986), 174–182.
5. I. I. Dikin, *Iterative solution of problems of linear and quadratic programming*, Soviet Math. Dokl. **8** (1967), 674–675.
6. ___, *On the speed of an iterative process*, Upravlyaemye Sistemi **12** (1974), 54–60.
7. N. K. Karmarkar, *A new polynomial-time algorithm for linear programming*, Combinatorica **4** (1984), 373–395.
8. N. Megiddo and M. Shub, *Boundary behavior of interior-point algorithms in linear programming*, Math. Oper. Res. **14** (1989), 97–146.
9. C. L. Monma and A. J. Morton, *Computational experiments with a dual affine variant of Karmarkar's method for linear programming*, Oper. Res. Lett. **6** (1987), 261–267.
10. R. D. C. Monteiro, I. Adler, and M. G. C. Resende, *A polynomial-time primal-dual affine scaling algorithm for linear and convex quadratic programming and its power series extension*, Technical Report, Department of Industrial Engineering and Operations Research, University of California, Berkeley, Calif., 1988.
11. J. A. Tomlin, *An experimental approach to Karmarkar's projective method for linear programming*, Technical Report, Ketron, Inc., Mountain View, CA, 1985.
12. J. A. Tomlin and J. S. Welch, *Implementing an interior-point method in a mathematical programming system*, Technical Report, Ketron, Inc., Mountain View, Calif., 1986.
13. R. J. Vanderbei, M. S. Meketon, and B. A. Freedman, *A modification of Karmarkar's linear programming algorithm*, Algorithmica **1** (1986), 395–407.

AT&T BELL LABORATORIES, MURRAY HILL, NEW JERSEY 07974
E-mail address, R. J. Vanderbei: rvdb@research.att.com
E-mail address, J. C. Lagarias: jcl@research.att.com

Contemporary Mathematics
Volume **114**, 1990

Phase 1 Search Directions for a Primal-Dual Interior Point Method for Linear Programming

IRVIN J. LUSTIG

ABSTRACT. A new method for obtaining an initial feasible interior point solution to a linear program and its dual is presented. Given a linear program, an auxiliary problem is created that uses a "big-M." By considering the search directions derived by solving the auxiliary problem as functions of the values of M, it is shown that as $M \to \infty$, the directions have limits and can be interpreted as Newton search directions on a certain system. Even though a proof of the theoretical convergence of the method is an open question, computational experience indicates that the method works well in practice.

1. Introduction

Since Karmarkar (1984) described his new polynomial-time algorithm for linear programming, many researchers have derived different variants of the algorithm. Most of these variants require that an initial interior feasible point be given. In order to find such a point, an auxiliary linear program is solved. An initial interior point for the auxiliary problem is easy to determine. The solution of this auxiliary program is found to be an initial interior point for the original linear program. Usually, the auxiliary program contains a large constant (a "big-M") that is required in order to guarantee the existence of a solution of the auxiliary problem. The use of such a constant can be unsettling in practice.

Megiddo (1986) described how one can take a primal and dual pair of linear programs and create a new primal-dual pair of auxiliary problems, with each problem having a big-M in the objective and the right-hand side. If the primal-dual interior point methods of Kojima et al. (1988), Monteiro and Adler (1989), or Lustig (1988a) are applied to the auxiliary problems, the search directions generated will depend on the value of M. In this paper,

1980 *Mathematics Subject Classification* (1985 *Revision*). Primary 90C05.
Key words and phrases. Linear programming, interior point methods, primal-dual algorithms, feasibility.

we show that as $M \to \infty$, the search directions have a limit that does not depend on the value of M. These directions can be interpreted as a Netwon direction on a certain system.

The presentation will use the notation of Lustig (1988a). As review, a summary of his method is presented in Section 2. Section 3 develops new search directions that find feasible points, and their properties are discussed in Section 4. Section 5 presents some computational experience that indicates that the method works well in practice.

2. The primal-dual algorithm

In this section, a review of the generic primal-dual algorithm is presented. The algorithm is based upon the work of Megiddo (1986), Kojima et al. (1988), and Monteiro and Adler (1989). The reader is referred to Lustig (1988a) for a more detailed discussion. We are interested in solving the linear program

$$(2.1) \qquad \begin{aligned} \text{minimize} \quad & c^T x \\ \text{subject to} \quad & Ax = b, \qquad x \geq 0 \end{aligned}$$

and its dual

$$(2.2) \qquad \begin{aligned} \text{maximize} \quad & b^T y \\ \text{subject to} \quad & A^T y + z = c, \qquad z \geq 0. \end{aligned}$$

Assume that $c \in \mathbf{R}^n$, $b \in \mathbf{R}^m$, and $A \in \mathbf{R}^{m \times n}$ are given. For the purpose of this discussion, assume that the linear programs (2.1) and (2.2) have optimal solutions. On iteration $k+1$, a feasible point $x^k > 0$ for the primal linear program (2.1) and a feasible point y^k and $z^k > 0$ for the dual linear program (2.2) are known. The generic primal-dual interior point algorithm generates points such that

$$(2.3a) \qquad x^{k+1} \leftarrow x^k - \alpha^k(p_x^1 - \mu^k p_x^2),$$

$$(2.3b) \qquad y^{k+1} \leftarrow y^k - \alpha^k(p_y^1 - \mu^k p_y^2),$$

and

$$(2.3c) \qquad z^{k+1} \leftarrow z^k - \alpha^k(p_z^1 - \mu^k p_z^2).$$

The search directions p_x^i, p_y^i, and p_z^i, $i = 1, 2$ are computed as follows:

$$(2.4a) \qquad p_y^1 = -(AXZ^{-1}A^T)^{-1}b,$$

$$(2.4b) \qquad p_z^1 = -A^T p_y^1,$$

$$(2.4c) \qquad p_x^1 = Xe - XZ^{-1}p_z^1,$$

$$(2.4d) \qquad p_y^2 = -(AXZ^{-1}A^T)^{-1}AZ^{-1}e,$$

$$(2.4e) \qquad p_z^2 = -A^T p_y^2,$$

$$(2.4f) \qquad p_x^2 = Z^{-1}e - XZ^{-1}p_z^2.$$

Here, X is a diagonal matrix with the components of x^k on the diagonal, Z is a diagonal matrix with the components of z^k on the diagonal, and $e = (1, 1, \dots, 1) \in \mathbf{R}^n$. On each iteration, the parameters α^k and μ^k are chosen so that $x^{k+1} > 0$ and $z^{k+1} > 0$. Lustig (1988a) has shown that only mild restrictions on the values of α^k and μ^k are necessary to guarantee convergence of the algorithm. However, the algorithm requires that initial *feasible* points x^0, y^0, and z^0 be given.

Note that the directions given above are written in two parts. The directions p_x^1, p_y^1, and p_z^1 are used to reduce the duality gap whereas the directions p_x^2, p_y^2, and p_z^2 center the iterates x^k and z^k. If we let $d^k = c^T x^k - b^T y^k$ represent the duality gap between the feasible primal-dual pair (x^k, y^k), the directions have the property that $c^T p_x^1 - b^T p_y^1 = d^k$ and $c^T p_x^2 - b^T p_y^2 = n$. The duality gap changes according to $d^{k+1} = d^k - \alpha^k(d^k - n\mu)$. If $\alpha^k > 0$ and $\mu^k < (d^k/n)$ hold, then the directions p_x^1, p_y^1, and p_z^1 reduce the duality gap. In addition, $d^k = e^T XZe$, so that the measure of complementarity also decreases.

3. Finding a feasible point

Megiddo (1986), Kojima et al. (1988), Monteiro and Adler (1989), and McShane et al. (1989) have suggested setting up the following pair of primal and dual linear programs to find the solution of a problem for which an initial interior feasible point is not known. The augmented primal system is
(3.1)

$$\text{minimize} \quad c^T x + M_P x_{n+1}$$

$$\text{subject to} \quad Ax + (b - Ax^0)x_{n+1} = b,$$

$$(A^T y^0 + z^0 - c)^T x + x_{n+2} = M_D, \qquad x, x_{n+1}, x_{n+2} \geq 0.$$

The dual of this system is

$$\text{maximize} \quad b^T y + M_D y_{m+1}$$

(3.2) $\qquad \text{subject to} \quad A^T y + (A^T y^0 + z^0 - c)y_{m+1} + z = c,$

$$(b - Ax^0)^T y + z_{n+1} = M_P,$$

$$y_{m+1} + z_{n+2} = 0, \qquad z, z_{n+1}, z_{n+2} \geq 0,$$

where $x^0 > 0$, $z^0 > 0$, and y^0 are arbitrary initial points. For large enough values of M_P and M_D, the optimal solution x to (3.1) will be the same as the optimal solution to (2.1). Similarly, the optimal vectors y and z to (3.2) will be the same as the optimal solution to (2.2). This solution will have the property that $x_{n+1} = 0$ and $y_{m+1} = 0$.

The generic primal-dual algorithm can be applied to the primal-dual pair of linear programs given by (3.1) and (3.2). An initial feasible solution is easily attained by using x^0, y^0, and z^0 and setting $x_{n+1}^0 = 1$ and $y_{m+1}^0 = -1$.

However, the selection of appropriate values of M_P and M_D is difficult in practice. McShane et al. (1989) show that the performance of their version of the primal-dual algorithm is sensitive to the choice of these values. Furthermore, choosing very large values of M_P and M_D can introduce numerical instabilities in an implementation of the algorithm. The purpose of deriving new directions is to eliminate M_P and M_D from the computations.

3.1. Derivation of limiting feasibility directions. When the primal-dual algorithm is applied to (3.1) and (3.2), the direction vectors obtained by applying equations (2.4) depend on the values of M_P and M_D. Our intent is to show that the contribution of M_P and M_D to these direction vectors gets small as the values of M_P and M_D get large. For ease of exposition, assume that $M = M_P = M_D$. Let $\bar{q}$ represent the direction vectors obtained when the primal-dual algorithm is applied to the artificial problems (3.1) and (3.2). For each $i = 1, 2$, it follows that $\bar{q}_x^i \in \mathbf{R}^{n+2}$, $\bar{q}_y^i \in \mathbf{R}^{m+1}$, and $\bar{q}_z^i \in \mathbf{R}^{n+2}$. The aim is to show that the limits

$$(3.3) \qquad (\bar{p}_x^i, \bar{p}_y^i, \bar{p}_z^i) = \lim_{M \to \infty} (\bar{q}_x^i, \bar{q}_y^i, \bar{q}_z^i), \qquad i = 1, 2$$

exist and to compute their values. Let $d_p = b - Ax^0$ and $d_D = A^T y^0 + z^0 - c$. Then

$$(3.4) \quad \begin{pmatrix} A & d_P & 0 \\ d_D^T & 0 & 1 \end{pmatrix} \begin{pmatrix} XZ^{-1} & & \\ & \frac{x_{n+1}}{z_{n+1}} & \\ & & \frac{x_{n+2}}{z_{n+2}} \end{pmatrix} \begin{pmatrix} A^T & d_D \\ d_P^T & 0 \\ 0 & 1 \end{pmatrix}$$
$$= \begin{pmatrix} AXZ^{-1}A^T + \frac{x_{n+1}}{z_{n+1}} d_P d_P^T & AXZ^{-1}d_D \\ d_D^T XZ^{-1}A & d_D^T XZ^{-1}d_D + \frac{x_{n+2}}{z_{n+2}} \end{pmatrix}.$$

For notational purposes, let

$$(3.5) \qquad \bar{q}_y^1 = \begin{pmatrix} \hat{q}_y \\ \hat{q}_{y_{m+1}} \end{pmatrix}$$

so that $\hat{q}_y$ represents a change for the dual variables y, and $\hat{q}_{y_{m+1}}$ represents a change for the artificial variable y_{m+1}. It is then necessary to solve the system of equations

$$(3.6) \quad \begin{pmatrix} AXZ^{-1}A^T + \frac{x_{n+1}}{z_{n+1}} d_P d_P^T & AXZ^{-1}d_D \\ d_D^T XZ^{-1}A & d_D^T XZ^{-1}d_D + \frac{x_{n+2}}{z_{n+2}} \end{pmatrix} \begin{pmatrix} \hat{q}_y \\ \hat{q}_{y_{m+1}} \end{pmatrix} = - \begin{pmatrix} b \\ M \end{pmatrix}$$

for $\bar{q}_y^1$. This is done by subtracting the proper multiples of the last row of this system from the other rows to eliminate the variable $\hat{q}_{y_{m+1}}$ from the first

m equations. The system of equations is then

(3.7)
$$\begin{pmatrix} AXZ^{-1}A^T + \frac{x_{n+1}}{z_{n+1}}d_P d_P^T - \frac{AXZ^{-1}d_D d_D^T XZ^{-1}A}{d_D^T XZ^{-1}d_D + (x_{n+2}/z_{n+2})} & 0 \\ d_D^T XZ^{-1}A & d_D^T XZ^{-1}d_D + \frac{x_{n+2}}{z_{n+2}} \end{pmatrix} \begin{pmatrix} \hat{q}_y \\ \hat{q}_{y_{m+1}} \end{pmatrix}$$

$$= - \begin{pmatrix} b - M\frac{AXZ^{-1}d_D}{d_D^T XZ^{-1}d_D + (x_{n+2}/z_{n+2})} \\ M \end{pmatrix}.$$

The vector $\hat{q}_y$ can now be determined by solving

(3.8)
$$\left(AXZ^{-1}A^T + \frac{x_{n+1}}{z_{n+1}}d_P d_P^T - \frac{AXZ^{-1}d_D d_D^T XZ^{-1}A}{d_D^T XZ^{-1}d_D + (x_{n+2}/z_{n+2})} \right) \hat{q}_y$$

$$= - \left(b - M\frac{AXZ^{-1}d_D}{d_D^T XZ^{-1}d_D + \frac{x_{n+2}}{z_{n+2}}} \right).$$

The matrix and right-hand side in equation (3.8) both depend on the value of M, since $x_{n+2} = M - d_D^T x$ and $z_{n+1} = M - d_P^T y$. As the value of M gets large, its contribution to the matrix and right-hand side of equation (3.8) diminishes. It follows that

(3.9)
$$\lim_{M \to \infty} b - M\frac{AXZ^{-1}d_D}{d_D^T XZ^{-1}d_D + \frac{x_{n+2}}{z_{n+2}}} = b - (z_{n+2})AXZ^{-1}d_D$$

and

(3.10)
$$\lim_{M \to \infty} \frac{AXZ^{-1}d_D d_D^T XZ^{-1}A}{d_D^T XZ^{-1}d_D + \frac{x_{n+2}}{z_{n+2}}} = 0.$$

Similarly,

(3.11)
$$\lim_{M \to \infty} \frac{x_{n+1}}{z_{n+1}}d_P d_P^T = 0.$$

This suggests that $\hat{p}_y = \lim_{M \to \infty} \hat{q}_y$ exists and is found by solving the system

(3.12)
$$(AXZ^{-1}A^T)\hat{p}_y = -(b - (z_{n+2})AXZ^{-1}d_D).$$

The value $\hat{q}_{y_{m+1}}$ is determined by solving

(3.13)
$$\left(d_D^T XZ^{-1}d_D + \frac{x_{n+2}}{z_{n+2}} \right) \hat{q}_{y_{m+1}} = -(M + d_D^T XZ^{-1}A\hat{q}_y).$$

Let $\hat{p}_{y_{m+1}} = \lim_{M \to \infty} \hat{q}_{y_{m+1}}$. It follows that

(3.14)
$$\hat{p}_{y_{m+1}} = -\lim_{M \to \infty} \frac{M + d_D^T XZ^{-1}A\hat{q}_y}{\frac{x_{n+2}}{z_{n+2}} + d_D^T XZ^{-1}d_D} = -z_{n+2}.$$

Therefore,

(3.15)
$$\hat{p}_y^1 = - \begin{pmatrix} (AXZ^{-1}A^T)^{-1}(b - (z_{n+2})AXZ^{-1}d_D) \\ z_{n+2} \end{pmatrix}.$$

Note that the contribution of $\bar{p}_y^1$ to the increase of y_{m+1} is in the correct direction, since $z_{n+2} = y_{m+1}$.

The vector $\bar{p}_z^1$ is found by writing

$$(3.16) \qquad \bar{p}_z^1 = - \begin{pmatrix} A^T & d_D \\ d_P^T & 0 \\ 0 & 1 \end{pmatrix} \bar{p}_y^1 = \begin{pmatrix} -A^T \hat{p}_y + z_{n+2} d_D \\ -d_P^T \hat{p}_y \\ z_{n+2} \end{pmatrix}.$$

Let $\hat{p}_z = -A^T \hat{p}_y + z_{n+2} d_D$. The vector $\bar{q}_x^1$ depends on M and is

$$(3.17) \qquad \bar{q}_x^1 = \begin{pmatrix} Xe - XZ^{-1}\hat{p}_z \\ x_{n+1} + \frac{x_{n+1}}{z_{n+1}} d_P^T \hat{p}_y \\ x_{n+2} - \frac{x_{n+2}}{z_{n+2}} z_{n+2} \end{pmatrix}.$$

It is easy to see that

$$(3.18) \qquad \bar{p}_x^1 = \lim_{M \to \infty} \bar{q}_x^1 = \begin{pmatrix} Xe - XZ^{-1}\hat{p}_z \\ x_{n+1} \\ 0 \end{pmatrix}.$$

Note that the direction of change of x_{n+1} is a descent direction for x_{n+1}.

A similar derivation shows that $(\bar{p}_x^2, \bar{p}_y^2, \bar{p}_z^2) = (p_x^2, p_y^2, p_z^2)$, as given by equations (2.4). Hence, the Phase 1 directions for centering *are the same* as the Phase 2 directions for centering. Feasibility of the primal point x^k or the dual point (y^k, z^k) does not change the centering direction. The variables x_{n+1} and y_{m+1} should not be centered, since their optimal value is zero. To summarize, the proposed algorithm changes x^k, y^k, and z^k according to

$$(3.19a) \qquad x^{k+1} \leftarrow x^k - \alpha^k (\tilde{p}_x^1 - \mu^k p_x^2),$$

$$(3.19b) \qquad y^{k+1} \leftarrow y^k - \alpha^k (\tilde{p}_y^1 - \mu^k p_y^2),$$

$$(3.19c) \qquad z^{k+1} \leftarrow z^k - \alpha^k (\tilde{p}_z^1 - \mu^k p_z^2),$$

$$(3.19d) \qquad x_{n+1}^{k+1} \leftarrow x_{n+1}^k - \alpha^k x_{n+1}^k,$$

$$(3.19e) \qquad y_{m+1}^{k+1} \leftarrow y_{m+1}^k - \alpha^k y_{m+1}^k,$$

where

$$(3.20a) \qquad \tilde{p}_y^1 = -(AXZ^{-1}A^T)^{-1}(b + (y_{m+1}^k)AXZ^{-1}d_D),$$

$$(3.20b) \qquad \tilde{p}_z^2 = -A^T \tilde{p}_y^1 - y_{m+1}^k d_D,$$

$$(3.20c) \qquad \tilde{p}_x^1 = Xe - XZ^{-1}\tilde{p}_z^1.$$

We will call these directions the "limiting feasibility directions." The algorithm always satisfies the primal constraints

$$(3.21) \qquad Ax + d_P x_{n+1} = b, \qquad x, x_{n+1} \geq 0,$$

and the dual constraints

$$(3.22) \qquad A^T y + d_D y_{m+1} + z = c, \qquad z \geq 0,$$

and it ignores the extra constraints in (3.1) and (3.2) with M_D and M_P on the right-hand side. If the search parameter α^k can be set to $\alpha^k = 1$ (independent of μ^k) on some iteration k with $x^{k+1} > 0$ and $z^{k+1} > 0$, then primal and dual interior feasible solutions have been found, and the algorithm can revert to its original form given by equations (2.3) and (2.4).

4. Properties of limiting feasibility directions

The directions given by equations (3.20) satisfy the following two properties, which are easy to verify:

$$(4.1a) \qquad A\tilde{p}_x^1 = -x_{n+1}d_P;$$

$$(4.1b) \qquad A^T\tilde{p}_y^1 + \tilde{p}_z^1 = -y_{m+1}d_D.$$

Equation (4.1a) indicates that the search direction for the primal variables x and the artificial variable x_{n+1} is in the null space of the equation for the equality constraints (3.21). Note that p_x^2 is in the null space of A. Equation (4.1b) indicates that the search direction for the dual variables y and z and the artificial variable y_{m+1} is in the null space of the constraints (3.22). The search directions also decrease the value of x_{n+1} and increase the value of y_{m+1}. (Recall that $x_{n+1}^0 = 1$ and $y_{m+1}^0 = -1$.)

The directions given by (3.20) also satisfy the set of equations

$$(4.2a) \qquad Z\tilde{p}_x^1 + X\tilde{p}_z^1 = XZe,$$

$$(4.2b) \qquad A\tilde{p}_x^1 + d_P x_{n+1} = 0,$$

$$(4.2c) \qquad A^T\tilde{p}_y^1 + \tilde{p}_z^1 + d_D y_{m+1} = 0.$$

This system indicates that the limiting feasibility directions are the Newton directions for the system

$$(4.3a) \qquad XZe = 0,$$

$$(4.3b) \qquad Ax + d_P(x_{n+1}) = b,$$

$$(4.3c) \qquad A^Ty + z + d_D(y_{m+1}) = c,$$

$$(4.3d) \qquad x_{n+1} = 0,$$

$$(4.3e) \qquad y_{m+1} = 0.$$

It is clear that an algorithm based on using the limiting feasibility directions will decrease the amount of infeasibility of the primal and dual linear programs (as measured by x_{n+1} and y_{m+1}) on each iteration. Hence, this monotonic decrease suffices to prove that the algorithm will converge. However, proving that the algorithm converges to primal and dual feasible points (or determining that either problem is infeasible) is difficult. The proof of convergence of an algorithm based on these new directions is currently an open problem.

4.1. Relationship to parametric programing. The algorithm has an inherently close relationship to parametric programming when one considers the first three equations in (4.3). Let $\theta = x_{n+1} = -y_{m+1}$. Since $x_{n+1}^0 = -y_{m+1}^0 = 1$, the algorithm maintains the equality of x_{n+1}^k and $-y_{m+1}^k$ on the kth iteration, provided that $d_P \neq 0$ and $d_D \neq 0$. Now, the set of equations

(4.4a) $$XZe = 0$$

(4.4b) $$Ax + d_P\theta = b$$

(4.4c) $$A^Ty + z - d_D\theta = c$$

(4.4d) $$x, z \geq 0$$

are the Karush-Kuhn-Tucker conditions for the parametric linear program,

(4.5)
$$\begin{aligned} \text{minimize} \quad & (c + \theta d_D)^T x \\ \text{subject to} \quad & Ax = b - \theta d_P, \qquad x \geq 0. \end{aligned}$$

A difficulty in showing that an algorithm using the limiting feasibility directions is convergent may be shown by considering the case when the current iterate is close to optimality for the parametric linear program for some value of θ. It must be shown that the algorithm does not stall when $e^T XZe$ is small and $\theta > 0$. (Note that if $Ax \neq b$ and $A^Ty + z \neq c$, then $e^T XZe \neq c^T x - b^T y$.) Alternatively, the parametric linear program could be solved for a fixed value of θ, and then an optimal solution to the original linear program could be found using an algorithm such as Dantzig's Self-Dual Algorithm (Lemke's Algorithm).

5. Computational experience

The algorithm described above has been implemented in conjunction with the generic primal-dual algorithm to solve large-scale linear programs. The details of the computational experiments are discussed by Lustig (1988b). The performance of the algorithm was measured in terms of iteration counts against that of McShane et al. (1989), which is a "big-M" method. Table 5.1 presents these results. The average savings was approximately 13%. Note that a feasible point for the primal and dual problem was usually discovered long before the algorithm terminated. The practical experience indicates that a convergence proof of an algorithm using the limiting feasibility directions should be possible.

Problem Name	Generic Primal-Dual			McShane et al.			Percentage of Iterations Saved
	Primal Phase 1	Dual Phase 1	Total Iterations	Primal Phase 1	Dual Phase 1	Total Iterations	
AFIRO	4	2	12	4	5	14	14
ADLITTLE	11	3	21	18	18	18	-17
SC205	6	7	17	20	9	20	15
SCAGR7	11	4	24	7	30	30	20
SHARE2B	9	8	21	6	16	18	-17
SHARE1B	9	11	30	11	86	87	66
SCORPION	10	3	19	20	5	20	5
SCAGR25	14	6	28	10	58	58	52
SCTAP1	6	3	21	7	8	27	22
BRANDY	28	22	30	28	28	28	-7
ISRAEL	19	9	33	9	31	34	3
SCFXM1	28	14	32	32	32	32	0
BANDM	19	9	26	32	31	32	19
E226	24	17	30	32	33	33	9
SCSDI	1	0	12	1	0	12	0
BEACONFD	11	8	15	19	20	20	25
SCRS8	24	16	32	57	57	57	44
SCFXM2	31	19	36	39	39	39	8
SCSD6	1	0	15	1	0	15	0
SHIP04S	15	0	18	20	20	22	18
FFFFF800	35	4	43				
SCFXM3	36	19	41	40	41	41	0
SCTAP2	6	3	22	5	24	25	12
SHIP04L	17	0	18	18	21	23	22
SHIP08S	15	0	19	21	24	24	21
SCTAP3	5	3	23	5	25	26	12
SHIP12S	20	0	22	27	28	28	21
25FV47	14	21	42	12	49	49	14
SCSD8	1	0	14	1	0	14	0
CZPROB	35	6	56	57	59	59	5
SHIP08L	20	0	22	22	23	24	8
SHIP12L	22	0	24	26	27	27	11

Problem FFFFF800 was not solved by McShane et al.

Table 5.1.

Comparisons of iteration counts

References

N. Karmarkar (1984), *A new polynomial-time algorithm for linear programming*, Combinatorica **4(4)**, 373–395.

M. Kojima, S. Mizuno, and A. Yoshise (1988), *A primal-dual interior point algorithm for linear programming*, Progress in Mathematical Programming (N. Negiddo, ed.), Springer-Verlag, New York, pp. 29–48.

I. J. Lustig (1988a), *A generic primal-dual interior point algorithm*, Technical Report SOR 88-3, Program in Statistics and Operations Research, Department of Civil Engineering and Operations Research, School of Engineering and Applied Science, Princeton University, Princeton, New Jersey.

_____, (1988b), *Feasibility issues in an interior point method for linear programming*, Technical Report SOR 88-9, Program in Statistics and Operations Research, Department of Civil Engineering and Operations Research, School of Engineering and Applied Science, Princeton University, Princeton, New Jersey, Math. Programming (to appear).

K. McShane, C. Monma, and D. Shanno (1989), *An implementation of a primal-dual interior point method for linear programming*, ORSA Journal on Computing **1**, 70–83.

N. Megiddo (1986), *Pathways to the optimal set in linear programming*, Progress in Mathematical Programming (N. Megiddo, ed.), Springer-Verlag, New York, 131–158.

R. C. Monteiro and I. Adler (1989), *Interior path following primal-dual algorithms, Part* I: *Linear programming*, Math. Programming **44**, 27–42.

PROGRAM IN STATISTICS AND OPERATIONS RESEARCH, DEPARTMENT OF CIVIL ENGINEERING AND OPERATIONS RESEARCH, SCHOOL OF ENGINEERING AND APPLIED SCIENCE, PRINCETON UNIVERSITY, PRINCETON, NEW JERSEY 08544

E-mail address: irv%basie@princeton.edu

Contemporary Mathematics
Volume **114**, 1990

Some Results Concerning Convergence of the Affine Scaling Algorithm

EARL R. BARNES

ABSTRACT. Several authors, for example [2] and [6], have advocated the use of a big-M technique to simplify the problem of finding a feasible starting point for the affine scaling algorithm. In this paper we point out a difficulty with convergence that can arise when this method is used. We propose a method for avoiding this difficulty. We also prove a convergence result for the affine scaling algorithm applied to the Phase I problem. Unlike earlier proofs, this one does not require the problem to be nondegenerate.

1. Introduction

Consider the linear programming problem

$$\text{minimize } c^T x$$

(1)
$$\text{subject to } Ax = b$$

$$x \geq 0$$

where c and x are n-dimensional column vectors, A is an $m \times n$ matrix, and b is an m-dimensional column vector. We assume that the feasible region for this problem has a nonempty interior and that the problem has a bounded solution. The affine scaling algorithm for solving this problem begins with a strictly positive vector x^0, satisfying $Ax^0 = b$, and iterates as follows.

If $x^k > 0$ satisfying $Ax^k = b$ is given, define

$$D_k = \text{diag}(x_1^k, \ldots, x_n^k)$$

and compute

$$\lambda_k = (AD_k^2 A^T)^{-1} AD_k^2 c.$$

Take

(2)
$$x^{k+1} = x^k - \frac{D_k^2(c - A^T \lambda_k)}{\|D_k(c - A^T \lambda_k)\|}.$$

1980 *Mathematics Subject Classification* (1985 *Revision*). Primary 90C05.

Observe that

$$c^T x^k - c^T x^{k+1} = (c - A^T \lambda_k)^T (x^k - x^{k+1}) = \|D_k(c - A^T \lambda_k)\|$$

(3)
$$= \left\{ \sum_{j=1}^{n} (x_j^k)^2 (c_j - a_j^T \lambda_k)^2 \right\}^{1/2}$$

where a_j denotes the jth column of A.

In practice there is generally no obvious way to select a feasible starting point x^0. The big M method proposes the following way around this difficulty.

Choose $x^0 > 0$ arbitrarily and solve the following modification of (1) for a very large value of M.

$$\text{minimize } c^T x + M x_{n+1}$$

(4)   subject to $Ax + (b - Ax^0)x_{n+1} = b$, $x \geq 0$, $x_{n+1} \geq 0$.

The point $x = x^0$, $x_{n+1} = 1$ provides a suitable starting point for applying the affine scaling algorithm to this problem.

For M very large, and x_{n+1} close to 1, the objective in (4) is dominated by the x_{n+1} term. In this case it is easy to see that the first steps of the affine scaling algorithm applied to the problem

$$\text{minimize } x_{n+1}$$

(5)   subject to $Ax + (b - Ax^0)x_{n+1} = b$, $x \geq 0$, $x_{n+1} \geq 0$,

starting at $x = x^0$, $x_{n+1} = 1$, behaves very much like the first steps of the affine scaling algorithm applied to (4), from the same starting point. Therefore, in order to analyze the behavior of the affine scaling algorithm for (4), it suffices to analyze its behavior for (5). This remark holds only for values of x_{n+1} close to 1. However, for some choices of x^0 we shall see that x_{n+1} remains close to 1 for many iterations of the affine scaling algorithm. This says that the affine scaling algorithm can spend a lot of time in Phase I if x^0 is not chosen properly. In Section 2 we describe a way of choosing x^0 to accelerate convergence in Phase I. It turns out that the rate of convergence depends on the size of $\|Ax^0 - b\|$, and this quantity must be decreased in order to accelerate convergence.

Another proposal for solving the Phase I problem has been offered by Adler et al. in [1]. These authors consider the dual of problem (1). When applied to problem (1) their method would choose a positive vector $\hat{x}^0$ and define

$$x^0 = \frac{\|b\|}{\|A\hat{x}^0\|} \hat{x}^0.$$

Given x^0 their method applies the big-M technique (5) to (1). Our analysis below shows that the rate of convergence depends on the size of $\|Ax^0 - b\|$, and increases as this quantity decreases. In the present case we have

$||Ax^0 - b|| \leq 2||b||$; however, it is not clear how to choose $\hat{x}^0$ to ensure that $||Ax^0 - b||$ is small when $||b||$ is large. So we take a different approach to choosing x^0. We use the method of successive projections to approximate a solution of $Ax = b$, $x > 0$. In this way we can always choose $||Ax^0 - b||$ small enough to achieve fast convergence in Phase I. The asymptotic behavior of this algorithm is discussed in [2], and more thoroughly in [3]. In [3] it is shown that in the nondegenerate case the asymptotic convergence rate depends only on the number of positive variables in the solution to which the algorithm is converging. If the algorithm converges to a solution having p positive variables the error between the current and optimal objective function values is reduced by a factor

$$(6) \qquad 1 - \frac{1}{\sqrt{n - p + \varepsilon_k}}$$

at the kth step. ε_k is a quantity that tends to 0 as $k \to \infty$.

2. The affine scaling updates

Let $a = b - Ax^0$ denote the artificial column in (5). When the affine scaling algorithm is applied to this problem the multiplier λ_k is given by

$$
\begin{aligned}
\lambda_k &= (x_{n+1}^k)^2 \{ AD_k^2 A^T + (x_{n+1}^k)^2 aa^T \}^{-1} a \\
&= (x_{n+1}^k)^2 \left\{ (AD_k^2 A^T)^{-1} - \frac{(x_{n+1}^k)^2 (AD_k^2 A^T)^{-1} aa^T (AD_k^2 A^T)^{-1}}{1 + (x_{n+1}^k)^2 a^T (AD_k^2 A^T)^{-1} a} \right\} a \\
&= (x_{n+1}^k)^2 \left\{ \frac{1}{1 + (x_{n+1}^k)^2 a^T (AD_k^2 A^T)^{-1} a} \right\} (AD_k^2 A^T)^{-1} a.
\end{aligned}
$$

To simplify the calculations that follow we set $\sigma_k = a^T (AD_k^2 A^T)^{-1} a$. A direct substitution into (3) gives

$$
\begin{aligned}
x_{n+1}^k - x_{n+1}^{k+1} &= \left\{ \sum_{j=1}^n (x_j^k)^2 (a_j^T \lambda_k)^2 + (x_{n+1}^k)^2 (1 - a^T \lambda_k)^2 \right\}^{1/2} \\
&= \left\{ \lambda_k^T \left(\sum_{j=1}^n (x_j^k)^2 a_j a_j^T \right) \lambda_k + (x_{n+1}^k)^2 (1 - a^T \lambda_k)^2 \right\}^{1/2} \\
&= \left\{ \lambda_k^T (AD_k^2 A^T) \lambda_k + \frac{(x_{n+1}^k)^2}{(1 + (x_{n+1}^k)^2 \sigma_k)^2} \right\}^{1/2}
\end{aligned}
$$

(continued)

$$= \left\{ \frac{(x_{n+1}^k)^2 a^T \lambda_k}{1 + (x_{n+1}^k)^2 \sigma_k} + \frac{(x_{n+1}^k)^2}{(1 + (x_{n+1}^k)^2 \sigma_k)^2} \right\}^{1/2}$$

$$= \frac{x_{n+1}^k}{\sqrt{1 + (x_{n+1}^k)^2 \sigma_k}}.$$

We rewrite this as

(7) $$x_{n+1}^{k+1} = \left(1 - \frac{1}{\sqrt{1 + (x_{n+1}^k)^2 \sigma_k}} \right) x_{n+1}^k.$$

Observe that we have found the exact behavior of the affine scaling algorithm applied to (5). Since $x_{n+1}^0 = 1$ and $\sigma_0 = (b - Ax^0)^T (AD_0^2 A^T)^{-1}(b - Ax^0)$ it is clear that if x^0 is chosen such that σ_0 is large, the algorithm makes only slight progress on the first iteration. Our investigation of the big M technique began when we observed several cases where many steps of the affine scaling algorithm were required to make a significant reduction in the objective function value. We encountered all of these problems at the IBM Watson Research Center. One of them (L27LAV) has only 146 constraints and 2656 variables. When x^0 was chosen to be the vector of all ones, more than 80 iterations were required to solve this problem. When x^0 was chosen by the method described in the next section, a solution was obtained in 7 iterations.

3. Choosing x^0

We will show how to choose x^0 such that $\sigma_0 \le n$. It will then follow that the error in our objective function can be reduced by a factor approximately equal to $1 - \frac{1}{\sqrt{n+1}}$ on the first iteration. In practice we have found that choosing x^0 to guarantee good progress on the first iteration is sufficient to guarantee good progress at each iteration.

Let $\varepsilon > 0$ be given. Consider the convex sets

$$S_1 = \{x | x \ge \varepsilon\}, \qquad S_2 = \{x | Ax = b\}.$$

Ideally, we would like to choose $x^0 \in S_1 \cap S_2$. This would result in $\sigma_0 = 0$. An algorithm for finding a point common to two convex sets has been proposed in [4]. It begins with a point $y^1 \in S_1$ and computes the point $\hat{y}^1 \in S_2$ nearest y^1. It then computes y^2 as the point in S_1 nearest $\hat{y}^1$. In general, given an iterate $y^k \in S_1$, $\hat{y}^k$ is defined to be the projection of y^k onto S_2 and y^{k+1} is the projection of $\hat{y}^k$ onto S_1. In mathematical terms we have

(8) $$\hat{y}^k = y^k + A^T (AA^T)^{-1}(b - Ay^k)$$

and

$$y_i^{k+1} = \max\{\hat{y}_i^k, \varepsilon\}, \qquad i = 1, \ldots, n,$$

with $y^1 \in S_1$ chosen arbitrarily.

Theorem 2 in [4] shows that the sequences $\{y^k\}$ and $\{\hat{y}^k\}$ converge to points y and $\hat{y}$ in S_1 and S_2, respectively, satisfying

(9) $$\|y - \hat{y}\| = \inf_{\substack{u \in S_1 \\ v \in S_2}} \|u - v\|.$$

Thus if $S_1 \cap S_2 \neq \varphi$, each of the sequences $\{y^k\}$, $\{\hat{y}^k\}$ converges to a point y satisfying $Ay = b$, $y \geq \varepsilon$. This suggests that we choose $x^0 = y^k$ for a reasonably large value of k. In practice we have never had to choose $k > 10$.

THEOREM 1. *Let $x^0 = y^k$ and define $D_0 = \operatorname{diag}(x_1^0, \ldots, x_n^0)$. Then for k sufficiently large we have*

(10) $$\sigma_0 = (b - Ax^0)^T (AD_0^2 A^T)^{-1} (b - Ax^0) \leq n.$$

PROOF. We have

$$\begin{aligned}
\sigma_0 &= (b - Ay^k)^T (AD_0^2 A^T)^{-1} (b - Ay^k) \\
&= (\hat{y}^k - y^k)^T A^T (AD_0^2 A^T)^{-1} A(\hat{y}^k - y^k) \\
&= [D_0^{-1}(\hat{y}^k - y^k)]^T D_0 A^T (AD_0^2 A^T)^{-1} AD_0 [D_0^{-1}(\hat{y}_k - y^k)].
\end{aligned}$$

Since $D_0 A^T (AD_0^2 A^T)^{-1} AD_0$ is a projection matrix we have

(11) $$\sigma_0 \leq \|D_0^{-1}(\hat{y}^k - y^k)\|^2 \leq \varepsilon^{-2} \|\hat{y}^k - y^k\|^2.$$

If $S_1 \cap S_2 \neq \varphi$ it follows from (9) that $\|\hat{y}^k - y^k\| \to 0$, so the theorem holds as soon as k is large enough that

$$\|\hat{y}^k - y^k\|^2 \leq n\varepsilon^2.$$

Suppose now that $S_1 \cap S_2 = \varphi$. Since the set of feasible solutions for (1) has a nonempty interior, there is a point $y > 0$ satisfying $Ay = b$. Let $\hat{y}$ denote the projection of y onto S_1. Thus

$$\hat{y}_i = \max\{y_i, \varepsilon\}, \qquad i = 1, \ldots, n.$$

Clearly $\|\hat{y} - y\|^2 < n\varepsilon^2$. Let $\delta > 0$ be chosen so small that $\|\hat{y} - y\|^2 + \delta \leq n\varepsilon^2$. It follows from (9) that

$$\|\hat{y}^k - y^k\|^2 \leq \|\hat{y} - y\|^2 + \delta \leq n\varepsilon^2$$

for k sufficiently large. Substituting this into (11) gives the conclusion of the theorem.

4. A convergence result

In [3] Dikin proves convergence of the affine scaling algorithm (2) for the case when (1) is nondegenerate. There are also convergence proofs in [2] and [6] for the case when (1) and its dual are nondegenerate. In this section we prove that when $x^0 > 0$ is chosen sufficiently close to a solution of $Ax = b$, the affine scaling algorithm applied to (5) converges at least as fast as a geometric progression with ratio $1 - 1/\sqrt{2}$. The proof requires no nondegeneracy assumptions. Moreover, our assumption about x^0 being close to a solution of $Ax = b$ does not make our convergence result a local one in the usual sense. As Figure 1(a) shows, it is possible to have x^0 quite close to S_2 while being a great distance from $S_1 \cap S_2$. Our convergence result depends solely on the size of σ_0, and we have seen in (10) that when $S_1 \cap S_2 \neq \varphi$, σ_0 can be made small just by taking x^0 close to S_2. In this section we assume $S_1 \cap S_2 \neq \varphi$.

It is a fortunate situation that our theorem does not require x^0 to be close to $S_1 \cap S_2$. For it is precisely in the case where we have $y^1 \in S_1$ close to S_2 and far from $S_1 \cap S_2$ that the method of successive projections has trouble finding a point near $S_1 \cap S_2$. Before we go further we explain why this is so.

As in Section 3, let y^k be a point in S_1 and let $\hat{y}^k$ be the point in S_2 nearest y^k. Let y be any point in $S_1 \cap S_2$. It follows from (8) and the condition $Ay = b$ that

$$(y - \hat{y}^k)^T (\hat{y}^k - y^k) = 0.$$

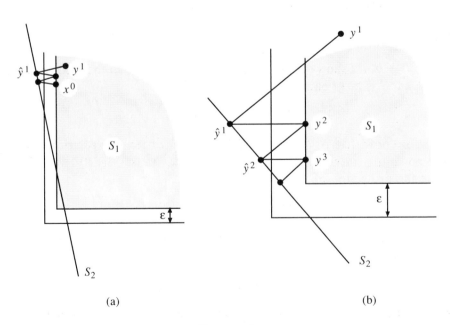

(a)   (b)

FIGURE 1

It follows that

$$\begin{aligned}
\|\hat{y}^k - y\|^2 &= \|(\hat{y}^k - y^k) + (y^k - y)\|^2 \\
&= \|\hat{y}^k - y^k\|^2 + 2(\hat{y}^k - y^k)^T(y^k - y) + \|y^k - y\|^2 \\
&= \|\hat{y}^k - y^k\|^2 + 2(\hat{y}^k - y^k)^T\{(\hat{y}^k - y) - (\hat{y}^k - y^k)\} + \|y^k - y\|^2 \\
&= \|y^k - y\|^2 - \|(\hat{y}^k - y^k)\|^2 \\
&= \left(1 - \frac{\|\hat{y}^k - y^k\|^2}{\|y^k - y\|^2}\right)\|y^k - y\|^2.
\end{aligned}$$

This equation shows that if the distance from y^k to S_2 is small compared to the distance from y^k to $S_1 \cap S_2$, then $\hat{y}^k$ is only slightly closer to $S_1 \cap S_2$ than y^k.

A similar analysis shows that

$$\|y^{k+1} - y\|^2 \leq \left(1 - \frac{\|y^{k+1} - \hat{y}^k\|^2}{\|\hat{y}^k - y\|^2}\right)\|\hat{y}^k - y\|^2$$

and there is very little slack in this inequality. It follows that when $\|\hat{y}^k - y^k\|$ is relatively small the sequences $\{y^k\}$, $\{\hat{y}^k\}$ converge very slowly to a point in $S_1 \cap S_2$. But in this case σ_0 is small by (11).

THEOREM 2. *If x^0 is chosen such that $\sigma_0 \leq 1$, the affine scaling algorithm applied to* (5) *converges at the rate of a geometric progression.*

PROOF. The affine scaling iterates (3) for problem (5) are given by

$$x_j^{k+1} = x_j^k\left(1 + \frac{x_{n+1}^k x_j^k a_j^T(AD_k^2 A^T)^{-1}a}{\sqrt{1 + (x_{n+1}^k)^2\sigma_k}}\right), \qquad j = 1, \dots, n,$$

and

(12) $$x_{n+1}^{k+1} = x_{n+1}^k\left(1 - \frac{1}{\sqrt{1 + (x_{n+1}^k)^2\sigma_k}}\right).$$

The main ingredient in our proof is the observation that for each k,

(13) $$|x_{n+1}^k x_j^k a_j^T(AD_k^2 A^T)^{-1}a| \leq 1, \qquad j = 1, \dots, n.$$

We will prove this condition by induction on k. First note that

$$\begin{aligned}
|x_{n+1}^k x_j^k a_j^T&(AD_k^2 A^T)^{-1}a| \\
&\leq \{(x_j^k)^2 a_j^T(AD_k^2 A^T)^{-1}a_j\}^{1/2}\{(x_{n+1}^k)^2 a^T(AD_k^2 A^T)^{-1}a\}^{1/2} \\
&= \{(x_j^k)^2 a_j^T(AD_k^T A^T)^{-1}a_j\}^{1/2}\{(x_{n+1}^k)^2\sigma_k\}^{1/2}.
\end{aligned}$$

The term $(x_j^k)^2 a_j^T(AD_k^2 A^T)^{-1}a_j$ is a diagonal element in the projection matrix

$$D_k A^T(AD_k^2 A^T)^{-1}AD_k$$

and is consequently ≤ 1. We therefore have

(14) $$|x_{n+1}^k x_j^k a_j^T (AD_k^2 A^T)^{-1} a| \leq \{(x_{n+1}^k)^2 \sigma_k\}^{1/2}$$

for each k. We have $x_{n+1}^0 = 1$ and $\sigma_0 \leq 1$ by hypothesis. Thus (13) holds for $k = 0$.

Assume now that $(x_{n+1}^k)^2 \sigma_k \leq 1$ so that (13) holds for some value of $k \geq 0$. Then, by (12),

$$x_j^{k+1} \geq x_j^k \left(1 - \frac{1}{\sqrt{1 + (x_{n+1}^k)^2 \sigma_k}} \right)$$

and consequently

(15) $$\frac{x_j^{k+1}}{x_{n+1}^{k+1}} \geq \frac{x_j^k}{x_{n+1}^k}, \qquad j = 1, \ldots, n.$$

If P and Q are $n \times n$ positive definite matrices we write $P \geq Q$ if $x^T P x \geq x^T Q x$ for each n-dimensional vector x. It is well known, and easy to show, that

$$P \geq Q \text{ implies } P^{-1} \leq Q^{-1}.$$

It follows from (15) that

$$\frac{1}{(x_{n+1}^{k+1})^2} AD_{k+1}^2 A^T = \frac{1}{(x_{n+1}^{k+1})^2} \sum_{j=1}^n (x_j^{k+1})^2 a_j a_j^T$$

$$\geq \frac{1}{(x_{n+1}^k)^2} \sum_{j=1}^n (x_j^k)^2 a_j a_j^T = \frac{1}{(x_{n+1}^k)^2} AD_k^2 A^T$$

so that

$$(x_{n+1}^{k+1})^2 (AD_{k+1}^2 A^T)^{-1} \leq (x_{n+1}^k)^2 (AD_k^2 A^T)^{-1}.$$

This implies that

(16) $$\begin{aligned} (x_{n+1}^{k+1})^2 \sigma_{k+1} &= (x_{n+1}^{k+1})^2 a^T (AD_{k+1}^2 A^T)^{-1} a \\ &\leq (x_{n+1}^k)^2 a^T (AD_k^2 A^T)^{-1} a = (x_{n+1}^k)^2 \sigma_k. \end{aligned}$$

If we now replace k by $k + 1$ in (14) and use the induction hypothesis, we see that (13) holds for each value of k. This implies that (16) also holds for each value of $k \geq 0$. Since $x_{n+1}^0 \sigma_0 \leq 1$ it follows from (12) that

$$x_{n+1}^{k+1} \leq \left(1 - \frac{1}{\sqrt{2}} \right) x_{n+1}^k \quad \text{for } k = 0, 1, \ldots.$$

This completes the proof of the theorem.

REFERENCES

1. I. Adler, N. Karmarkar, M. G. C. Resende, and G. Veiga, *An implementation of Karmarkar's algorithm for linear programming*, Report No. ORC 86-8, May 1987, Operations Research Center, University of California, Berkeley, CA.
2. E. R. Barnes, *A variation on Karmarkar's algorithm for solving linear programming problems*, Math. Programming **36** (1986), 174–182.
3. I. I. Dikin, *On the speed of an iterative process*, Uprawlyaemye Sistemi **12** (1974), 54–60.
4. L. G. Gubin, B. T. Polyak, and E. V. Raik, *The method of projections for finding the common point of convex sets*, U.S.S.R. Comput. Math. and Math. Phys. **6** (1967), 1–24.
5. R. J. Vanderbei and J. C. Lagarias, *I. I. Dikin's convergence result for the affine scaling algorithm*, this volume, pp. 109–119.
6. R. J. Vanderbei, M. S. Meketon, and B. A. Freedman, *A modification of Karmarkar's linear programming algorithm*, Algorithmica **1** (1986), 395–407.

SCHOOL OF INDUSTRIAL AND SYSTEMS ENGINEERING, GEORGIA INSTITUTE OF TECHNOLOGY, ATLANTA, GEORGIA 30332-0205

E-mail address: ebarnes @gtri01.gatech.edu

Contemporary Mathematics
Volume **114**, 1990

Dual Ellipsoids and Degeneracy in the Projective Algorithm for Linear Programming

KURT M. ANSTREICHER

ABSTRACT. We consider the construction of dual ellipsoids in the standard-form projective algorithm for linear programming. As originally suggested by Todd and further developed by Ye, these ellipsoids potentially allow for the removal of primal variables by proving that they must be zero in all optimal solutions. When the linear program is nondegenerate, Ye has shown that in fact the solution basis can eventually be identified exactly in this manner. However, Ye's argument does not generalize in the presence of primal degeneracy. We show here, under a mild assumption, that in all cases the dual ellipsoids converge to lie in the affine hull of the set of dual optimal solutions. This observation suggests a Lagrangian strengthening of the dual ellipsoid construction, which could significantly improve performance in the degenerate case.

1. Introduction

An important issue in the use of interior algorithms for linear programming is the problem of ultimately obtaining an exact, optimal, basic solution. Although interior algorithms appear to have the capability to obtain near-optimal solutions in a small number of iterations, the additional "clean-up" required to obtain a provably optimal solution could easily ruin any apparent advantage over the simplex method. Not surprisingly, several researchers have suggested (at least in talks) heuristic schemes for eliminating variables "on the fly" when solving linear programs using interior algorithms. Moreover, at least one theoretically rigorous approach has been devised. It was first shown in Todd [1988a] that ellipsoids that contain all dual optimal solutions can easily be generated in the standard-form variant of the projective algorithm. Ye [1987] then recast these ellipsoids in the space of dual slacks and developed a simple, closed-form test that can prove that a primal variable must be zero in all optimal solutions. Ye also showed that when the linear

---

1980 *Mathematics Subject Classification.* Primary 90C05.
Key words and phrases. Linear programming, Karmarkar's algorithm, projective algorithm, degeneracy.

program to be solved is nondegenerate, the ellipsoid-based test will eventually
identify the solution basis exactly. Moreover, this approach is computationally practicable in that all variables can be tested at the expense of roughly
one iteration of the algorithm.

Unfortunately, Ye's proof that the ellipsoid construction will eventually
identify the solution basis does not appear to easily generalize in the presence of primal degeneracy. In fact, preliminary computational experiments
on degenerate problems (Todd [1988b]) have shown that the dual ellipsoids
may perform very poorly as the primal iterates near optimality. In this paper we consider the dual ellipsoid construction with a specific view on the
effect of primal degeneracy. We show, under a mild assumption, that in all
cases the dual ellipsoids converge to lie in the affine hull of the set of dual
optimal solutions. This limiting behavior is generally not sufficient for a
variable-elimination procedure based on the dual ellipsoids to work. However, an important consequence of the limiting behavior is that failure of the
procedure is ultimately caused by the lack of enforcement of nonnegativity
constraints on dual slack variables. Motivated by this observation, we suggest
a Lagrangian strengthening of the ellipsoid construction that enforces nonnegativity constraints. The Lagrangian procedure is computationally practicable
and could significantly improve performance in the degenerate case.

2. The projective algorithm

We briefly describe here the "standard-form variant" of the projective algorithm for linear programming, as devised in Anstreicher [1986], Gay [1987],
Gonzaga [1985], Steger [1985], and Ye and Kojima [1987]. For a more detailed description, including a proof of the main convergence result, consult
any of the above references. Consider a standard form linear program

$$\text{LP} \quad z^* = \min \tilde{c}^\top \tilde{x},$$
$$\tilde{A}\tilde{x} = b,$$
$$\tilde{x} \geq 0,$$

where $\tilde{x} \in \Re^{n-1}$, and $\tilde{A}$ is a $m \times (n-1)$ matrix. We assume that $\{\tilde{x} \in \Re^{n-1} | \tilde{A}\tilde{x} = 0, \tilde{c}^\top \tilde{x} \leq 0\} = \{0\}$, and that $e \in \Re^{n-1}$ is feasible for LP.
(Throughout the paper, e will denote a vector of varying dimension, all of
whose components are equal to one.) We also assume that a valid lower
bound $z^0 \leq z^*$ is known; an algorithm for generating such a z^0 is given in
Anstreicher [1986].

Following the methodology of Anstreicher [1986], we consider a fractional
linear program on the simplex, equivalent to LP:

$$\text{FLP} \quad z^* = \min \frac{c^\top x}{d^\top x},$$
$$Ax = 0,$$
$$x \in S,$$

where $x \in \Re^n$, $c^\top = [\hat{c}^\top, 0]$, $d^\top = [0^\top, 1]$, $A = [\tilde{A}, -b]$, and $S \subset \Re^n$ is the simplex $\{x \geq 0 | e^\top x = n\}$. On each iteration $k \geq 0$ of the projective algorithm for LP, we have an $x^k > 0$, feasible for FLP, and a valid lower bound $z^k \leq z^*$. ($x^0 = e$, and z^0 is assumed known.) Fix k, and let D be the diagonal matrix with $D_{ii} = x_i^k$, $i = 1, \ldots, n$. Using the projective transformation $T(\cdot): S \to S$,

$$y = T(x) = \left(\frac{n}{e^\top D^{-1} x}\right) D^{-1} x,$$

$$x = T^{-1}(y) = \left(\frac{n}{e^\top D y}\right) D y,$$

FLP is equivalent to the "transformed" problem

$$\overline{\text{FLP}} \quad \min \frac{\overline{c}^\top y}{\overline{d}^\top y},$$

$$\overline{A} y = 0,$$

$$y \in S,$$

where $\overline{A} = AD$, $\overline{c} = Dc$, and $\overline{d} = Dd$. Since $T(x^k) = e$, a step starting at e in $\overline{\text{FLP}}$ corresponds to a step starting at x^k in FLP.

We now describe the step taken in $\overline{\text{FLP}}$ on a given iteration k. For any $t \in \Re^n$ we use the notation t_q and t_p to denote the projections of t onto the subspaces $\{y | \overline{A} y = 0\}$ and $\{y | \overline{A} y = 0, e^\top y = 0\}$, respectively. First, the lower bound z^k is updated to a value $\overline{z} \geq z^k$. Define a function $\overline{\theta}_2(\cdot): \Re \to \Re$ by

$$\overline{\theta}_2(z) = \min_{j=1,\ldots,n} (\overline{c}_q - z \overline{d}_q)_j.$$

If $\overline{\theta}_2(z^k) \leq 0$ let $\overline{z} = z^k$. Otherwise there is a unique $\overline{z} > z^k$ with $\overline{\theta}_2(\overline{z}) = 0$, and in fact $\overline{z}$ may be computed via a simple "ratio test." Once $\overline{z}$ is determined, the step in $\overline{\text{FLP}}$ is of the form $y(\alpha) = e - (\alpha/\|g\|)g$, where $g = \overline{c}_p - \overline{z} \overline{d}_p$. In practice the step length α is chosen by an approximate line search of the potential function $f(\overline{c} - \overline{z}\overline{d}, y) = n \ln((\overline{c} - \overline{z}\overline{d})^\top y) - \sum_{i=1}^n \ln(y_i)$. Following the choice of α we set $x^{k+1} = T^{-1}(y(\alpha))$, $z^{k+1} = \overline{z}$, and go to iteration $k + 1$.

Let $v^k = c^\top x^k / d^\top x^k$ denote the objective value in FLP on iteration k. The basic convergence result for the algorithm is that for all $k \geq 0$, α may be chosen so that

(2.1) $$(v^k - z^k) < \frac{\exp(-0.3k/n)}{d^\top x^k}(v^0 - z^0),$$

where it can be shown that there is a $\delta > 0$ such that $d^\top x^k = x_n^k \geq \delta$, $k \geq 0$. Equation (2.1) follows from the fact that for any fixed k, and $y = T(x)$, $f(\overline{c} - \overline{z}\overline{d}, y)$ differs from $f(c - z^{k+1}d, x)$ by a constant that is independent of x. Moreover, it can be shown directly that there is always a choice of α so that $f(\overline{c} - \overline{z}\overline{d}, y(\alpha)) - f(\overline{c} - \overline{z}\overline{d}, e) < -0.3$. (For details see Anstreicher [1986].)

3. Dual ellipsoids

Consider an iteration k of the projective algorithm on which the lower bound is increased: $z^{k+1} > z^k$. As described in the previous section, it is then the case (by construction) that $\overline{c}_q - z^{k+1}\overline{d}_q \geq 0$. Let $c^k = c - z^{k+1}d = (\tilde{c}^\top, -z^{k+1})^\top$, and let $u^k \in \mathfrak{R}^m$ be such that $\overline{c}_q - z^{k+1}\overline{d}_q = \overline{c}_q^k = \overline{c}^k - \overline{A}^\top u^k$. Then $\overline{c}_q^k \geq 0$, implying $c^k - A^\top u^k \geq 0$, and the latter is exactly $\tilde{c} - \tilde{A}^\top u^k \geq 0$, $b^\top u^k \geq z^{k+1}$. Thus the process of updating the lower bound implicitly generates a feasible solution to the dual of LP, with dual objective value at least z^{k+1}. This is the fundamental duality result of Todd and Burrell [1986] as extended to the standard form algorithm in Gay [1987] and Ye and Kojima [1987].

We next describe the dual ellipsoid constructions of Todd [1988a] and Ye [1987]. Let u be any optimal solution to the dual of LP. Then certainly $\tilde{A}^\top u \leq \tilde{c}$ and $b^\top u \geq z^{k+1}$, so $A^\top u \leq c^k$. Letting $s = c^k - A^\top u$, we then have $Ds \geq 0$. Then $\|Ds\|^2 \leq |e^\top Ds|^2 = |(c^k - A^\top u)^\top x^k|^2 = \gamma_k^2$, where $\gamma_k = c^{k^\top} x^k$. Thus all dual slack vectors s corresponding to optimal dual solutions u are contained in the ellipsoid

$$S^k = \{s | s^\top D^2 s \leq \gamma_k^2, s = c^k - A^\top u, u \in \mathfrak{R}^m\}.$$

This is the dual slack ellipsoid of Ye [1987]. It is then straightforward to show that

$$(3.1) \quad \begin{aligned} S^k &= \{s | (s - s^k)^\top D^2 (s - s^k) \leq \epsilon_k^2, s = c^k - A^\top u, u \in \mathfrak{R}^m\} \\ &= \{s | (u - u^k)^\top A D^2 A^\top (u - u^k) \leq \epsilon_k^2, s = c^k - A^\top u, u \in \mathfrak{R}^m\}, \end{aligned}$$

where $s^k = c^k - A^\top u^k$ and $\epsilon_k^2 = \gamma_k^2 - \|Ds^k\|^2$. (Note that $Ds^k = \overline{c}_q^k$.) The third formula for S^k follows naturally from the derivation of

$$U^k = \{u | (u - u^k)^\top A D^2 A^\top (u - u^k) \leq \epsilon_k^2\}$$

in Todd [1988a]. Todd showed directly that the ellipsoid U^k contains all dual optimal solutions, and clearly S^k is the image of U^k under the affine mapping $s = c^k - A^\top u$. Todd also characterized the volume of U^k, and under a nondegeneracy assumption showed that the volumes of U^k converge to zero at least geometrically fast. A related result of Ye is that the logarithm of the volume of the set $\{s | s^\top D^2 s \leq \gamma_k^2\}$ differs from $f(c^k, x^k)$ by a constant, and therefore reducing the potential function $f(c^k, \cdot)$ by a constant produces a constant factor reduction in the volume of $\{s | s^\top D^2 s \leq \gamma_k^2\}$.

Note, however, that the fact that the volume of the latter set approaches zero certainly does not imply that S^k collapses to a point.

The primary motivation for considering the ellipsoids S^k is that if $s_i > 0$ for all $s \in S^k$, then $x_i^* = 0$ in all optimal solutions of FLP. A procedure for eliminating variables in FLP can then be based on computing the minimum value of s_i over $s \in S^k$. We consider this process, first suggested in Todd [1988a] and more fully developed in Ye [1987], in the next section. Our purpose here is to develop a simple, partial characterization of the limiting behavior of U^k. This characterization is essentially qualitative, but provides considerable insight into what can go wrong with a variable-elimination procedure based on S^k when LP is degenerate. In the next section we introduce a strengthened variable-elimination procedure based on these observations.

Henceforth we assume that $z^k < z^*$ for all $k \geq 0$, so that the lower bound is updated infinitely many times. Let $\mathscr{K} = \{k \geq 0 | z^{k+1} > z^k\}$. Then for all $k \in \mathscr{K}$,

$$
(3.2) \quad \begin{aligned}
U^k &= \{u | (u - u^k)^\top A D^2 A^\top (u - u^k) \leq \epsilon_k^2\} \\
&= \{u | \|D A^\top (u^k - u)\|^2 \leq \epsilon_k^2\} \\
&= \left\{u | \sum_{i=1}^{n} (x_i^k)^2 |A_i^\top (u^k - u)|^2 \leq \epsilon_k^2\right\},
\end{aligned}
$$

where A_i denotes the ith column of A. Note that since x_n^k is bounded away from zero, $A_n = -b$, $b^\top u^k \to z^*$, and $\epsilon_k \to 0$, it is obviously the case that in the limit U^k, $k \in \mathscr{K}$ lie in the flat $\{u | b^\top u = z^*\}$, in the sense that for any fixed u with $b^\top u \neq z^*$, there is a $k(u)$ such that $u \notin U^k$ for all $k \in \mathscr{K}$, $k \geq k(u)$. In fact, a stronger result holds. We show below, under a mild assumption, that in the limit U^k, $k \in \mathscr{K}$ lie in the affine hull of the set of dual optimal solutions—an affine space of dimension generally much less than $m - 1$.

If T is any set of vectors in $\mathfrak{R}^n$, we use $[T]$ to denote the linear span of the elements of T. For any $t \in \mathfrak{R}^n$, $t \geq 0$, let $\beta(t) = \{j | t_j > 0\}$, $\alpha(t) = \{j | t_j = 0\}$. A feasible solution x of FLP is *nondegenerate* if $[\{A_j | j \in \beta(x)\}]$ has dimension m. An optimal solution x^* of FLP is *maximal* if the dimension of $[\{A_j | j \in \beta(x^*)\}]$ is maximal among all optimal solutions. Note then that any nondegenerate optimal solution is immediately maximal, but a maximal optimal solution could be degenerate. For $z \leq z^*$, let $u(z) = \{u | \tilde{A}^\top u \leq \tilde{c}, b^\top u \geq z\}$. Then $u(z^*)$ is the set of optimal solutions to the dual of LP, and $u^k \in u(z^{k+1})$ for every $k \in \mathscr{K}$. Let U^* denote the affine hull of $u(z^*)$, and let u^* be any point in the relative interior of $u(z^*)$; if $u(z^*)$ is a singleton, then $\{u^*\} = U^* = u(z^*)$. Finally let $s^* = (c - z^* d) - A^\top u^* = ((\tilde{c} - \tilde{A}^\top u^*)^\top, 0)^\top$.

LEMMA 3.1. *Let x^* be any maximal optimal solution of* FLP. *Then* $[\{A_j|j \in \alpha(s^*)\}] = [\{A_j|j \in \beta(x^*)\}]$.

PROOF. Immediately we have $\beta(x^*) \subset \alpha(s^*)$, by the complementary slackness condition for x^* and s^*. Since u^* is in the relative interior of $u(z^*)$, $\alpha(s^*) = \{j|(c - z^*d)_j - A_j^\top u = 0 \ \forall u \in u(z^*)\}$. By the Tucker strong complementarity theorem for LP, there is then an x^{**}, optimal for FLP, with $x_j^{**} > 0$ for all $j \in \alpha(s^*)$. But x^* is a maximal optimal solution, so $[\{A_j|j \in \beta(x^*)\}] = [\{A_j|j \in \beta(x^{**})\}] = [\{A_j|j \in \alpha(s^*)\}]$.   Q.E.D.

THEOREM 3.2. *Assume that $x^k \to x^*$, where x^* is a maximal optimal solution of* FLP. *Then for every $u \notin U^*$, there is a $k(u)$ so that $u \notin U^k$ for all $k \in \mathcal{K}$, $k \geq k(u)$.*

PROOF. Let $V^* = [\{A_j|j \in \alpha(s^*)\}]^\perp$, so that $U^* = u^* + V^*$. For any $u \in \mathfrak{R}^m$, there is then a unique decomposition $u = u^* + v(u) + w(u)$, where $v(u) \in V^*$, and $w(u) \perp V^*$. Clearly, $u \in U^*$ if and only if $w(u) = 0$. Since u^k, $k \in \mathcal{K}$ approach $u(z^*)$, it must be that $w(u^k) \to 0$, $k \in \mathcal{K}$. Now fix $u \notin U^*$, so $w(u) \neq 0$. Then $w(u^k - u) = w(u^k) - w(u) \to w(u)$, $k \in \mathcal{K}$. Since $V^{*\perp} = [\{A_j|j \in \beta(x^*)\}]$, and $x^k \to x^*$, there must be a $j \in \beta(x^*)$ so that $A_j^\top(u^k - u)$ is bounded away from zero for sufficiently large $k \in \mathcal{K}$. But x_j^k is also bounded away from zero, and $\epsilon_k \to 0$, so (3.2) implies that for large enough $k \in \mathcal{K}$, $u \notin U^k$.   Q.E.D.

Although Theorem 3.2 is quite simple, it has nontrivial consequences when one considers the performance of a variable-elimination scheme based on S^k. Note that if x^* is nondegenerate, then immediately $U^* = \{u^*\}$, and U^k, $k \in \mathcal{K}$ converge to u^*. It follows that all i such that $s_i^* > 0$ can eventually be identified using S^k, which is of course the desired result. When all optimal solutions of FLP are degenerate, however, $u(z^*)$ is not a singleton, and U^k, $k \in \mathcal{K}$ cannot possibly converge to a point. However, Theorem 3.2 implies that if $s_i > 0$ for all dual optimal u, then eventually any $u \in U^k$ having $s_i = 0$ are in fact dual infeasible. This observation motivates a strengthening of the variable-elimination procedure to be developed in the next section.

Finally, we note that some assumption similar to the condition that x^* be maximal appears to be necessary for Theorem 3.2 to hold. For example, consider the standard form linear program LP with $m = 3$, $n = 5$, and

$$\tilde{A} = \begin{pmatrix} 1 & 0 & 0 & 0 \\ 0 & 1 & -1 & 0 \\ 0 & 0 & 1 & 1 \end{pmatrix}, \qquad b = \begin{pmatrix} 1 \\ 0 \\ 1 \end{pmatrix}, \qquad \tilde{c} = \begin{pmatrix} 1 \\ 0 \\ 0 \\ 0 \\ 0 \end{pmatrix}.$$

Optimal solutions of LP are then of the form $(1, \theta, \theta, 1 - \theta)^\top$ for $0 \leq \theta \leq 1$, and the unique dual optimal solution is $u^* = (1, 0, 0)^\top$. However, if in FLP $x^k \to x^*$, where x^* is the degenerate optimal solution $(2.5, 0, 0, 0, 2.5)$, then in the limit U^k, $k \in \mathscr{K}$ can only be restricted to lie in the affine space $u^* + [\{(0, 1, 0)^\top\}]$. In all cases, if x^* is in the relative interior of the set of optimal solutions of FLP, then x^* is maximal. Hence, the maximality condition seems quite mild for the projective algorithm, which (in continuous form, at least) naturally converges to the "central" optimal solution.

4. Eliminating variables

As mentioned in the previous section, the ellipsoids S^k are primarily of interest because of their potential use in identifying variables that must be zero in an optimal solution of FLP. (Ye refers to such a variable-elimination scheme as "building down" to an optimal basis.) For any fixed $k \in \mathscr{K}$ and $i < n$, the natural problem to consider is the "ellipsoid optimization problem"

$$\text{EOP} \quad \min s_i$$

$$\frac{1}{2}(s - s^k)^\top D^2 (s - s^k) \le \frac{\epsilon_k^2}{2}$$

$$c^k - A^\top u = s.$$

THEOREM 4.1. (Ye [1987]). *Let s' denote the solution of* EOP, *and let $\mu' \in \mathfrak{R}$ and $\nu' \in \mathfrak{R}^n$ denote the corresponding* KKT *multipliers for the constraints* $0.5(s - s^k)^\top D^2 (s - s^k) - 0.5\epsilon_k^2 \le 0$ *and* $c^k - A^\top u - s = 0$, *respectively. Then*

$$s_i' = s_i^k - \epsilon_k \sqrt{q_i}, \qquad \mu' = \frac{\sqrt{q_i}}{\epsilon_k}, \qquad \nu' = I_i - D^2 q,$$

where I_i is the ith column of an $n \times n$ identity matrix, $\eta = (AD^2A^\top)^{-1}A_i$, and $q = A^\top \eta$.

PROOF. A straightforward application of the *KKT* optimality conditions for EOP. Q.E.D.

Since in the nondegenerate case each q_i is bounded (see Todd [1988a], Ye [1987]), Theorem 4.1 implies that in the nondegenerate case any i with $s_i^* > 0$ will eventually be identified via $s_i' > 0$ in the solution of EOP. Assuming that $AD^2A^\top$ has been factorized, the dominant effort in obtaining s_i' is the computation of q_i, which would generally be accomplished (in $O(m^2)$ operations) by solving the system $AD^2A^\top \eta = A_i$ for η, and then letting $q_i = A_i^\top \eta$. Thus q (and ν') can be obtained in an additional $O(mn)$ operation as a by-product of the computation of q_i.

Unfortunately, in the degenerate case there is little reason to expect that at x^k nears optimality, S^k will be able to identify *any* variables that must be zero, even when the degree of degeneracy (i.e., the dimension of U^*) is small. However, as indicated above, the ultimate downfalling of the use of

S^k is the lack of enforcement of nonnegativity constraints on s in EOP. Of course, explicitly adding the constraints $s \geq 0$ to EOP would result in a problem as difficult as the original linear program LP. However, a Lagrangian construction that explicitly enforces $s \geq 0$, while relaxing all other constraints, provides a computationally feasible strengthening of EOP. Specifically, consider the Lagrangian optimization problem

$$\text{LOP} \quad \max L(\mu, \nu),$$
$$A\nu = 0,$$
$$\mu \geq 0,$$

where $L(\mu, \nu)$ is the Lagrangian dual function

$$(4.1) \qquad L(\mu, \nu) = \min_{s \geq 0} s_i + \frac{\mu}{2}((s - s^k)^\top D^2 (s - s^k) - \epsilon_k^2) + \nu^\top (c^k - s).$$

Note that for fixed μ and ν, the objective in the problem defining $L(\mu, \nu)$ is a separable, convex quadratic. Thus, evaluating $L(\mu, \nu)$ can be accomplished in $O(n)$ arithmetic operations for any fixed μ and ν. In fact, a straightforward analysis obtains

$$L(\mu, \nu') = \nu'^\top c^k + \frac{\mu}{2}(\gamma_k^2 - 2\epsilon_k^2) - \frac{\mu}{2} \sum_{j=1}^{n} (x_j^k)^2 \left[\left(s_j^k - \frac{q_j}{\mu} \right)^+ \right]^2,$$

where for any scalar a, $a^+ = a$ if $a \geq 0$, and 0 otherwise. Furthermore, standard Lagrangian duality results (Bazaraa and Shetty [1979, Chapter 6]) imply that for any $\mu \geq 0$ and ν with $A\nu = 0$, $L(\mu, \nu)$ is a valid lower bound on the problem obtained by adding the constraints $s \geq 0$ to EOP. Finally, the original solution value s_i' in EOP corresponds exactly to evaluating $L(\mu', \nu')$ *ignoring* the constraints $s \geq 0$ in (4.1). Thus, enforcing the nonnegativity constraints in (4.1) will always produce $L(\mu', \nu') \geq s_i'$. If in fact $L(\mu', \nu') > 0$, then $s_i > 0$ in all dual optimal solutions, and the variable x_i could be eliminated from FLP. If on the other hand $L(\mu', \nu') \leq 0$, the lower bound on s_i could be further strengthened by performing "dual ascent" in the problem LOP. It is unlikely that modifying ν' would be worth the computational effort of satisfying the constraint $A\nu = 0$, but ascent on the scalar μ could certainly be worthwhile. In fact, whenever $L(\mu', \nu') > s_i'$, standard Lagrangian duality results imply that μ should be decreased from μ' so as to maximize $L(\cdot, \nu')$ in LOP.

REFERENCES

K. M. Anstreicher [1986], *A monotonic projective algorithm for fractional linear programming*, Algorithmica **1**, 483–498.

M. S. Bazaraa and C. M. Shetty [1979], *Nonlinear programming—Theory and algorithms*, John Wiley and Sons, New York, NY.

D. M. Gay [1987], *A variant of Karmarkar's linear programming algorithm for problems in standard form*, Math. Programming **37** (1), 81–90.

C. Gonzaga [1985], *A conical projection algorithm for linear programming*, Department of Electrical Engineering and Computer Science, University of California, Berkeley, CA.

A. E. Steger [1985], *An extension of Karmarkar's algorithm for bounded linear programming problems*, M. S. thesis, State University of New York, Stony Brook, NY.

M. J. Todd [1988a], *Improved bounds and containing ellipsoids in Karmarkar's linear programming algorithm*, Math. Oper. Res. **13** (4), 650–659.

——, [1988b], private communication.

M. J. Todd and B. P. Burrell [1986], *An extension of Karmarkar's algorithm for linear programming using dual variables*, Algorithmica **1**, 409–424.

Y. Ye [1987], *A 'build-down' simplex-Karmarkar method for linear programming*, Department of Engineering Economic Systems, Stanford University, Stanford, CA.

Y. Ye and M. Kojima [1987], *Recovering optimal dual solutions in Karmarkar's polynomial algorithm for linear programming*, Math. Programming **39**, 305–317.

DEPARTMENT OF OPERATIONS RESEARCH, YALE UNIVERSITY, NEW HAVEN, CONNECTICUT 06520

E-mail address: anskurm@yalevm

Contemporary Mathematics
Volume **114**, 1990

A Note on Limiting Behavior of the Projective and the Affine Rescaling Algorithms

MIROSLAV D. AŠIĆ, VERA V. KOVAČEVIĆ-VUJČIĆ,
AND MIRJANA D. RADOSAVLJEVIĆ-NIKOLIĆ

ABSTRACT. In this paper we show that both the projective (Karmarkar [**6**]) and the affine rescaling algorithms (Dikin [**4**], Barnes [**3**]) have a unique asymptotic direction of convergence to the optimum, provided that some nondegeneracy assumptions are satisfied. The degenerate case is still open, although a similar behavior was observed in many numerical experiments.

1. Introduction

The asymptotic behavior of various rescaling algorithms has been studied by many researchers. Both the discrete and the continuous variants were investigated and a variety of techniques applied. We mention here a few results that are closely related to the present work.

Megiddo and Shub in [**7**] consider the asymptotic behavior of continuous trajectories defined by the affine rescaling vector field. Under nondegeneracy conditions they show that every interior orbit is tangent to the direction uniquely determined by the componentwise inverse of the reduced cost vector at the optimal vertex. Adler and Monteiro in [**1**] show that a similar result holds also in the presence of primal and dual degeneracies. Witzgall, Boggs, and Domich present in [**12**] an alternative proof based on g-center manifolds.

Shub in [**10**] analyzes both continuous and discrete trajectories defined by the projective rescaling vector field. Under nondegeneracy assumptions he proves that the continuous trajectories are tangent at the optimum to the same direction as in [**7**]. In the discrete case he obtains an estimate of the rate of convergence of the generated sequence of points to the optimal solution.

1980 *Mathematics Subject Classification* (1985 *Revision*). Primary 90C05.
The authors' research supported in part by Republicka Zajednica za Nauku SR Srbije.
The third author died on March 11, 1990. The other authors dedicate this paper to her memory.

We shall discuss this result in more detail in Section 2. Finally, Monteiro [8] obtains limiting results for the tangential directions to continuous trajectories induced by the projective rescaling vector field without nondegeneracy assumptions.

In this paper we analyze discrete trajectories induced by both the projective and the affine rescaling algorithms with the small step size ($\alpha \leq \frac{1}{3}$ for the projective algorithm and $\alpha \leq \frac{1}{2}$ for the affine algorithm). Under nondegeneracy assumptions we prove results analogous to those in [7] and [10]. The degenerate case remains open, although numerical experience indicates that results similar to [1] and [8] hold.

2. The projective rescaling algorithm

Following [6], the projective rescaling algorithm is stated with respect to the linear programming problem in Karmarkar's standard form:

$$
\begin{aligned}
&\text{minimize} \quad c^T x \\
&\text{subject to} \quad Ax = 0 \\
&\qquad\qquad\quad e^T x = 1 \\
&\qquad\qquad\quad x \geq 0,
\end{aligned}
$$

(1)

where $A = [a_{ij}]_{m \times n}$, $x, c \in R^n$, and $e = (1, \ldots, 1)^T \in R^n$. We also assume that rank $A = m$, $Ae = 0$, and that the optimal value of the objective function is zero.

Given $\alpha \in (0, 1)$ the algorithm generates a sequence (x^k) defined by

$$
x^0 = \frac{1}{n} e, \qquad x^{k+1} = \frac{D_k b^k}{e^T D_k b^k}, \qquad k = 0, 1, \ldots,
$$

where $D_k = \text{diag}(x_1^k, \ldots, x_n^k)$,

$$
b^k = \frac{1}{n} e - \frac{\alpha}{\sqrt{n(n-1)}} \frac{c_p(x^k)}{\|c_p(x^k)\|},
$$

and $c_p(x^k)$ is the projection of the vector $D_k c$ onto the null space of the matrix

$$
\begin{bmatrix} AD_k \\ e^T \end{bmatrix}.
$$

Assuming that problem (1) has the unique nondegenerate solution, the following theorem can be proved:

THEOREM 1. *Let (x^k) be the sequence generated by the projective rescaling algorithm with $\alpha \leq \frac{1}{3}$. Suppose that $\overline{x} = (\overline{x}_1, \ldots, \overline{x}_{m+1}, 0, \ldots, 0)^T$, $\overline{x}_1 > 0, \ldots, \overline{x}_{m+1} > 0$ is the unique optimal solution to problem (1) and let $\overline{c} = (0, \ldots, 0, \overline{c}_{m+2}, \ldots, \overline{c}_n)^T$ be the reduced cost vector associated with the* *optimal basis* B. *Then* $\bar{c}_j > 0$, $j = m + 2, \ldots, n$, *and moreover,*

(i)
$$\lim_{k \to \infty} \frac{\bar{c}_j x_j^k}{\bar{c}^T x^k} = \frac{1}{n - m - 1}, \qquad j = m + 2, \ldots, n,$$

(ii)
$$\lim_{k \to \infty} \frac{c_p(x^k)}{c^T x^k} = \left(\underbrace{-\frac{1}{n}, \ldots, -\frac{1}{n}}_{m+1}, \frac{1}{n - m - 1} - \frac{1}{n}, \ldots, \frac{1}{n - m - 1} - \frac{1}{n} \right)^T,$$

(iii)
$$\lim_{k \to \infty} \frac{c^T x^{k+1}}{c^T x^k} = \frac{\sqrt{(n - 1)(n - m - 1)(m + 1)} - \alpha(m + 1)}{\sqrt{(n - 1)(n - m - 1)(m + 1)} + \alpha(n - m - 1)}.$$

The proof of Theorem 1 follows directly from the proof of Theorem 3 in [2]. Theorem 1 implies the following:

THEOREM 2. *Suppose that the assumptions of Theorem 1 are satisfied. Then*
$$\lim_{k \to \infty} \frac{x_N^{k+1} - x_N^k}{c^T x^{k+1} - c^T x^k} = \frac{1}{n - m - 1} \left(\frac{1}{\bar{c}_{m+2}}, \ldots, \frac{1}{\bar{c}_n} \right)^T,$$

where x_N^{k+1} and x_N^k are nonbasic parts of x^{k+1} and x^k, respectively.

PROOF. Let us first note that by the definition of x^{k+1} for $j = 1, \ldots, n$ we have

(2)
$$\frac{x_j^{k+1} - x_j^k}{c^T x^{k+1} - c^T x^k} = \frac{x_j^{k+1} - x_j^k}{\bar{c}^T x^{k+1} - \bar{c}^T x^k} = \frac{\dfrac{b_j^k x_j^k}{e^T D_k b^k} - x_j^k}{\dfrac{\bar{c}^T D_k b^k}{e^T D_k b^k} - \bar{c}^T x^k}$$
$$= \frac{b_j^k - e^T D_k b^k}{\dfrac{\bar{c}^T D_k b^k}{\bar{c}^T x^k} - e^T D_k b^k} \frac{x_j^k}{\bar{c}^T x^k}.$$

Using (ii) of Theorem 1 we get
$$\lim_{k \to \infty} \frac{c_p(x^k)}{\|c_p(x^k)\|} = \frac{1}{\sqrt{\dfrac{1}{n - m - 1} - \dfrac{1}{n}}}$$
$$\left(-\frac{1}{n}, \ldots, -\frac{1}{n}, \frac{1}{n - m - 1} - \frac{1}{n}, \ldots, \frac{1}{n - m - 1} - \frac{1}{n} \right)^T.$$

Now by the definition of b^k it follows that
(3)
$$\lim_{k \to \infty} b_j^k = \frac{1}{n} + \frac{\alpha}{\sqrt{n(n - 1)}} \frac{1}{n} \frac{1}{\sqrt{\dfrac{1}{n - m - 1} - \dfrac{1}{n}}} = u, \qquad j = 1, \ldots, m + 1,$$

$$\lim_{k \to \infty} b_j^k = \frac{1}{n} - \frac{\alpha}{\sqrt{n(n - 1)}} \sqrt{\frac{1}{n - m - 1} - \frac{1}{n}} = v, \qquad j = m + 2, \ldots, n.$$

On the other hand,

$$\frac{\bar{c}^T D_k b^k}{\bar{c}^T x^k} = \sum_{j=1}^n \frac{\bar{c}_j x_j^k}{\bar{c}^T x^k} b_j^k = \sum_{j=m+2}^n \frac{\bar{c}_j x_j^k}{\bar{c}^T x^k} b_j^k,$$

and by (i) of Theorem 1 and (3) we get

$$(4) \qquad \lim_{k \to \infty} \frac{\bar{c}^T D_k b^k}{\bar{c}^T x^k} = v.$$

We also have

$$(5) \qquad \lim_{k \to \infty} e^T D_k b^k = \sum_{j=1}^{m+1} \bar{x}_j u = u.$$

Using (3), (4), (5), and (i) of Theorem 1, from (2) we finally get

$$\lim_{k \to \infty} \frac{x_j^{k+1} - x_j^k}{c^T x^{k+1} - c^T x^k} = \frac{1}{n-m-1} \frac{1}{\bar{c}_j}, \qquad j = m+2, \dots, n. \quad \square$$

REMARK. The limit obtained in Theorem 2 uniquely determines the limit of the basic part of the asymptotic direction.

In [10] Shub shows that in the nondegenerate case there exist constants K_1, $K_2 > 0$ and a β,

$$0 < \beta \le \frac{1}{\sqrt{(m+1)(n-m-1)/n} + \alpha(n-m-1)/n}$$

such that

$$K_1(1 - \alpha\beta)^k \le \|x_N^k\| \le K_2(1 - \alpha\beta)^k,$$

which is an estimate of the rate of convergence of x_N^k to $\bar{x}_N$ and is not directly related to Theorem 2. However, it seems that Theorem 1 can be used to derive stronger results of this type.

3. The affine rescaling algorithm

The affine rescaling algorithm applies to the linear programming problem in standard form

$$(6) \qquad \begin{aligned} \text{minimize} \quad & c^T x \\ \text{subject to} \quad & Ax = b \\ & x \ge 0, \end{aligned}$$

where $A = [a_{ij}]_{m \times n}$, $x, c \in R^n$, $b \in R^m$. We again assume that rank $A = m$ and $Ae = b$, where $e = (1, \dots, 1)^T \in R^n$.

For the given $\alpha \in (0, 1)$ the algorithm generates the sequence (x^k) defined by

$$x^0 = e, \qquad x^{k+1} = D_k b^k, \qquad k = 0, 1, \dots,$$

where

$$D_k = \operatorname{diag}(x_1^k, \ldots, x_n^k), \qquad b^k = e - \alpha \frac{c_p(x^k)}{\|c_p(x^k)\|},$$

and $c_p(x^k)$ is the projection of the vector $D_k c$ onto the null space of the matrix AD_k.

Like Barnes in [3] we shall assume that (6) has no degenerate basic feasible solutions and that its dual has no degenerate basic solutions. Then the following theorem on the limiting behavior of the sequence (x^k) can be proved:

THEOREM 3. *Let* (x^k) *be the sequence generated by the affine rescaling algorithm with* $\alpha \le \frac{1}{2}$. *Suppose that* $\bar{x} = (\bar{x}_1, \ldots, \bar{x}_m, 0, \ldots, 0)^T$, $\bar{x}_1 > 0, \ldots, \bar{x}_m > 0$ *is the (unique) optimal solution to problem* (6), *and let* $\bar{c} = (0, \ldots, 0, \bar{c}_{m+1}, \ldots, \bar{c}_n)^T$ *be the reduced cost vector associated with the optimal basis* B. *Then* $\bar{c}_j > 0$, $j = m+1, \ldots, n$, *and furthermore,*

(i)
$$\lim_{k \to \infty} \frac{\bar{c}_j x_j^k}{\bar{c}^T x^k} = \frac{1}{n-m}, \qquad j = m+1, \ldots, n,$$

(ii)
$$\lim_{k \to \infty} \frac{c_p(x^k)}{c^T x^k - c^T \bar{x}} = \Big(\underbrace{0, \ldots, 0}_{m}, \frac{1}{n-m}, \ldots, \frac{1}{n-m} \Big)^T,$$

(iii)
$$\lim_{k \to \infty} \frac{c^T x^{k+1} - c^T \bar{x}}{c^T x^k - c^T \bar{x}} = 1 - \frac{\alpha}{\sqrt{n-m}}.$$

The proof of Theorem 3 follows from Theorem 3 in [9].

REMARK. Assumptions on primal and dual nondegeneracy are needed here only to establish the convergence of the sequence (x^k). Some new results on global convergence of the affine rescaling method indicate that these assumptions might be relaxed (Tsuchiya [11]). The arguments about the limiting behavior of the sequence (x^k) in the proof of Theorem 3 are based only on the assumption that the limit point $\bar{x}$ is the unique nondegenerate solution to problem (6).

As a consequence of Theorem 3 we have the following:

THEOREM 4. *Let the assumptions of Theorem 3 be satisfied. Then*

$$\lim_{k \to \infty} \frac{x_N^{k+1} - x_N^k}{c^T x^{k+1} - c^T x^k} = \frac{1}{n-m} \Big(\frac{1}{\bar{c}_{m+1}}, \ldots, \frac{1}{\bar{c}_n} \Big)^T.$$

PROOF. By the definition of the affine rescaling algorithm it follows that for $j = 1, \ldots, n$,

(7)
$$\frac{x_j^{k+1} - x_j^k}{c^T x^{k+1} - c^T x^k} = \frac{x_j^{k+1} - x_j^k}{\bar{c}^T x^{k+1} - \bar{c}^T x^k} = \frac{b_j^k - 1}{\frac{\bar{c}^T D_k b^k}{\bar{c}^T x^k} - 1} \frac{x_j^k}{\bar{c}^T x^k}.$$

Using (ii) of Theorem 3 we get

$$\lim_{k\to\infty} \frac{c_p(x^k)}{\|c_p(x^k)\|} = \left(0, \ldots, 0, \sqrt{\frac{1}{n-m}}, \ldots, \sqrt{\frac{1}{n-m}}\right)^T$$

and thus

$$(8) \qquad \lim_{k\to\infty} b_j^k = 1 - \alpha\sqrt{\frac{1}{n-m}} = w, \qquad j = m+1, \ldots, n.$$

Now, similarly as in Theorem 2, we conclude

$$\lim_{k\to\infty} \frac{\bar{c}^T D_k b^k}{\bar{c}^T x^k} = w,$$

which together with (7), (8), and (i) of Theorem 3 implies

$$\lim_{k\to\infty} \frac{x_j^{k+1} - x_j^k}{c^T x^{k+1} - c^T x^k} = \frac{1}{n-m}\frac{1}{\bar{c}_j}, \qquad j = m+1, \ldots, n. \quad \square$$

In the conclusion let us point out that Theorems 1 and 3 establish the existence of the limit

$$(9) \qquad \lim_{k\to\infty} \frac{x_N^k - \bar{x}_N}{\|x_N^k - \bar{x}_N\|},$$

while Theorems 2 and 4 establish the limit

$$(10) \qquad \lim_{k\to\infty} \frac{x_N^{k+1} - x_N^k}{\|x_N^{k+1} - x_N^k\|}$$

for both algorithms. The existence of the latter implies under some mild and obvious assumptions the existence of the former, but not vice versa, as can be seen from simple examples. In a sense, one might say that Theorems 2 and 4 give more information on the asymptotic behavior of the search direction than do Theorems 1 and 3.

Acknowledgment

The authors wish to thank a referee for several useful comments and especially for pointing out the relationship between the limits (9) and (10). The distinction between them is reminiscent of Stolz's Theorem (see, e.g., [5]) and even a proof can be given that goes along similar lines as that of Stolz's Theorem.

References

1. I. Adler and R. D. C. Monteiro, *Limiting behaviour of the affine scaling continuous trajectories for linear programming problems*, this volume, pp. 189–211.
2. M. D. Ašić, V. V. Kovačević-Vujčić, and M. D. Radosavljević-Nikolić, *Asymptotic behaviour of Karmarkar's method for linear programming*, Math. Programming **46** (1990), 1–18.

3. E. R. Barnes, *A variation on Karmarkar's algorithm for solving linear programming problems*, Math. Programming **36** (1986), 174–182.
4. I. I. Dikin, *O shodimosti odnovo iteracionovo procesa*, Upravliaemye sistemy, IM, IK, SO AN SSSR **12** (1974), 54–60.
5. G. M. Fihtengolc, *Kurs Diferencialnovo i Integralnovo Iscislenia* I (Nauka, Moskva, 1966).
6. N. Karmarkar, *A new polynomial-time algorithm for linear programming*, Combinatorica **4(4)** (1984), 373–395.
7. N. Megiddo and M. Shub, *Boundary behaviour of interior point algorithms in linear programming*, Math. Oper. Res. **14** (1989), 97–146.
8. R. D. C. Monteiro, *Convergence and boundary behaviour of the projective scaling trajectories for linear programming*, J. Complexity (to appear).
9. M. D. Radosavljević-Nikolić, *Asymptotic behaviour of the affine variant of Karmarkar's algorithm for linear programming*, XX Conference on Mathematical Optimization, Humboldt Universität zu Berlin (April 4–9, 1988).
10. M. Shub, *On the asymptotic behaviour of the projective rescaling algorithm for linear programming*, J. Complexity **3** (1987), 258–269.
11. T. Tsuchiya, *Global convergence of the affine scaling methods for degenerate linear programming problems*, Technical Report, The Institute of Statistical Mathematics (Tokyo, Japan, 1990).
12. C. Witzgall, P. Boggs, and P. Domich, *On the convergence behavior of trajectories for linear programming*, this volume, pp. 161–187.

DEPARTMENT OF MATHEMATICS, THE OHIO STATE UNIVERSITY, NEWARK, OHIO 43055
E-mail address: tj5351@ohstvma.ircc.ohio-state.edu

FACULTY OF ORGANIZATIONAL SCIENCES, BELGRADE UNIVERSITY, BELGRADE, YUGOSLAVIA

FACULTY OF ORGANIZATIONAL SCIENCES, BELGRADE UNIVERSITY, BELGRADE, YUGOSLAVIA

- 3 -
Trajectories
of Interior-Point Methods

Contemporary Mathematics
Volume **114**, 1990

On the Convergence Behavior of Trajectories
for Linear Programming

CHRISTOPH WITZGALL, PAUL T. BOGGS,
AND PAUL D. DOMICH

ABSTRACT. The convergence behavior of center trajectories arising from the use of the logarithmic barrier function in linear programming is examined based on the concept of analytic center of a system of linear constraints. Main results are the convergence of A-trajectories to g-centers of the optimal face and the convergence of their tangential directions. Both results hold in the presence of primal and dual degeneracies. g-center varieties are introduced extending the concept of analytic g-centers.

Introduction

In the early 1960s, Huard [13], [14] suggested a "method of centers" for minimizing a multivariate convex function $f(u)$ subject to convex inequalities $g_i(u) \leq 0$, $i = 1, \ldots, n$. Starting at a point $u^{(0)}$, a subsequent trial point $u^{(1)}$ is chosen as a suitable "center" of the system

$$g_0(u) = f(u) - f(u^{(0)}) \leq 0, \qquad g_i(u) \leq 0, \qquad i = 1, \ldots, n$$

of convex constraints. The process is then repeated, defining $u^{(2)}$, with $u^{(1)}$ in the role of $u^{(0)}$, and so on. Accumulation points of the resulting sequence of centers are optimal solutions. This method was soon recognized (see Fiacco and McCormick [11]) as essentially equivalent to the approach based on sequential unconstrained minimization techniques (SUMT) proposed earlier by Fiacco and McCormick [9], [10] for solving nonlinear programming problems. Here a "barrier function," such as the logarithmic barrier function, is used to transform a constrained minimization problem into a sequence of

1980 *Mathematics Subject Classification* (1985 *Revision*). Primary 90C05, 65K05, 34A99.

Contribution of the National Institute of Standards and Technology and not subject to copyright in the United States. This research was supported in part by ONR Contract N-0014-87-F0053.

unconstrained ones:

$$minimize \ f(u) - \pi_k \sum_{i=1}^{n} \ln(-g_i(u)), \qquad k = 1, 2, \ldots,$$

with penalty parameters π_k selected to converge to zero. With this particular choice of the barrier function, the SUMT method relates to the method of centers based on the "analytic center" in the parlance of Sonnevend [24]. Indeed, the points specified by the method of centers lie on the same "center trajectory" as those obtained by SUMT.

Linear programs constitute a very special case of the convex minimization problem, and it was only following the proposal of Karmarkar's method [15] that the usefulness for linear programming of such "interior point methods" was generally recognized, sparking an intense research effort. Karmarkar's polynomial complexity bound for linear programming, for instance, was improved by the proposal by Renegar [21] of a version of the method of centers.

Many post-Karmarkar methods took their departure from trajectory-following procedures that go back to Murray [19], Murray and Wright [20], and the classic text by Fiacco and McCormick [12]. Bayer and Lagarias [4], [5], [6] in their pathbreaking work identified and examined two such sets of trajectories, "A-trajectories" and "P-trajectories," and their associated direction fields. A-trajectories are affine invariant and correspond to SUMT with a logarithmic barrier function modified by the addition of a linear function. The associated field direction is often called "dual affine direction" or "affine scaling direction," and is utilized in a class of methods (e.g., Megiddo [16], Adler, Karmarkar, Resende, and Veiga [2], Barnes [3], Vanderbei, Meketon, and Freedman [26], [7], with forerunner Dikin [8]) for which a substantial body of computational experience has been accumulated. Karmarkar's method, on the other hand, relates to the P-trajectories.

A main result of this paper is the continuity of the analytic "g-centers" of a system of linear constraints with respect to its right-hand side. This implies that A-trajectories converge to the corresponding g-centers of the optimal face of feasible solutions (see Megiddo [17], Adler and Monteiro [1]). Our proof utilizes a lemma that bounds g-centers away from the boundary. This lemma is also used to derive other convergence properties.

A-trajectories have limit tangents. Moreover, it was frequently observed and is easily proved for nondegenerate optima that, if the optimum is unique, all A-trajectories have the same limit tangent. Adler and Monteiro [1] prove these facts in full generality. We give a different proof that utilizes "g-varieties." Megiddo and Shub [18] show continuity of an affine vector field on the boundary of the feasible region. This result was derived for a different purpose and does not imply convergence of tangential directions since the limits may vanish.

We also work with a more general form of linear program than is commonly used. This is to ensure that the dual of that linear program falls again under this formulation and thus permits relating primal and dual "g-centers." A broader and more detailed exposition of the material presented in this paper will be contained in a National Institute of Standards Technical Report by the authors [27].

1. Systems of linear inequalities and equations

Consider a system of n linear inequalities and l linear equations in R^m:

(1.1)
$$A^T u \le b, \qquad u \in R^m,$$
$$E^T u = f.$$

Here A denotes an $m \times n$ matrix, E an $m \times l$ matrix, while b and f are column vectors of n and l components, respectively.

Constraints that are either equations in the first place, or are inequalities that are binding for all feasible solutions, will be called

(1.2)   *tight*

within a solvable constraint system (1.1). The relative interior of the solution polyhedron consists of all feasible points u for which all nontight inequalities are nonbinding, that is, for which all

(1.3)   *residuals* $r_i(u) = b_i - A_i^T u$

of nontight inequalities are positive.

It is clear that replacing tight inequalities (1.2) by their corresponding equations in system (1.1) does not alter the solution set. We call this modification

(1.4)   *regularization.*

In a regularized system, none of the inequalities $A_i^T u \le b_i$ are tight. Note that only solvable systems can be regularized, as otherwise tightness is not defined. Thus solvability is understood whenever we refer to regularized systems.

Each face F of the solution polyhedron consists of the solutions to the

(1.5)   *face system:*
$$A_G^T u \le b_G,$$
$$A_F^T u = b_F,$$
$$E^T u = f,$$

which arises from the original system by replacing the inequalities that are binding for all points in F by their corresponding equations. In (1.5), we denote the matrix and right-hand sides of the remaining inequalities by A_G and b_G, respectively. Similarly, A_F and b_F represent the equations that were inequalities previously. Each such face system is automatically regularized (1.4).

We are now prepared to define analytic g-centers for bounded systems of linear constraints, that is, systems with bounded solution sets. With each linear constraint system we associate the

(1.6)   *logarithmic barrier function* $B(u) = \sum_{i \in S} \ln r_i(u)$,

where S is the subset of indices i belonging to nontight inequalities. If all inequalities are tight, then we define $B(u) = 0$ for all feasible solutions. Gradient and hessian matrix of $B(u)$ are

(1.7) $\nabla_u B(u) = -A_S R_S^{-1}(u)e, \qquad e^T = (1, 1, \ldots, 1),$

$\nabla_{uu}^2 B(u) = -A_S R_S^{-2}(u)A_S^T,$

where A_S is the submatrix of A consisting of the columns that correspond to nontight constraints $A_i^T u \le b_i$, $i \in S$, and $R_S(u)$ is the diagonal matrix of positive residuals of the same constraints and a square submatrix of the diagonal matrix of all residuals (1.3)

$$R(u) = \text{diag}(r_1(u), r_2(u), \ldots, r_n(u)).$$

The expression given in (1.7) implies that the hessian of the barrier function is negative semidefinite with none of its null-directions parallel to the solution set. On the relative interior of that polyhedron the logarithmic barrier function is therefore strictly concave. Subtracting an arbitrary linear term $g^T u$, $g \in R^n$ from the logarithmic barrier function $B(u)$ maintains strict concavity. The modified barrier function $B(u) - g^T u$ thus has a unique maximizer $u^h(g)$, the

(1.8)   *analytic g-center* or, simply, g-*center*

of the constraint system. In the case $g = 0$ we arrive at the classical center definition of Huard and Sonnevend. In the full-dimensional case, the condition for the g-center is simply that $\nabla_u B(u) = g$. Otherwise, it takes the form

(1.9) $\nabla_u B(u) = -A_S R_S^{-1}(u)e = g + [A_T, E]w,$

where the submatrix A_T of A corresponds to the tight inequalities. If the system of constraints has been regularized (1.4), then the condition simplifies to

(1.10) $\nabla_u B(u) = -A R^{-1}(u)e = g + Ew.$

We denote the analytic g-center of the face system associated with a face F of the solution polyhedron by

(1.11) $u_F^h(g).$

The g-center associated with a vertex of the solution polyhedron is, of course, that vertex itself.

Different vectors g may lead to the same g-center. Let L denote the linear subspace parallel to the affine hull H of a solution set. Then two g-centers $u^h(g^{(1)})$ and $u^h(g^{(2)})$ of the same constraint system coincide if and only if the vectors $g^{(1)}$, $g^{(2)}$ have the same orthogonal projection onto L. Also, for every relative interior solution u of a linear constraint system there exists a unique vector $g \in L$ for which $u = u^h(g)$. These considerations lead to the definition of higher dimensional analogues to analytic g-centers of a linear constraint system (1.1). We select in R^m, parallel to the affine hull H of the solutions set of (1.1) but otherwise arbitrary, a k-dimensional linear manifold, which we call a

(1.12)   *template*.

Along with such a template we consider the set of all its parallel displacements in R^m. If the system of linear equations

$$(1.13) \qquad C^T u = d$$

describes the template, then any parallel displacement of the template can be described by a system that differs from (1.13) only by the choice of the right-hand sides d. Adjoining one of those linear systems to the original system yields an expanded system

$$(1.14) \qquad \begin{aligned} A^T u &\leq b, \\ E^T u &= f, \\ C^T u &= d. \end{aligned}$$

We now vary the right-hand sides d such that the corresponding parallel displacements of the template meet the relative interior of the original solution polyhedron. The locus of all g-centers of such systems forms an $\hat{m} - k$-dimensional algebraic manifold or variety, the

(1.15)   *g-center variety*

for the template (1.13). Here $\hat{m}$ denotes the dimension of the original solution polyhedron. The variety is characterized by the existence of Lagrange multipliers w, z (compare (1.10)) such that

$$(1.16) \qquad -AR^{-1}(u)e = g + Ew + Cz.$$

Consider the template displacement passing through the g-center $u^h(g)$ of the original system (1.1). We call the corresponding expanded system (1.14) a

(1.17)   *g-center restriction*

of the system (1.1) by the template (1.13). There is a unique g-center restriction for each template.

(1.18) *The g-center restriction has the same g-center* $u^h(g)$ *as the original linear constraint system* (1.1). *Every g-center variety contains the g-center* $u^h(g)$.

PROOF. The g-center condition $-AR^{-1}(u^h(g))e = g + Ew$ (1.10), which has to hold because of the optimality of $u^h(g)$ for system (1.1), implies condition (1.16) with zero Lagrange multipliers z. □

The connection of g-centers with the Legendre transform (see for instance Rockafellar [22] for its definition) of the logarithmic barrier function has been developed by Bayer and Lagarias [6].

(1.19) *For a regularized constraint system* (1.1), *the mapping* $g \rightarrow u^h(g)$ *provides a 1-1 correspondence between the linear subspace* $\{g: E^T g = 0\}$ $(= R^m$ *in the full-dimensional case*) *and the relative interior of the solution polyhedron. Under this mapping, g-center varieties are the inverse images of flats. The set of all g-center varieties is therefore closed under intersection.*

2. Bounding centers away from the boundary

We will now establish a lower bound on the distance from a center to the boundary relative to the length of the corresponding chord. In what follows we will describe several applications of this result. In particular, it will be used in the next section to prove that centers vary continuously with right-hand sides, even if there is loss of dimension.

In this section, we will require linear constraint systems of the form (1.1) to be bounded, that is, to have bounded solution polyhedra. We call two different points on the relative boundary of a bounded solution polyhedron P

(2.1)   *antipodal*

if they are collinear with g-center $u^h(g) \in P$ of the corresponding constraint system.

(2.2) LEMMA. *If the g-center* $u^h(g)$ (1.8) *of a bounded system of linear constraints* (1.1) *is expressed as the weighted mean of two antipodal points* v *and* z,

$$u^h = \lambda v + \mu z, \qquad \lambda + \mu = 1, \qquad \lambda, \mu \geq 0,$$

then

$$\lambda, \mu \geq \frac{1}{n + \|g\| \, \|v - z\|} \geq \frac{1}{n + \|g\|\delta},$$

where n is the number of inequalities and δ is the diameter of the solution polyhedron P of the constraint system.

PROOF. Any line through two antipodal points defines a 1-dimensional g-center restriction (1.17). We refer to that linear constraint system as the

(2.3)   *antipodal restriction system*

generated by the antipodal line. The lemma thus holds in general if it holds for every such restriction system. To prove the latter, consider the unit vector $a = (v - z)/\|v - z\|$ in the direction of the antipodal line. The mapping $u = T(s) = u^h(g) + as$, $T: R^1 \to R^m$, with $\|a\| = 1$ maps an interval onto the diameter bounded by the two antipodal points while preserving euclidean lengths between points. The mapping transforms the inequalities of the restriction system (2.3) into the constraints $A_i^T(u^h(g) + as) \le b_i$. Of these constraints, those with $A_i^T a = 0$ do not depend on s, are trivially true, and may therefore be deleted. All constraints resulting from tight inequalities and equations are in that category. Otherwise, if $A_i^T a > 0$, then $s \le r_i(u^h(g))/A_i^T a$, and if $A_i^T a < 0$, then $s \ge r_i(u^h(g))/A_i^T a$. The mapping $T(s)$ thus transforms the restriction system (2.3) into a linear inequality system of the following form in the 1-dimensional variable s:

$$
(2.4) \qquad
\begin{array}{llll}
-s \le p_i, & 0 < p_1 \le p_i, & i = 1, 2, \ldots, n_l, & n_l \ge 1, \\
+s \le q_i, & 0 < q_1 \le q_i, & i = 1, 2, \ldots, n_r, & n_r \ge 1.
\end{array}
$$

It is well known and readily seen that analytic centers are invariant under such affine transformations. In particular, the g-center $u^h(g)$ maps—inversely—into the $\overline{g}$-center $s^h(\overline{g}) = 0$ of system (2.4), where

$$
\overline{g} = a^T g, \qquad |\overline{g}| \le \|g\|.
$$

The lemma will thus follow if

(2.5) *for any number $\overline{g} \in R^1$ the $\overline{g}$-center of the 1-dimensional system of $\hat{n}$ inequalities (2.4) is of the form*

$$
s^h(\overline{g}) = 0 = \lambda(-p_1) + \mu q_1,
$$

where, with $\overline{\delta} = q_1 + p_1 = \|v - z\| \le \delta$ and $\hat{n} = n_l + n_r \le n$,

$$
\lambda, \mu \ge \frac{1}{\hat{n} + |\overline{g}|\overline{\delta}}.
$$

PROOF OF (2.5). Since $-p_1$, q_1 are the only pair of antipodal points, the claim here is that $\underline{s} \le s^h(\overline{g}) \le \overline{s}$, where

$$
\underline{s} = \left(1 - \frac{1}{\hat{n} + |\overline{g}|\overline{\delta}}\right)(-p_1) + \left(\frac{1}{\hat{n} + |\overline{g}|\overline{\delta}}\right) q_1,
$$

$$
\overline{s} = \left(\frac{1}{\hat{n} + |\overline{g}|\overline{\delta}}\right)(-p_1) + \left(1 - \frac{1}{\hat{n} + |\overline{g}|\overline{\delta}}\right) q_1.
$$

In order to verify these estimates, let $m(s)$ denote the derivative of the logarithmic barrier function for system (2.4):

$$
m(s) = \sum_{i=1}^{n_l} \frac{1}{p_i + s} - \sum_{i=1}^{n_r} \frac{1}{q_i - s}.
$$

Suppose $-p_1 < s < \underline{s}$. Then

$$p_1 + s < p_1 + \underline{s} = \frac{p_1 + q_1}{\hat{n} + |\overline{g}|\overline{\delta}} = \frac{\overline{\delta}}{\hat{n} + |\overline{g}|\overline{\delta}},$$

and, recalling $q_1 \le q_i$, $i = 1, \ldots, n_r$,

$$q_i - s > q_1 - \underline{s} = \frac{(\hat{n} - 1 + |\overline{g}|\overline{\delta})(p_1 + q_1)}{\hat{n} + |\overline{g}|\overline{\delta}} = \frac{(\hat{n} - 1 + |\overline{g}|\overline{\delta})\overline{\delta}}{\hat{n} + |\overline{g}|\overline{\delta}}.$$

It follows that $m(s) > |\overline{g}|$; indeed

$$m(s) > \frac{1}{p_1 + \underline{s}} - \frac{n_r}{q_1 - \underline{s}}$$

$$= \frac{\hat{n} + |\overline{g}|\overline{\delta}}{\overline{\delta}} - \frac{n_r(\hat{n} + |\overline{g}|\overline{\delta})}{(\hat{n} - 1 + |\overline{g}|\overline{\delta})\overline{\delta}} = \left(\frac{\hat{n} + |\overline{g}|\overline{\delta}}{\hat{n} - 1 + |\overline{g}|\overline{\delta}} \right) \left(\frac{n_l - 1 + |\overline{g}|\overline{\delta}}{\overline{\delta}} \right) \ge |\overline{g}|.$$

Since $m(s^h(\overline{g})) = \overline{g}$ characterizes the $\overline{g}$-center, it follows from the above result that $s^h(\overline{g})$ cannot lie in the interval $-p_1 < s < \underline{s}$, whence $\underline{s} \le s^h(\overline{g})$. The analogous verification of $s^h(\overline{g}) \le \overline{s}$ completes the proof of (2.5) and therefore of Lemma (2.2). $\square$

Lemma (2.2) is sharp if $g = 0$ and the constraint system is a minimal representation of a regular d-simplex ($n = d + 1$). In this case, $\lambda = 1/n$ for a diameter through a vertex. This yields Karmarkar's [15] observation that the circumscribed radius of a d-simplex equals d times the inscribed one.

3. Convergence of analytic g-centers

A sequence of linear constraint systems,

$$(3.1) \qquad A^T u \le b^{(k)}, \qquad E^T u = f^{(k)}, \qquad k = 1, 2, \ldots,$$

is

$$(3.2) \qquad\qquad\qquad \textit{convergent}$$

if the sequences $\{b^{(k)}\}_{k=1,2,\ldots}$ and $\{f^{(k)}\}_{k=1,2,\ldots}$ of right-hand sides converge to limits $b^{(0)}$ and $f^{(0)}$, respectively. If the individual systems in sequence (3.1) are bounded, that is, if their solution polyhedra P^k are bounded, then every sequence $\{u^{(k)}\}_{k=1,2,\ldots}$ of solutions $u^{(k)} \in P^k$ has an accumulation point. Any such accumulation point will solve the limit system,

$$(3.3) \qquad\qquad A^T u \le b^{(0)}, \qquad E^T u = f^{(0)},$$

which is readily seen to have itself a bounded solution polyhedron $P^{(0)}$.

We call a convergent sequence (3.2) of linear constraint systems

$$(3.4) \qquad\qquad\qquad \textit{regularly convergent}$$

if corresponding inequalities $A_i^T u \leq b_i$ are either tight (1.2) for all integers k or nontight for all integers k, in other words, if the tightness pattern is independent of k. Since there are only finitely many such tightness patterns— each characterized by a subset $T \subseteq [1, \ldots, n]$ of indices for which the corresponding inequalities are tight for all k—it follows that

(3.5) *every convergent sequence* (3.2) *of linear constraint systems is the union of finitely many regularly convergent subsequences and a finite subsequence.*

We also observe

LEMMA (3.6). *Every vertex* $v^{(0)}$ *of the solution polyhedron* $P^{(0)}$ *of the limit system* (3.3) *is the limit of a sequence of points* $u^{(k)}$ *that belong to the relative boundary of the solution polyhedron* $P^{(k)}$ *of the kth constraint system in the sequence* (3.1). (*Each point* $u^{(k)}$ *may in fact be chosen as a vertex of* $P^{(k)}$.)

PROOF. Let $v^{(0)}$ be a vertex of the limit polyhedron $P^{(0)}$. Choose an objective function $c^T u$ such that $v^{(0)}$ is the unique solution of the corresponding linear program. Then it follows by well-known arguments that $v^{(0)}$ is the limit of optimal vertices to the linear programs that are obtained for the same objective function from the linear constraint systems (3.1). □

We are now ready to formulate our main result.

(3.7) THEOREM. *The g-centers of a convergent sequence* (3.1) *of bounded linear constraint systems converge towards the g-center of their limit system* (3.3).

PROOF. The sequence of centers $u^{(k)}$ possesses accumulation points, all of which solve the limit system. Let $\hat{u}$ be such an accumulation point. We show first that

(3.8) $\hat{u}$ *lies in the relative interior of the limit polyhedron.*

By Lemma (3.6) there exists a sequence of points $v^{(k)}$, each in the relative boundary of $P^{(k)}$, which converge towards $v^{(0)}$ as $k \to \infty$. Let $z^{(k)}$ be the antipodal point of $v^{(k)}$ with respect to the center $u^{(k)}$. Then, by Lemma (2.2),

$$u^{(k)} = \lambda_k v^{(k)} + \mu_k z^{(k)}, \qquad \lambda_k + \mu_k = 1, \qquad \lambda_k, \mu_k \geq \frac{1}{n + \|g\|\delta},$$

where n is the number of columns of matrix A, and δ is an upper bound on the diameters of all solution polyhedra $P^{(k)}$ in the sequence. Such an upper bound exists since otherwise the solutions of the limit system would not be bounded.

The points $z^{(k)}$ have an accumulation point $\hat{z}$ that solves the limit system (3.3). Consider the coefficients λ_k for a subsequence such that $z^{(k)}$ converges to $\hat{z}$. Since these coefficients are contained in the compact interval

$$\left[\frac{1}{n + \|g\|\delta}, 1 - \frac{1}{n + \|g\|\delta} \right],$$

they have an accumulation point λ_0 in that interval. With $\mu_0 = 1 - \lambda_0$,

$$\lambda_0, \mu_0 \geq \frac{1}{n + \|g\|\delta},$$

and

$$\hat{u} = \lambda_0 v^{(0)} + \mu_0 \hat{z}.$$

If the constraint $A_i^T u \leq b_i^{(0)}$ is binding at $\hat{u}$, $A_i^T \hat{u} = b_i^{(0)}$, then $A_i^T v^{(0)} = b_i^{(0)}$ must hold also, since $\lambda_0 > 0$. Thus the constraint $A_i^T u \leq b_i^{(0)}$ is binding at any vertex $v^{(0)}$ of $P^{(0)}$ and must therefore be tight. In other words, those constraints of the limit system (3.3) that are binding at $\hat{u}$ are the tight constraints in (3.3). This establishes (3.8).

According to (3.5) it suffices to prove the theorem for regularly convergent sequences of systems (3.1), which may be assumed regularized (1.4) with a common equation matrix E. In this case, the individual g-centers are characterized by condition (1.10), which becomes

$$(3.9) \qquad -AR^{-1}(u^{(k)})e = g + Ew^{(k)}.$$

We partition the matrix A into the portions A_T, A_S representing those inequalities that are tight or nontight, respectively, in the limit system (3.3). Condition (3.9) can then be written as

$$(3.10) \qquad -A_T R_T^{-1}(u^{(k)})e_T - A_S R_S^{-1}(u^{(k)})e_S = g + Ew^{(k)},$$

with R_T, R_S the corresponding square diagonal submatrices of the matrix R of residuals and e_T, e_S simply vectors of suitable length, all components of which have value 1. As $u^{(k)} \to \hat{u}$ for a suitable subsequence of (3.1), the second term in (3.1) converges to

$$A_S R_S^{-1}(\hat{u})e_S$$

since by (3.8) the residuals $r_i(\hat{u})$ are positive for $i \in S$, so that their reciprocals are finite and continuous at $\hat{u}$ as a function of u. It follows in view of (3.10) that the sequence of vectors

$$-A_T y^{(k)} - g - Ew^{(k)}, \qquad y^{(k)} = R_T^{-1}(u^{(k)})e_T$$

converges. Since all vectors of this convergent sequence are members of the linear subspace—a closed set—

$$M = \{-A_T y - g - Ew : y \in R^{|T|}, w \in R^l\},$$

so is the limit, which thus can be expressed in the form

$$-A_T \hat{y} - g - E\hat{w}.$$

Note that the vectors $\hat{y}$, $\hat{w}$ are, in general, not limits of the sequences $\{y^{(k)}\}$ and $\{w^{(k)}\}$. Nevertheless, these vectors serve as Lagrange multipliers in the relation

$$-A_S R_S^{-1}(\hat{u})e_T = g + A_T \hat{y} + E\hat{w},$$

which is the center condition (1.10), establishing any accumulation point $\hat{u}$ of the sequence of g-centers $u^{(k)}$ of the systems (3.1) as the unique center $u^{(0)}$ of the limit system (3.3). □

4. g-center trajectories

We consider linear programs of the form

$$(4.1) \qquad \begin{array}{ll} \textit{minimize} & c^T u \quad \textit{for } u \in R^m, \\ \textit{subject to} & A^T u \leq b, \\ & E^T u = f. \end{array}$$

Here c is an m-dimensional objective vector, A is an $m \times n$ matrix, E an $m \times l$ matrix. Vectors b and f are right-hand sides of suitable length. If at all, then the objective function $c^T u$ assumes its minimum, the

$$(4.2) \qquad \textit{optimal value } t^*,$$

on an entire face of the polyhedron of feasible solutions, the

$$(4.3) \qquad \textit{optimal face}$$

of the linear program. The boundedness of the optimal face is a natural regularity condition for what follows. It implies that $\text{rank}([A, E]) = m$ and the existence of optimal vertices. On occasion we will also require that the objective function actually varies over the feasible region, that is, that the linear program is nonconstant. If the constraint system of the linear program is regularized (1.4), then the linear program is nonconstant if and only if $\text{rank}(E) < \text{rank}([E, c])$.

Consider then a nonconstant linear program (4.1) with bounded optimal face. Its optimal solutions and their value are not affected if we append an

$$(4.4) \qquad \textit{objective constraint } c^T u \leq t, \qquad t \geq t^*.$$

Note that the resulting

$$(4.5) \qquad \textit{clipped}$$

constraint system is bounded if the optimal face is bounded. Hence there exists a g-center $u^h(g, t)$ (1.8), which uniquely maximizes the appended and modified barrier function

$$(4.6) \quad B(u) + \ln(t - c^T u) - g^T t = \sum_{i \in S} \ln(b_i - A_i^T u) + \ln(t - c^T u) - g^T t,$$

where S denotes the set of indices associated with nontight inequalities in the linear program (4.1). As t varies from $+\infty$ to t^*, the corresponding centers $u^h(g, t)$ will describe a curve that leads from the g-center of the original constraint system towards optimal solutions, the

$$(4.7) \qquad \textit{(analytic) g-center trajectory.}$$

Following Bayer and Lagarias [5], g-center trajectories are also referred to as

(4.8)   *A-trajectories.*

For $g = 0$, the

(4.9)   *(analytic) center trajectory or central trajectory*

is obtained. Different selections of the vector g may produce the same trajectory, possibly in different parametrizations, in particular, if those selections differ by a multiple of the objective vector c (compare Section 1).

At each trajectory point, the g-center condition (1.10) takes the form

$$(4.10) \qquad -AR^{-1}(u^h(g, t))e - \frac{c}{t - c^T u^h(g, t)} = g + Ew(t),$$

assuming that the constraints of the linear program have been regularized (1.4). In this formula, the expression

$$h(g, t) = t - c^T u^h(g, t)$$

is of particular interest and will be called the

(4.11)   *g-center deficiency.*

For each parameter value $t > t^*$, the objective value $c^T u^h(g, t)$ of the associated point $u^h(g, t)$ on the center trajectory will be closer to the optimal value t^*. This is simply because the g-center $u^h(g, t)$ lies in the relative interior of the clipped (4.5) polyhedron and because the objective constraint (4.4) is not tight. In other words, the g-center deficiency $h(t, t)$ is always positive for $t > t^*$ and vanishes for $t = t^*$. As $0 < h(g, t) = t - c^T u^h(g, t) < t - t^*$ for $t > t^*$, the objective values along the center trajectory converge to the optimal value. This proves convergence of the center trajectory to an optimal solution of the linear program (4.1) provided that the optimal solution is unique. Conversely, $h(g, t) \to 0$ implies $t - t^* \to 0$. This follows from the following lemma and the fact that the diameter $\delta(t)$ of polyhedron $P(t)$ does not increase as $t \to t^*$.

(4.12) LEMMA. *If n denotes the number of inequalities in the linear program (4.1), then*

$$0 < \frac{t - t^*}{n + 1 + \|g\|\delta(t)} \leq h(g, t) < t - t^* \quad for \quad t > t^*,$$

where $\delta(t)$ denotes the diameter of the clipped (4.5) polyhedron.

PROOF. The second relation is at issue. Let u^* be an optimal solution and consider the straight ray emanating from u^* through $u^h(g, t)$. Since $t^* = c^T u^* < c^T u^h(g, t) < t$, there exists beyond $u^h(g, t)$ on this ray a point w with $c^T w = t$. Denote the clipped polyhedron by $P(t)$, and let v be the last point in $P(t)$ on the straight-line segment from $u^h(g, t)$ to w. Clearly $c^T v \leq t$. The points u^* and v are antipodal (2.1) with respect to the g-center $u^h(g, t)$ of the polyhedron $P(t)$, and there exist coefficients λ, μ with

$$u^h(g, t) = \lambda u^* + \mu v, \qquad \lambda + \mu = 1, \qquad \lambda, \mu \geq 0.$$

From this:

$$c^T u^h(g, t) = \lambda c^T u^* + \mu c^T v \leq \lambda t^* + \mu t.$$

By Lemma (2.2),

$$\lambda \geq \frac{1}{n + 1 + \|g\|\delta(t)},$$

so that

$$h(g, t) = t - c^T u^h(g, t) \geq (1 - \mu)t - \lambda t^* = \lambda(t - t^*) \geq \frac{t - t^*}{n + 1 + \|g\|\delta(t)}. \qquad \square$$

It will be shown in Section 6, (6.7), that the center deficiency $h(g, t)$ varies monotonically with the parameter t.

The above result may be seen as a poor man's complexity gauge: consider the "method of g-centers" with starting point $u^{(0)}$, $c^T u^{(0)} > t^*$ and the recursion $t^{(k)} = c^T u^{(k)}$, $u^{(k+1)} = u^h(g, t^{(k)})$, $k = 0, 1, \ldots$. Since

$$c^T u^h(g, t) - t^* = t - t^* - h(g, t) \leq t - t^* - \frac{t - t^*}{n + 1 + \|g\|\delta(t)}$$

$$= (t - t^*) \left(1 - \frac{1}{n + 1 + \|g\|\delta(t)} \right)$$

by Lemma (4.12), we find

(4.13) THEOREM. *The values $t^{(k)}$ in the method of g-centers converge at least linearly to the objective value t^*.*

As an immediate consequence of Theorem (3.7) we note

(4.14) THEOREM. *As t decreases to the objective value t^*, the g-center trajectory (4.7) converges towards the g-center of the optimal face (4.3).*

This theorem is of interest only if the optimal face is not a vertex, that is, in the case of dual degeneracy. Indeed, we did establish convergence earlier in this section for the case that the optimal solution is unique. Moreover, if the optimal face is a facet, that is, a face of codimension 1, then the theorem can again be proved directly, as we will see later in this section.

Several similar definitions describe the same trajectory. For instance, the

(4.15) *objective cross section*

through trajectory point $u^h(g, t)$ is represented by the linear constraint system that results when the equation $c^T u = c^T u^h(g, t)$ is appended to the constraints of the linear program (4.1). The objective cross section is obviously a g-center restriction (1.17), and the g-center of the cross section will coincide with $u^h(g, t)$, that is, with the intersection of the cross section with the g-center trajectory. This shows that

(4.16) *the g-center trajectory is a portion of an algebraic curve stretching backwards beyond the g-center of the feasible region, or to infinity if the feasible region is not bounded. This full trajectory is also the g-center variety (1.15) for the system of constraints in linear program (4.1) with the objective (hyper) plane as its template (1.12).*

Suppose the linear program (4.1) has a full-dimensional feasible region and an optimal facet, that is, an optimal face of one dimension less. Then the constraint system in (4.1) contains one or more inequalities that are of the form $-c^T u \leq -t^*$ up to a positive multiplier. Now consider the g-center trajectory characterized in terms of g-centers of objective cross sections (4.16). The g-center trajectory of (4.1) is then seen as a portion of the g-center trajectory of the linear program that results if the inequalities that are thus equivalent to $-c^T u \leq -t^*$ are deleted. In addition, the hyperplane through the optimal facet is an objective cross section of the resulting reduced program, its g-center the last point on the g-center trajectory of the full program (4.1). This reasoning—which works, however, only for facets—provides an alternate proof that the g-center trajectory converges to the g-center of the optimal face (see Theorem (4.14)).

5. Dual variables or shadows

This section will feature another application of Theorem (3.7). It will be used to show that Lagrange multipliers associated with g-trajectories converge to a common limit, the analytic center of the optimal face for a

dual

to the linear program (4.1). That dual linear program takes the form

(5.1)
$$\begin{aligned} &\textit{minimize} \quad b^T x + f^T w \quad \textit{for } x \in R^n, w \in R^l, \\ &\textit{subject to} \quad Ax + Ew = -c, \\ & \quad x \geq 0. \end{aligned}$$

Here A, E, c, b, f are as in (4.1): A an $m \times n$ matrix, E an $m \times l$ matrix. The m vector c, the objective vector of that linear program, now serves as right-hand side of the equation constraints. The n vector b and l vector f together define the objective function. Note that if u and x, w are feasible solutions of (4.1) and (5.1), respectively, then

(5.2) $-c^T u = x^T A^T u + w^T E^T u \leq x^T b + w^T f = b^T x + f^T w$,

that is, the objective values of one program bound the negatives of the objective values of the other from above. By the duality theorem of linear programming, the

(5.3)   *duality gap* $c^T u + b^T x + f^T w \geq 0$

vanishes if both feasible solutions are optimal. In other words, the optimal values t^*, s^* of the two programs are opposites: $s^* = -t^*$.

For analyses of program (5.1)—particularly if the matrix E is empty—it is frequently required that there exist solutions with $x > 0$. Note that this is the condition that the constraints of that program are regularized. Boundedness of the optimal face of a linear program translates into regularization of the constraints of its dual. More precisely:

(5.4)   (i) *If the linear program* (4.1) *has an optimal face that is bounded, then the constraints of its dual* (5.1) *are regularized* (1.4), *that is, there exists a dual feasible solution with* $x > 0$.

(ii) *If the constraints of the linear program* (4.1) *are regularized and the columns of matrix* E *are linearly independent, then the linear program* (5.1) *possesses an optimal face that is bounded, provided there are feasible solutions at all.*

(iii) *If the linear program* (4.1) *has a bounded optimal face and is regularized, and if the matrix* E *is of maximum rank, then its dual has a bounded optimal face and is regularized, too.*

The proof of this statement is based on fundamental facts of the theory of linear inequalities and can be found in the authors' technical report [27]. An optimal solution to a linear program is

(5.5)   *strictly optimal*

if it lies in the relative interior of the optimal face. Note that the relative interior of a single point is that point itself; every unique vertex optimum is therefore strictly optimal. See, for instance, Schrijver [23] for the following result:

(5.6)   *Any two strictly optimal solutions* u^*, $\begin{bmatrix} x^* \\ w^* \end{bmatrix}$ *of linear programs* (4.1) *and* (5.1), *respectively, are*

                        strictly complementary,

that is,

$$x_i^* > 0 \quad \text{if and only if} \quad b_i - A_i^T u^* = 0.$$

In what follows, we assume that the linear program (4.1) has optimal solutions and that the optimal face is bounded. We then associate with each point $u^h(g, t)$ on a given g-trajectory a feasible solution of a perturbation of the dual program (5.1). To this end we recall the condition (4.10) that a point $u^h(g, t)$ of the g-center trajectory must satisfy. Multiplying that condition by $h(g, t) = t - c^T u^h(g, t) > 0$ yields—with an obvious redefinition of the Lagrange multipliers w—

$$h(g, t)AR^{-1}(u^h(g, t))e + Ew(g, t) = -c - h(g, t)g.$$

The positive n vector

$$x = x(g, t) = h(g, t)R^{-1}(u^h(g, t))e$$

forms, together with the l vector $w = w(g, t)$, a feasible solution to the perturbed linear program

(5.7)
$$\begin{aligned}
minimize \quad & b^T x + f^T \quad \text{for } x \in R^n, w \in R^l, \\
subject \; to \quad & Ax + Ew = -c - h(g, t)g, \\
& x \geq 0.
\end{aligned}$$

We call that solution the

(5.8)
$$g\text{-shadow} \quad \begin{bmatrix} x(g, t) \\ w(g, t) \end{bmatrix}$$

of the g-center trajectory point $u = u^h(g, t), t > t^*$.

Note that for $g = 0$, the linear program (5.7) is in fact identical to the dual (5.1) of the original linear program. This means that the shadow of any point on the analytic center trajectory is automatically dual feasible. Not only that, but

(5.9) THEOREM. *The shadow of a point on the analytic center trajectory of a regularized linear program with bounded optimal face lies on the analytic center trajectory of its dual linear program, provided the latter is nonconstant. (If the dual linear program is constant, then the shadow will always coincide with the analytic center of the dual constraint system.) The parameter t of the original trajectory point $u^h(t)$ relates to the parameter s of its shadow*

$$\begin{bmatrix} x(t) \\ w(t) \end{bmatrix}$$

on the dual trajectory via their respective center deficiencies (4.11):

$$h^D(s) = s - b^T x(t) - f^T w(t) = t - c^T u^h(t) = h(t).$$

The point and its shadow move together towards respective optimal solutions along their respective center trajectories.

For related results see, for instance, Bayer and Lagarias [5] and Todd and Burrell [25]. We omit the proof of this theorem and proceed to extend it to general g-trajectories. Consider the linear program

(5.10)
$$\begin{aligned}
minimize \quad & (c + h(g, t)g)^T u \quad \text{for } u \in R^m \\
subject \; to \quad & A^T u \leq b, \\
& E^T u = f,
\end{aligned}$$

whose dual is the perturbed program (5.7). The program arises from the original linear program by perturbation of the objective function.

The following four propositions (5.11)–(5.15) hold under the proviso that $u^h(g, t)$ is sufficiently close to the trajectory limit $u^h(g, t^*)$; that is,

$$h(g, t)\|g\| < \varepsilon,$$

for some $\varepsilon > 0$ associated with program (4.1). This conditions will assure that the perturbed dual program (5.7) is still regularized, and that therefore by (5.4) the perturbed primal program (5.10) still has a bounded optimal face. The latter linear program thus has center trajectories.

(5.11)   *The points $u^h(g, t)$ on the g-center trajectory of the original linear program (4.1) lie on the analytic center trajectory of the perturbed program (5.10),*

$$u^h(g, t) = \hat{u}^h(\hat{t}), \qquad \hat{t} = (cz + h(g, t)g)^T u^h(g, t) + h(g, t).$$

PROOF. Parameter $\hat{t}$ has been chosen such that the center deficiencies (4.11) agree for the two interpretations:

(5.12) $\hat{h}(\hat{t}) = \hat{t} - (c + h(g, t)g)^T u^h(g, t) = h(g, t).$

(t thus follows by a straightforward transformation of (4.10) that $u^h(g, t)$ satisfies

$$-AR^{-1}(u^h(g, t))e - \frac{c + h(g, t)g}{\hat{t} - (c + h(g, t)g)^T u^h(g, t)} = Ew(g, t). \qquad \square$$

The following is also a direct consequence of (5.12).

(5.13)   *The g-shadow (5.8) of the g-trajectory point $u = h^h(g, t)$ with respect to the original linear program (4.1) coincides with the shadow of the same point $u = \hat{u}^h(\hat{t})$ but is interpreted as a point on the analytic center trajectory of the perturbed linear program (5.10),*

$$x(g, t) = \hat{x}(\hat{t}), \qquad w(g, t) = \hat{w}(\hat{t}).$$

By Theorem (5.9), the above shadow lies on the analytic center trajectory of the perturbed dual program (5.7). Let $s(t)$ be the parameter of the shadow with respect to that trajectory. It follows that

(5.14)   *The g-shadow (5.8) of the point $u = u^h(g, t)$ is the analytic center of the—perturbed and clipped—constraint system*

$$Ax + Ew = -c - h(g, t)g,$$
$$x \geq 0,$$
$$b^T x + f^T w \leq s(t).$$

The key observation now is that, with s^* the optimal value of the unperturbed dual program (5.1),

(5.15) $s(t) \to s^*$ as $h(g, t) \to 0.$

PROOF. Theorem (5.9) relates the center deficiencies of point $\hat{u}^h(\hat{t})$ on the analytic center trajectory of (5.10) and of its shadow on the analytic center trajectory of (5.7): $\hat{h}^D(s(t)) = \hat{h}(\hat{t})$. Denote by $s^*(t)$ the optimal value of the perturbed dual program (5.7). Then Lemma (4.12), applied to the perturbed dual program (5.7), and (5.12) yield

$$s(t) - s^*(t) \le \hat{h}^D(s(t))(m + 1) = \hat{h}(\hat{t})(m + 1) = h(g, t)(m + 1).$$

Thus $|s(t) - s^*(t)| \to 0$ as $h(g, t) \to 0$. By the well-known continuity of the optimal value of a linear program with respect to perturbations of its right-hand side, $|s^*(t) - s^*| \to 0$ (see for instance [27]). Thus

$$|s(t) - s^*| \le |s(t) - s^*(t)| + |s^*(t) - s^*| \to 0 \quad as \ h(g, t) \to 0. \quad \square$$

The following extends Theorem (5.9) and also states that all A-trajectories have the same shadow limit.

(5.16) THEOREM. *For* $t \to t^*$ *the g-shadows* (5.8)

$$\begin{bmatrix} x(g, t) \\ w(g, t) \end{bmatrix}$$

of the points $u^h(g, t)$ *on any g-trajectory converge to a strictly optimal* (5.5) *solution*

$$\begin{bmatrix} x^* \\ w^* \end{bmatrix}$$

of the dual linear program (5.1). *That solution is the analytic center of the optimal face of the dual linear program. It is therefore independent of the choice of* g. *It is also strictly complementary* (5.6) *to the analytic g-center of the optimal face of the primal linear program* (4.1)

PROOF. By (5.15), and since $t \to t^*$ implies $h(g, t) \to 0$, the constraint system in (5.14) has the limit system

$$\begin{aligned} Ax + Ew &= -c, \\ x &\ge 0, \\ b^T x + f^T w &\le s^*, \end{aligned}$$

so that the theorem follows by Theorem (3.7) and by observation (5.6). $\square$

6. The basic ODE's

In this section, we will derive and analyze ordinary differential equations (compare Bayer and Lagarias [5]) whose solution curves are precisely the A-trajectories. We will not treat the general case, but rather consider in this section the special case in which there are no tight inequalities and no equations, and in which the polyhedron of feasible solutions is therefore full dimensional. Adler and Monteiro [1] address the second important special

case, the classic case of primal linear programs, in which the solution polyhedron is the intersection of a linear manifold with the nonnegative orthant,
and none of the nonnegativity conditions are tight.

As in preceding sections, we will assume that the optimal face of the underlying linear program

(6.1)
$$\begin{aligned} minimize \quad & c^T u \quad for \ u \in R^m, \\ subject \ to \quad & A^T u \le b, \end{aligned}$$

is nonempty and bounded. As pointed out in Section 4, this implies that the
matrix A assumes its full rank m. We also require $c \ne 0$ so that the linear
program will be nonconstant.

A first differential equation describing a g-center trajectory is derived by
recalling that g-centers $u^h(g, t)$ are characterized by

$$\nabla_u L(u, t) = g, \qquad L(u, t) = B(u) + \ln(t - c^T u);$$

that is, the appended logarithmic and modified barrier function $L(u, t)$ (4.6)
is maximized as a function of u for any value of t with $t^* < t < \infty$. Thus

(6.2)
$$\nabla_u L(u^h(g, t), t) = g$$

with

$$\nabla_u L(u, t) = -AR^{-1}(u)e - \frac{c}{t - c^T u}$$

holds for all points

$$u(t) = u^h(g, t)$$

of the g-center trajectory. We differentiate (6.2) with respect to t to obtain

(6.3)
$$\nabla^2_{uu} L(u, t)u' + \nabla^2_{ut} L(u, t) = 0,$$

where

$$\nabla^2_{uu} L(u, t) = -AR^{-2}(u)A^T - \frac{cc^T}{(t - c^T u)^2}, \qquad \nabla^2_{ut} L(u, t) = \frac{c}{(t - c^T u)^2}.$$

This is an implicit differential equation for the trajectory $u(t) = u^h(g, t)$.
As matrices A and R have full rank, matrix $AR^{-2}(u)A^T$ is nonsingular.
The above implicit differential equation can be made therefore explicit:

$$u' = -\nabla^2_{uu} L(u, t)^{-1} \nabla^2_{ut} L(u, t).$$

In particular, we find the expressions

$$\begin{aligned} u' &= \left(AR^{-2}(u)A^T + \frac{cc^T}{(t - c^T u)^2} \right)^{-1} \frac{c}{(t - c^T u)^2} \\ &= \frac{(AR^{-2}(u)A^T)^{-1} c}{(t - c^T u)^2 + c^T(AR^{-2}A^T)^{-1} c} \end{aligned}$$

and, by changing the trajectory parameter,

(6.5)
$$u' = (AR^{-2}(u)A^T)^{-1} c.$$

Every interior feasible point $u^{(0)}$ can serve as initial condition for any parameter t_0 with $t^* < t_0 \leq \infty$. Indeed, we have $u^h(g, t_0) = u^{(0)}$ for $g = \nabla_u B(u^{(0)}) - c/(t_0 - c^T u^{(0)})$.

Note that the optimum is approached for decreasing curve parameter t. Indeed, as $c \neq 0$,

$$(6.6) \qquad\qquad c^T u' > 0$$

for $t > t^*$ follows immediately from, say, (6.5). Moreover,

(6.7) *the center deficiency $h(g, t) = t - c^T u^h(g, t)$ (4.11) decreases monotonically as t decreases to t^*; that is, it increases with t.*

PROOF. By (6.4)

$$c^T u' = \frac{c^T (AR^{-2}(u)A^T)^{-1}c}{(t - c^T u)^2 + c^T (AR^{-2}A^T)^{-1}c} < 1.$$

Thus $h'(g, t) = 1 - c^T u'(t) > 0$. □

The above forms of differential equations are only defined for $t > t^*$, that is, as long as the trajectory stays in the interior of the feasible region. Indeed some entries of the diagonal matrix R^{-2} grow infinitely large as the boundary is approached, suggesting that the differential equation becomes ill conditioned in the proximity of the optimal solution. In the case of a unique optimum, however, we will show that the condition of the differential equation does not deteriorate and that multiplication by $(t - c^T u)^2$ yields equivalent forms of differential equations (6.4) that assume a limit as t approaches t^*. These equivalent forms are

$$(6.8) \qquad u' = (AX^2 A^T + cc^T)^{-1}c = \frac{(AX^2 A^T)^{-1}c}{1 + c^T (AX^2 A^T)^{-1}c}$$

as well as, with yet another trajectory parameter,

$$(6.9) \qquad\qquad u' = (AX^2 A^T)^{-1}c,$$

where matrix

$$X = X(u) = (t - c^T u)R^{-1}(u)$$

was seen in Section 5 to converge for $u \to u^*$. Indeed, the shadow $x = Xe$ (5.8) was seen to converge towards a strictly optimal Lagrange multiplier vector x^* (Theorem (5.16)). Beyond this observation we have

(6.10) THEOREM. *If the optimal solution u^* to a linear program is unique, then all A-trajectories, including the center trajectory, have a common limit tangent $(u')^*$.*

PROOF. For simplicity we assume again that the feasible region of the linear program is full dimensional. Since matrix X converges, so does $AX^2 A^T$. The rank of its limit $A(X^*)^2 A^T$ is at issue. Its rank equals that of AX^*. Let submatrix A_F characterize the inequalities that are binding for the unique solution u^*. Since the corresponding equations $A_F^T u = b_F$ have that unique solution, matrix A_F has full rank m. By (5.6), the shadow limit x^* is strictly complementary. In other words, $x_F^* > 0$, and $x_G^* = 0$ for the remainder of x^*. Splitting the limit matrix X^*, whose diagonal is formed by x^*, into the corresponding portions X_F^*, $X_G^* = 0$, yields

$$\text{rank}(AX^*) = \text{rank}(A_F X_F^*) = m,$$

since both A_F and X_F^* have full rank. Matrix AX^2A^T is thus invertible at the limit, guaranteeing the existence of the limit $u'(g, t^*)$. Since the shadow limit x^* does not depend on g, neither does $(u')^* = u'(g, t^*)$. □

Note that so far the existence of a limit tangent has been established only for unique optima. In the following section, we will prove such existence in the general case.

In form (6.8) of the differential equation the term cc^T is added to the matrix to be inverted. This addition may improve the numerical condition of the differential equation; it does not, however, improve the limit rank of the matrix in question:

(6.11) $$\text{rank}(A(X^*)^2A^T + cc^T) = \text{rank}(A(X^*)^2A^T).$$

PROOF. The statement is equivalent to $\text{rank}(A_F) = \text{rank}([A_F, c])$, where submatrix A_F again corresponds to the portion of the inequality system that is binding at a strictly optimal (5.5) solution u^*. Rank equality then follows from $c = -A_F x_F^* = -Ax^*$. □

7. Convergence proofs using center varieties

As an immediate consequence of Theorem (3.7) we note

(7.1) THEOREM. *The definition of g-center varieties* (1.15) *of a system* (1.1) *of linear constraints can be extended continuously to the boundary of the solution polytope.*

In what follows, we assume that this extension has been made.

In Section 4, we introduced objective cross sections (4.15), that is, intersections of the polyhedron of feasible solutions of linear program (4.1) with the hyperplane described by $c^T u = t$, $t \geq t^*$. These objective cross sections were used to derive an alternate description of the g-center trajectory: locus of the g-centers of the objective cross sections or, more precisely, the g-centers of the constraint systems arising from adding the cross section equation to the constraint system of the linear program. As was pointed out also in Section 4 in case an entire facet would be optimal, this alternate description makes it immediately clear that the center trajectory has a natural extension beyond the optimal point; hence, there exist limit point and limit tangent. In this section, we will use center varieties (1.15) in order to extend that line of argument to general multiple optima. We will again assume that the

system of linear constraints in linear program (4.1) is regularized (1.4), that
the program has optimal solutions, and that the optimal face is bounded.

We first extend the definition of the objective cross section to lower
dimensions. Consider the system of equations

$$(7.2) \qquad\qquad A_F^T u = b_F, \qquad E^T u = f,$$

which arise from the equations and those inequalities of the aforementioned
linear program that are binding for its entire optimal face. Equations (7.2)
thus describe the affine hull of the optimal face

$$\{u : A_F^T u = b_F,\ A_G^T u \leq b_G,\ E^T u = f\}.$$

We use the affine hull of the optimal face—with its description (7.2)—as a
template (1.12) to generate a g-center variety (1.15), the

$$(7.3) \qquad\qquad\qquad \textit{objective } g\textit{-center variety.}$$

We refer to intersections of parallel displacements of the affine hull of the
optimal face with the feasible region as

$$(7.4) \qquad\qquad\qquad \textit{objective sections}$$

of linear program (4.1), generalizing the concept of objective cross section
(4.15). Indeed, if the optimal face is an entire facet, the objective sections are
the objective cross sections, and the objective g-center variety (7.3) coincides
with the g-center trajectory. In general, the objective g-center variety is an
algebraic manifold of dimension $\hat{m} - k$, where $\hat{m}$ is the dimension of the
feasible region and k is the dimension of the optimal face. If the optimal
solution is a unique vertex, then $k = 0$ and the objective sections (7.3) are
the points in the feasible region. In that case, the objective g-center variety
coincides with the latter. Objective varieties thus are nontrivial only if there
are multiple optimal solutions. Note that

(7.5) *the objective g-center variety (7.3) contains the g-center trajectory*
(4.7). The optimal limit point $u^h(g, t^*)$ of the g-center trajectory is the
intersection of the objective g-center variety with the optimal face.

PROOF. Let $u = u^h(g, t)$ be a point on the center trajectory, and recall
that $u(t)$ is the g-center of its objective cross section (4.15). The objective
section (7.4) through u is a g-center restriction (1.17) of the above objective
cross section. By (1.8), u is the g-center of that objective section and lies
therefore on the objective g-center variety. The g-center trajectory is thus
located on the objective g-center variety, and so is the optimal limit point
$u^* = u^h(g, t^*)$. As to the second statement, we note that by its definition
as a template any objective section has a unique intersection with its corresponding g-center variety. The optimal face is the objective section through
the limit point u^*. □

The point of objective g-center varieties is that they are occasionally ex-
tendable beyond the optimal face:

(7.6) *The objective g-center variety* (7.3) *of the linear program* (4.1) *is contained in the g-center variety of the same dimension that arises from the same template after the constraints that bind the optimal face are deleted from the linear program.*

PROOF. The objective g-center variety is characterized by

$$-AR^{-1}(u)e = g + Ew + A_F z_F,$$

where w, z_F are suitable Lagrange multipliers (see (1.16)). Splitting matrix A into two parts, A_F and A_G, depending on whether or not the corresponding inequalities bind the optimal face, and denoting by R_F and R_G the corresponding square portions of the matrix of residuals R, we find

$$(7.7) \qquad -A_G R_G^{-1}(u)e_G = g + Ew + A_F \hat{z}_F, \qquad \hat{z}_F = z_F + R_F^{-1}(u)e_F.$$

This relation characterizes a g-center variety with template $A_F^T u = b_F$, $E^T u = f$ for the constraint system that arises from the constraint system of the linear program by deleting the inequalities $A_F^T u \leq b_F$. In the relative interior of the original feasible region, the objective g-center variety coincides with the g-center variety defined in a larger region by (7.7). □

The Lagrange conditions (7.7) can be used to derive equations for the objective g-center variety. To this end, consider any matrix B that is complementary orthogonal to the matrix $[E, A_F]$, that is, any matrix whose columns span the null space of $[E, A_F]^T$: $[E, A_F]^T B = 0$. It thus follows immediately from (7.7):

(7.8) *The objective g-center variety* (7.3) *has equation* $-B^T A_G R_G^{-1}(u)e_G = B^T g$.

Based on the above observations, we modify the differential equation for the A-trajectories so it still holds in the limit even if the optimal solution to the linear program is not unique. For simplicity of notation we will again assume that a full-dimensional linear program (6.1) is given. Multiplying the implicit differential equation (6.3) $(t - c^T u)^2$ yields

$$(7.9) \qquad\qquad (AX^2(u)A^T + cc^T)u' = c.$$

Here $X(u) = (t - c^T u)^2 R^{-1}(u)$ is the diagonal matrix whose entries are the components of the g-shadow (5.8) (or Lagrange multipliers) of the trajectory point $u = u^h(g, t)$. As long as the trajectory stays in the interior, $x = X(u)e > 0$, all diagonal components are positive. At the limit $u = u^*$, $x^* = X(u^*)e$ is a strictly optimal solution of the linear program (5.1) and as such is strictly complementary to the strictly optimal solution u^* (see (5.6)). Thus

$$x_F^* > 0, \quad x_G^* = 0.$$

Splitting the diagonal matrix $X(u)$ correspondingly into portions $X_F(u)$, $X_G(u)$, it follows that, in the limit,

$$\text{rank}(AX(u^*)) = \text{rank}(A_F X_F(u^*)) = \text{rank}(A_F) = m - k$$

and, consequently, using also (6.11),

$$(7.10) \qquad \text{rank}(AX^2(u^*)A^T + cc^T) = \text{rank}(AX^2(u^*)A^T) = m - k,$$

where k again denotes the dimension of the optimal face. If $k > 0$, there is rank loss in the limit, and the implicit differential equation (7.9) then does not determine u' uniquely. In what follows, we will derive additional conditions for u' that will fully determine the limit direction u'.

The A-trajectories lie on their respective objective g-varieties characterized by

$$-AR^{-1}(u)e = g + A_F z_F.$$

Since $\text{rank}(A_F) = m - k$, there exists an $m \times k$ matrix B of rank k that is complementary orthogonal to A_F. As E is now considered empty

$$-B^T AR^{-1}(u)e = -B^T A_G R_G^{-1}(u)e_G = B^T g.$$

We have seen earlier (7.8) that this represents an equation system for the objective g-center variety (7.3). This observation provides a geometric interpretation for the following analysis. Differentiating the above relation with respect to t and changing sign yields

$$(7.11) \qquad B^T AR^{-2}(u)A^T u' = B^T A_G R_G^{-2}(u)A_G^T u' = 0.$$

Since the matrix $R_G(u)$ does not contain any of the residuals that vanish for $u = u^*$, $R_G^{-1}(u^*)$ exists. Thus

$$\text{rank}(B^T A_G R_G^{-2} A_G^T) = \text{rank}(B^T A_G R_G^{-2} A_G^T B) = \text{rank}(B^T A_G R_G^{-1}) = \text{rank}(B^T A_G).$$

We proceed to show that

$$(7.12) \qquad \text{rank}(B^T A_G) = k.$$

To this end we consider an $(m - k) \times m$ submatrix A_H consisting of a maximal set of linearly independent columns of A_F. Thus $\text{rank}(A_H) = \text{rank}(A_F) = m - k$ and the $m \times m$ matrix $Q = [A_H, B]$ is nonsingular by definition of B. Thus $Q^T A$ still has rank m and is of the form

$$\begin{bmatrix} A_H^T A_F & A_H^T A_G \\ 0 & B^T A_G \end{bmatrix}.$$

This requires (7.12) since the top portion has at most rank $m - k$. Now consider the two conditions

$$(7.13) \qquad \begin{aligned} (AX^2(u)A^T + cc^T)u' &= c, \\ B^T A_G R_G^{-2}(u)A_G^T u' &= 0, \end{aligned}$$

and their limit

$$(7.14) \qquad \begin{aligned} (A_F X_F^2(u^*)A_F^T + cc^T)(u')^* &= c, \\ B^T A_G R_G^{-2}(u^*)A_G^T (u')^* &= 0. \end{aligned}$$

By their derivation, conditions (7.13) are a consequence of the implicit differential equation (7.9) and must therefore hold for all $u = u^h(g, t)$, $t > t^*$. They determine u' uniquely; in fact, the first of the two conditions by itself does. If the limit system (7.14) has a unique solution $(u')^*$— and that suggests itself in view of (7.10) and (7.12)—then that solution, by well-known facts about convergent sequences of linear equation systems, the limit of the solutions $u' = u'(g, t)$ to (7,9). This will prove

(7.15) THEOREM. *Conditions* (7.13) *provide an implicit differential equation for A-trajectories that possesses a limit as $t \to t^*$. The limit conditions* (7.14) *have a unique solution* $(u')^*$ *to which the tangential directions u' of the respective A-trajectories converge*

PROOF. As mentioned above, all that remains to be shown is that the linear equations (7.14) have a unique solution. The first condition is solvable if, for any m vector v, $v^T(A_F^T X_F^2 A_F^T + cc^T) = 0$ implies $v^T c = 0$. But this is plain since $v^T(A_F^T X_F^2 A_F^T + cc^T) = 0$ implies

$$v^T A_F^T X_F^2 A_F^T v + v^T cc^T v = \|v^T A_F X_F\|^2 + \|v^T c\|^2 = 0.$$

This also shows that, for any solution v of the homogeneous system associated with the first condition in (7.14), $v^T A_F X_F = 0$ and, by the positivity of the diagonal elements of X_F, $v^T A_F = 0$. Since the columns of matrix B form by definition a basis of all vectors orthogonal to A_F, there exists a vector z such that $v = Bz$. Every solution w of the first condition is therefore of the form $w = w^{(0)} - Bz$, with $w^{(0)}$ an arbitrary fixed solution. Substitution into the second condition yields

$$B^T A_G R_G^{-2} A_G^T Bz = B^T A_G R_G^{-2} A_G^T w^{(0)}.$$

These equations for z are readily shown to be solvable by an argument analogous to the one used above for the first condition. (They are also the normal equations for a linear regression problem, and are therefore known to be solvable, as are the equations of the first condition when the relation $c = -A_F X_F e_F$ is taken into account.) We have thus established the existence of a solution of (7.14). To show uniqueness, we establish zero as the only solution of the homogeneous system

$$(A_F X_F^2 A_F^T + cc^T)v = 0,$$
$$B^T A_G R_G^{-2} A_G^T v = 0.$$

We have already seen that the first condition in the above homogeneous system implies both $c^T v = 0$ and $A_F^T v = 0$, as well as the existence of a vector z with $v = Bz$. Multiplying the second homogeneous condition by z^T yields $v^T A_G R_G^{-2} A_G v = 0$. From this, $R_G^{-1} A_G^T v = 0$, and subsequently, $A_G^T v = 0$.

Taken together, we have $c^T v = 0$, $A^T v = 0$. Since the linear program in question (6.1) has a bounded optimal face, this implies $v = 0$. □

Acknowledgments

We thank Mark Hartmann for helpful comments, the editor, and an anonymous referee for their diligence.

REFERENCES

1. I. Adler and R. D. C. Monteiro, *The limiting behavior of the affine scaling trajectories for linear programming problems*, Technical Report ESRC 88-9, Engineering Systems Research Center, University of California at Berkeley, April 1988.

2. I. Adler, M. G. C. Resende, G. Veiga, and N. Karmarkar, *An implementation of Karmarkar's algorithm for linear programming*, Math. Programming **44** (1989), 297–335.

3. E. R. A. Barnes, *A variation on Karmarkar's algorithm for solving linear programming problems*, Math. Programming **36** (1986), 174–182.

4. D. A. Bayer and J. C. Lagarias, *The nonlinear geometry of linear programming*: I. *Affine and projective scaling trajectories*, II. *Legendre transform coordinates*, III. *Central trajectories*. IV, *Karmarkar's linear programming algorithm and Newton's method*, internal memoranda, AT&T Bell Laboratories, Murray Hill, N.J. 1986.

5. D. A. Bayer and J. C. Lagarias, *The nonlinear geometry of linear programming*: I, *affine and projective scaling trajectories*, Trans. Amer. Math. Soc. **314** (1989), 499–526.

6. D. A. Bayer and J. C. Lagarias, *The nonlinear geometry of linear programming.II, Legendre transform coordinates and central trajectories*, Trans. Amer. Math. Soc. **314** (1989), 527–581.

7. P. T. Boggs, P. D. Domich, J. R. Donaldson and C. Witzgall, *Algorithmic enhancements to the method of centers for linear programming*, ORSA J. Comput. **1** (1989), 159–171.

8. I. I. Dikin, *Iterative solution of problems of linear and quadratic programming*, Soviet Math. Dokl. **8** (1967), 674–675.

9. A. V. Fiacco and G. P. McCormick, *Programming under nonlinear constraints by unconstrained minimization: A primal-dual method*, Research Paper RAC-TP-96, The Research Analysis Corporation, 1963.

10. A. V. Fiacco, and G. P. McCormick, *The sequential unconstrained minimization technique for nonlinear programming, a primal-dual method*, Management Sci. **10** (1964), 601–617.

11. A. V. Fiacco and G. P. McCormick, *The sequential unconstrained minimization technique (SUMT) without parameters*, Oper. Res. **15** (1967), 820–827.

12. A. V. Fiacco and G. P. McCormick, *Nonlinear programming: Sequential unconstrained minimization techniques*, John Wiley and Sons, Inc., New York, 1968.

13. P. Huard, *Résolution des p. m. à constraintes non linéaries par la méthode des centres*, Tech. Rep. Note E. D. F. no. HR 5.690, Électricité De France, May 1964.

14. P. Huard, *Resolution of mathematical programming with nonlinear constraints by the method of centres*, Nonlinear Programming (J. Abadie, ed.), North-Holland, Amsterdam, 1967, pp. 209–219.

15. N. Karmarkar, *A new polynomial-time algorithm for linear programming*, Combinatorica **4** (1984), 373–395.

16. N. Megiddo, *A variation of Karmarkar's algorithm*, Technical report, IBM Almaden Research Center, San Jose, Calif., 1985.

17. N. Megiddo, *Pathways to the optimal set in linear programming*, Progress in Mathematical Programming: Interior-point Algorithms and Related Methods (N.Megiddo, ed.), papers from the Conference held in Pacific Grove, California, March 1–4, 1987, Springer-Verlag, New York, 1989, pp. 131–158.

18. N. Megiddo and M. Shub, *Boundary behavior of interior point algorithms in linear programming*, Math. Oper. Res. **14** (1989), 97–146.

19. W. Murray, *An algorithm for constrained minimization*, Optimization; Symposium of the Institute of Mathematics and Its Applications (R. Fletcher, ed.), University of Keele, 1968, Academic Press, London, 1969, pp. 247–258.

20. W. Murray and M. H. Wright, *Numerical aspects of trajectory-following algorithms for nonlinearly constrained optimization*, Computers and Mathematical Programming: Proceedings of the Bicentennial Conference on Mathematical Programming 1976 (W. W. White, ed.), NBS Special Publication 502, National Bureau of Standards, Gaithersburg, Md., 1978, pp. 194–204.
21. J. Renegar, *A polynomial-time algorithm, based on Newton's method for linear programming*, Math. Programming **40** (1988), 59–93.
22. R. T. Rockafellar, *Convex analysis*, Princeton Univ. Press, Princeton, N. J., 1969.
23. A. Schrijver, *Theory of linear and integer programming*, John Wiley and Sons, Inc., New York, 1986.
24. G. Sonnevend, *An "analytic center" for polyhedrons and new classes of global algorithms for linear (smooth, convex) programming*, Proc. 12th IFIP Conference on System Modeling and Optimization, Budapest, 1985.
25. M. J. Todd and B. P. Burrell, *An extension of Karmarkar's algorithm for linear programming using dual variables*, Algorithmica 1, **4** (1986), 409–424.
26. R. J. Vanderbei, M. S. Meketon, and B. A. Freedman, *A modification of Karmarkar's linear programming algorithm*, Algorithmica 1, **4** (1986), 395–407.
27. C. Witzgall, P. T. Boggs and P. D. Domich, *On center trajectories and their relatives in linear programming*, NIST Technical Report (in preparation; first draft, April 1988).

APPLIED AND COMPUTATIONAL MATHEMATICS DIVISION, NATIONAL INSTITUTE OF STANDARDS AND TECHNOLOGY, GAITHERSBURG, MARYLAND 20899
E-mail address: Witzgall@CAM.NIST.GOV

APPLIED AND COMPUTATIONAL MATHEMATICS DIVISION, NATIONAL INSTITUTE OF STANDARDS AND TECHNOLOGY, GAITHERSBURG, MARYLAND 20899
E-mail address: Boggs@CAM.NIST.GOV

APPLIED AND COMPUTATIONAL MATHEMATICS DIVISION, NATIONAL INSTITUTE OF STANDARDS AND TECHNOLOGY, BOULDER, COLORADO 80303-3328
E-mail address: Domich@CAM.NIST.GOV

Contemporary Mathematics
Volume **114**, 1990

Limiting Behavior of the Affine Scaling Continuous Trajectories for Linear Programming Problems

ILAN ADLER AND RENATO D. C. MONTEIRO

ABSTRACT. We consider the continuous trajectories of the vector field induced by the primal affine scaling algorithm as applied to linear programming problems in standard form. By characterizing these trajectories as solutions of certain parametrized logarithmic barrier families of problems, we show that these trajectories tend to an optimal solution that in general depends on the starting point. By considering the trajectories that arise from the Lagrangian multipliers of the above-mentioned logarithmic barrier families of problems, we show that the trajectories of the dual estimates associated with the affine scaling trajectories converge to the so-called "centered" optimal solution of the dual problem. We also present results related to asymptotic direction of the affine scaling trajectories. We briefly discuss how to apply our results to linear programs formulated in formats different from the standard form. Finally, we extend the result to the primal–dual affine scaling algorithm.

1. Introduction

The primal affine scaling (PAS) algorithm for solving linear programming problems was first presented by Dikin [4] and later was (independently) reintroduced by Barnes [2] and Vanderbei, Meketon, and Freedman [13] as a variant of the projective interior point algorithm for linear programming that was presented in the seminal work of Karmarkar [6]. The PAS algorithm is designed for linear programming problems in standard form. A similar algorithm, the so-called dual affine scaling (DAS) algorithm designed for problems in inequality form, was implemented by Adler, Karmarkar, Resende, and Veiga [1]. Both variants are currently the most experimented interior

1980 *Mathematics Subject Classification* (1985 *Revision*). Primary 90C05; Secondary 90C25.
Key words and phrases. Interior-point methods, linear programming, Karmarkar's algorithm, logarithmic barrier function, affine scaling algorithms, continuous trajectories for linear programming.

point algorithms for linear programming and exhibit promising results. (See [1], [2], and Monma and Morton [10].)

A third variant, a primal–dual affine scaling (PDAS) algorithm, was presented and analyzed by Monteiro, Adler, and Resende [11]. Our main interest in this paper is to analyze the limiting behavior of the continuous trajectories of the three major variants of the affine scaling algorithms. Karmarkar [7] and Bayer and Lagarias [3] discussed some properties of the continuous PAS trajectories. Megiddo and Shub [9] analyzed the limiting behavior of these trajectories near the optimal vertex under the assumption of primal and dual nondegeneracy. Megiddo [8] presented and proved some limiting behavior of the trajectories associated with the PDAS algorithm. By presenting a weighted primal affine scaling algorithm and characterizing its infinitesimal trajectories as solutions of certain parametrized logarithmic barrier families of problems, we extend the above results and present some new results related to all three affine scaling variants with no nondegeneracy assumptions. The logarithmic barrier function method is formally studied in Fiacco and McCormick [5] in the context of nonlinear optimization.

The paper is organized as follows. In Section 2, we define a weighted PAS algorithm and present a characterization of its infinitesimal trajectories as solutions of certain parametrized logarithmic barrier families of problems. By considering the optimality conditions associated with these families of problems, we automatically obtain dual trajectories associated with the weighted PAS trajectories. We also discuss the effects of the initial conditions on the derived trajectories. In Section 3, we present the limiting behavior of the weighted PAS trajectories and its associated dual trajectories. We show in particular that, while the (weighted) PAS trajectories tend to an optimal solution of the linear program under consideration that depends on the starting point, the dual trajectories tend to the so-called (weighted) "center" of the optimal face of the dual problem. In Section 4, we apply the results in Section 3 to show that the (weighted) dual estimates also tend to the (weighted) "center" of the optimal face of the dual problem as x traverses a (weighted) PAS trajectory converging to an optimal solution of the linear program under consideration. In Section 5, we present some results related to the asymptotic direction of the primal and dual trajectories. In Sections 6 and 7, we briefly discuss how to apply the results of the preceding sections (2–5) to the dual affine scaling (DAS) and the primal–dual affine scaling (PDAS) trajectories, respectively. We conclude by offering some remarks in Section 8.

2. Characterization of the weighted PAS trajectories

In this section, we introduce our terminology and refresh few notions that are probably familiar to the reader. We define a family of weighted affine scaling vector fields, which will be the object of our study, and then characterize their trajectories as solutions of certain parametrized logarithmic barrier families of problems. This characterization will allow us to draw conclusions about the limiting behavior of these trajectories as will be described in the next section.

We start by briefly reviewing the notion of a vector field. Let $E \in \mathbf{R}^n$ be a finite dimensional affine space. Let $T(E)$ denote the subspace determined by translating the affine space E to the origin of $\mathbf{R}^n$; that is, $T(E)$ is the set of vectors tangent to the affine space E. A vector field in an open subset O of E is a mapping $\Phi : O \to T(E)$. The trajectory of the vector field Φ passing through the point $x^o \in O$ at time $t^o \in \mathbf{R}$ is determined by the solution curve of the following differential equation:

$$\dot{x}(t) = \Phi(x(t)),$$
$$x(t^o) = x^o.$$

In what follows, we will be interested in a specific family of vector fields, namely, the weighted primal affine scaling vector fields. Before describing this family of vector fields, we need to introduce our terminology. Consider the linear programming problem in standard form

$$\text{(P)} \quad \min c^T x,$$
$$\text{s.t.} \ Ax = b,$$
$$x \geq 0$$

and its dual problem

$$\text{(D)} \quad \max b^T y,$$
$$\text{s.t.} \ A^T y + z = c,$$
$$z \geq 0$$

where A is an $m \times n$ matrix and b, c are vectors of length m and n, respectively. The following notation will be used throughout this paper. Let

$$S_A = \{x \in \mathbf{R}^n ; \ Ax = b\},$$
$$S_F = \{x ; x \in S_A, \ x \geq 0\},$$
$$S_I = \{x ; x \in S_A, \ x > 0\},$$
$$T_A = \{(y, z) \in \mathbf{R}^m \times \mathbf{R}^n ; \ A^T y + z = c\},$$
$$T_F = \{(y, z) ; (y, z) \in T_A, \ z \geq 0\},$$
$$T_I = \{(y, z) ; (y, z) \in T_A, \ z > 0\}.$$

The sets S_F and T_F are the feasible sets of problems (P) and (D), respectively; S_A and T_A are the affine hulls of S_F and T_F, respectively; and S_I and T_I are the relative interior of S_F and T_F, respectively. The lower-case letter e will denote the vector of all ones whose dimension is dictated by the appropriate context. If x is a lower-case letter that denotes a vector $x = (x_1, \ldots, x_n)^T$, then a capital letter will denote the diagonal matrix with the components of the vector on the diagonal; i.e., $X = \text{diag}(x_1, \ldots, x_n)$. Also, $\mathbf{R}_+^n$ will denote the set of n-vectors with all components strictly positive. If x and z are two n-vectors, we define their product xz to be the vector $XZe = (x_1 z_1, \ldots, x_n z_n)^T$. The inverse of x under this operation is denoted by x^{-1} and is given by $x^{-1} \equiv X^{-1} e$. In this way, expressions like x^{-1} and $x^{-2} z$ are defined if all the components of x are nonzero. No confusion should arise between the expressions xz and $x^T z$ where the latter just denotes the inner product of x and z. Given an $m \times n$ matrix A and a subset B of the index set $\{1, \ldots, n\}$, we denote by A_B the submatrix of A associated with the index set B.

We impose the following assumptions on problem (P):

ASSUMPTION 2.1.

(a) The set S_I is nonempty.
(b) The set of optimal solutions of problem (P) is nonempty and bounded.
(c) Rank$(A) = m$.

We should point out that Assumption (c) can be discarded in our development. However, we keep it since it will simplify the arguments considerably. Assumption (b) can be verified in several alternative ways, as the following proposition shows.

PROPOSITION 2.1. *Assume that the set S_I is nonempty. Then the following conditions are equivalent*:

(a) *For all (or some) $w = (w_1, \ldots, w_n) \in \mathbf{R}^n_+$ and $\mu > 0$, the problem*
$\min\{c^T x - \mu \sum_{j=1}^n w_j \ln x_j \, ; \, x \in S_I\}$ *has a (unique) global solution.*
(b) *The set T_I is nonempty.*
(c) *The set of optimal solutions of problem (P) is nonempty and bounded.*
(d) *For all (or some) $\bar{x} \in S_F$, the set $\{x \in S_F \, ; \, c^T x \leq c^T \bar{x}\}$ is bounded.*

We refer the reader to [5] and [8] for arguments that lead to the proof of Proposition 2.1.

Next, we introduce and motivate briefly the weighted PAS algorithms. Assume that a vector $w \in \mathbf{R}^n_+$ is given. Given an interior point $\bar{x}$ for problem (P), that is, $\bar{x} \in S_I$, the algorithm computes a search direction $\Delta x \equiv \Delta x(\bar{x})$ as follows. Consider the linear scaling transformation $\Psi_w : \mathbf{R}^n \to \mathbf{R}^n$ defined as $\Psi_w(x) = D^{-1} x$, $x \in \mathbf{R}^n$, where $D \equiv W^{-1/2} \overline{X}$. Observe that $\Psi_w(\bar{x}) = w^{1/2}$. In the transformed space, problem (P) becomes

$$\min (Dc)^T v \, ,$$
$$\text{s. t. } ADv = b \, ,$$
$$v \geq 0.$$

The search direction d_v in the transformed space is obtained by projecting the gradient vector Dc orthogonally onto the linear subspace $\{v \, ; \, ADv = 0\}$. Specifically, d_v is the solution of the following problem.

(2.1)
$$\min \tfrac{1}{2} \|Dc - v\|^2 \, ,$$
$$\text{s. t. } ADv = 0 \, ,$$

which yields the explicit solution

$$d_v = \left[I - DA^T \left(AD^2 A^T \right)^{-1} AD \right] Dc$$

together with the quantities

(2.2)
$$y^E(\bar{x}) = (AD^2 A^T)^{-1} AD^2 c,$$
$$z^E(\bar{x}) = c - A^T (AD^2 A^T)^{-1} AD^2 c$$

where $y^E(\bar{x})$ is the Lagrangian multiplier associated with the constraint of (2.1) and $z^E(\bar{x}) = c - A^T y^E(\bar{x})$. We shall call $y^E(\bar{x})$ and $z^E(\bar{x})$ the "dual estimates" at the point $\bar{x}$ for reasons that will become apparent in Section 5. Finally, the search direction Δx in the original space is given as

(2.3)
$$\Delta x = Dd_v = D \left[I - DA^T \left(AD^2 A^T \right)^{-1} AD \right] Dc.$$

The next iterate $\hat{x}$ is then computed as $\hat{x} = \bar{x} - \alpha \Delta x$ where $\alpha = \alpha(\bar{x}) > 0$ is an appropriate step size that guarantees that $\hat{x} > 0$. We refer to the algorithm outlined above as the w-PAS algorithm.

Note that the w-PAS algorithm as introduced in [4], [2], and [13] uses $w \equiv e$ (that is, $D = X$). We refer to this algorithm as the PAS algorithm. Our generalization will be useful in extending the results to the primal-dual affine scaling method as will be discussed in Section 7.

We next describe a family of affine scaling vector fields that are induced by the algorithm described above. Let $E = \{x; Ax = b\}$ and $O \equiv S_I \subseteq E$. Observe that $T(E) = \{x; Ax = 0\}$. The weighted primal affine scaling vector field with weight $w \in \mathbf{R}^n_+$, or for brevity, w-PAS vector field, is the vector field $\Phi_w : O \to T(E)$ defined as follows:

$$\Phi_w(x) \equiv D \left[I - DA^T \left(AD^2 A^T \right)^{-1} AD \right] Dc, \qquad x \in O \equiv S_I$$

where $D = W^{-1/2} X$. We are interested in studying the behavior of the trajectories of Φ_w, that is, the solution curves of the differential equation

(2.4)
$$\dot{q}(t) = \Phi_w(q(t)).$$

In particular, since problem (P) is in minimization form, we will be interested in the behavior of $q(t)$ as t decreases, that is, as $c^T q(t)$ monotonically decreases. With this aim, we consider solution curves of the following reparametrized differential equation:

(2.5)
$$\dot{x}(\mu) = \frac{1}{\mu^2} \Phi_w(x(\mu)).$$

It turns out that by studying the behavior of the trajectories of (2.5), we obtain information about the trajectories of (2.4) as the following result shows.

PROPOSITION 2.2. *Let $x^o \in S_I$ and $t^o > 0$ be given. Then $q(t)$ is a solution of (2.4) satisfying $q(t^o) = x^o$ if and only if $x(\mu) \equiv q(t^o + (t^o)^{-1} - \mu^{-1})$ is a solution of (2.5) satisfying $x(t^o) = x^o$.*

The proof of Proposition 2.2 is immediate. Note that the behavior, as t approaches $-\infty$, of a solution curve of (2.4) passing through $x^o \in S_I$ is completely determined by the behavior of a solution curve of (2.5) as μ approaches 0.

We now turn our efforts towards characterizing the solution curves of (2.5) as a path of solutions of a logarithmic barrier family of problems. Consider the following family of problems parametrized by the penalty parameter $\mu > 0$.

(2.6)
$$(P_\mu) \quad \min c^T x - \mu \left[p^T x + \sum_{j=1}^{n} w_j \ln x_j \right],$$
$$\text{s. t. } Ax = b,$$
$$x > 0$$

where $p \in \mathbf{R}^n$ is an arbitrary given vector. Since the objective function of problem (P_μ) is a strictly convex function, it follows that the global solution of (P_μ), if it exists, is completely characterized by the following Karush–Kuhn–Tucker stationary condition:

(2.7)
$$c - A^T y - \mu w x^{-1} = \mu p,$$
$$Ax = b, \qquad x > 0$$

where $y \in \mathbf{R}^m$ is the Lagrangian multiplier associated with the equality constraint of problem (P_μ). Let $I(p)$ denote the set of parameters $\mu > 0$ such that problem (P_μ) (and hence system (2.7)) has a solution. Note that by Proposition 2.1, problem (P_μ) has a global solution if and only if the set $Y(p, \mu) \equiv \{y; A^T y < c - \mu p\} \neq \varnothing$. Hence, $I(p) = \{\mu > 0; Y(p, \mu) \neq \varnothing\}$. The next result gives the form of the set $I(p)$.

PROPOSITION 2.3. *The set $I(p) = (0, d_p)$ for some $d_p > 0$. Moreover, $I(0) = (0, \infty)$. Also, if the set S_F is bounded then $d_p = \infty$ for all $p \in \mathbf{R}^n$.*

PROOF. It is straightforward to see that the set $L(p) \equiv \{\mu; Y(p, \mu) \neq \varnothing\}$ is convex and open. By Assumption 2.1.b and Proposition 2.1, we have that $Y(p, 0) \neq \varnothing$. These two observations show that $L(p)$ is an open interval such that $0 \in L(p)$. Since $I(p) = L(p) \cap \mathbf{R}_+$, the first statement follows. Obviously, if $p = 0$ then $d_p = \infty$.

When the set S_F is bounded, it follows that the set of optimal solutions is nonempty and bounded for any objective function, and hence, by Proposition 2.1, $Y(p, \mu) \neq \varnothing$ for all $\mu \in \mathbf{R}$. Thus, in this case $I(p) = (0, \infty)$ for all $p \in \mathbf{R}^n$. $\square$

In order to simplify the notation in what follows, let $z = c - A^T y$ for $y \in \mathbf{R}^m$. Then system (2.7) can be rewritten in an equivalent way as

(2.8a) $$z - \mu x^{-1} w - \mu p = 0,$$

(2.8b) $$Ax - b = 0, \qquad x > 0,$$

(2.8c) $$A^T y + z - c = 0.$$

Let $(x_p(\mu), y_p(\mu), z_p(\mu))$ denote the solution of system (2.8). It can be easily verified that the path $\mu \in I(p) \to (x_p(\mu), y_p(\mu), z_p(\mu))$ has derivatives of all order. We next show that $x_p(\mu)$, $\mu \in I(p)$ is a solution of the differential equation (2.5), for all $p \in \mathbf{R}^n$. Differentiating system (2.8), we obtain

(2.9) $$\dot{z}_p(\mu) + \mu x_p^{-2}(\mu)\dot{x}_p(\mu)w - x_p^{-1}(\mu)w - p = 0,$$

(2.10) $$A\dot{x}_p(\mu) = 0,$$

(2.11) $$A^T \dot{y}_p(\mu) + \dot{z}_p(\mu) = 0.$$

Equations (2.8.a) and (2.9) then imply that

(2.12) $$\dot{z}_p(\mu) + \mu x_p^{-2}(\mu)\dot{x}_p(\mu)w = \frac{z_p(\mu)}{\mu}.$$

In order to simplify the formulas below, we drop the subscript p and the indication of variable μ. Solving for $\dot{x}$, $\dot{y}$, and $\dot{z}$ using relations (2.10), (2.11), and (2.12) and letting $D = W^{-1/2}X$, we obtain

(2.13)
$$\begin{aligned}
\dot{x} &= \frac{1}{\mu^2} D \left[I - DA^T \left(AD^2 A^T \right)^{-1} AD \right] Dz \\
&= \frac{1}{\mu^2} D \left[I - DA^T \left(AD^2 A^T \right)^{-1} AD \right] Dc \\
&= \frac{1}{\mu^2} \Phi_w(x),
\end{aligned}$$

(2.14)
$$\begin{aligned}
\dot{y} &= -\frac{1}{\mu}(AD^2 A^T)^{-1} AD^2 z \\
&= \frac{1}{\mu}[y - (AD^2 A^T)^{-1} AD^2 c],
\end{aligned}$$

(2.15)
$$\begin{aligned}
\dot{z} &= \frac{1}{\mu} A^T (AD^2 A^T)^{-1} AD^2 z \\
&= \frac{1}{\mu}[A^T (AD^2 A^T)^{-1} AD^2 c - A^T y] \\
&= \frac{1}{\mu}[z + A^T (AD^2 A^T)^{-1} AD^2 c - c]
\end{aligned}$$

where the alternative formulas above were obtained by using relation (2.8.c). Expression (2.13) says that $x_p(\mu)$ is a solution of the differential equation (2.5). We summarize the above discussion in the following result.

THEOREM 2.1. *Let $(x_p(\mu), y_p(\mu), z_p(\mu))$ denote the path of solutions for the parametrized system of equations* (2.8). *Then $x_p(\mu)$ is a solution of the differential equation* (2.5).

Recall that our primary interest is to study the behavior of $x_p(\mu)$ passing through some given $x^o \in S_I$. However, system (2.8) as developed by considering the optimal solutions $x_p(\mu)$ for the family of logarithmic barrier problems (P_μ) involves the Lagrangian multiplier $y_p(\mu)$ and its slack $z_p(\mu)$. Thus, we should also study the effect of the initialization on these trajectories. Observe that for every given $\mu_o > 0$ and $(y^o, z^o) \in T_A$, if we set $p = z^o/\mu_o - w(x^o)^{-1}$ then it follows that $\mu_o \in I(p)$ and $(x_p(\mu_o), y_p(\mu_o), z_p(\mu_o)) = (x^o, y^o, z^o)$. By Theorem 2.1, it then follows that $x_p(\mu)$ is a solution of (2.5) satisfying $x_p(\mu_o) = x^o$. In order to study the dependence on the initial conditions $\mu_o > 0$, $x^o \in S_I$, and $(y^o, z^o) \in T_A$, we denote the trajectories $x_p(\mu)$, $y_p(\mu)$, and $z_p(\mu)$ with $p = z^o/\mu_o - w(x^o)^{-1}$ by $x(\mu; \mu_o, x^o, y^o, z^o)$, $y(\mu; \mu_o, x^o, y^o, z^o)$, and $z(\mu; \mu_o, x^o, y^o, z^o)$, respectively. We refer to $x_p(.)$ as a "weighted primal affine scaling" trajectory, or for brevity, w-PAS trajectory, and to $(y_p(.), z_p(.))$ as its associated "dual trajectory." When $w = e$, we refer to $x_p(.)$ as a PAS trajectory. As should be expected, the initial condition $(y^o, z^o) \in T_A$ does not affect the w-PAS trajectory $x(\mu; \mu_o, x^o, y^o, z^o)$. This fact is stated in the following result, which also shows the behavior of the w-PAS trajectory on the initial condition $\mu_o > 0$.

PROPOSITION 2.4. *Given $x^o \in S_I$, (y^o, z^o), $(y^1, z^1) \in T_A$, and $\mu_o, \mu_1 > 0$ then for all μ such that $0 < \mu \le \mu_1$, we have*

$$(2.16) \qquad x(\mu; \mu_1, x^o, y^1, z^1) = x(\rho(\mu); \mu_o, x^o, y^o, z^o)$$

where $\rho(\mu) \equiv (1/\mu + 1/\mu_o - 1/\mu_1)^{-1}$.

PROOF. Let $x(\mu)$ and $\tilde{x}(\mu)$ denote the expression on the left and right side of equality (2.16). By the observation following Theorem 2.1, it follows that $x(\mu)$ is a solution of (2.5) satisfying $x(\mu_1) = x^o$. On the other hand, by differentiating $\tilde{x}(\mu)$, one can easily verify that $\tilde{x}(\mu)$ is a solution of (2.5) satisfying $\tilde{x}(\mu_1) = x^o$. From the theory of differential equations, there exists a unique solution of (2.5) with initial condition (x^o, μ_1). Hence, $x(\mu) = \tilde{x}(\mu)$ for all $\mu \in (0, \mu_1]$. This completes the proof of the proposition. $\square$

We next show how the initial conditions $\mu_0 > 0$ and $(y^o, z^o) \in T_A$ affect the dual trajectories.

PROPOSITION 2.5. *Given $x^o \in S_I$, (y^o, z^o), $(y^1, z^1) \in T_A$, and $\mu_o, \mu_1 > 0$ then for all μ such that $0 < \mu \le \mu_1$, we have*
(2.17)

$$y(\mu; \mu_1, x^o, y^1, z^1) = \frac{\mu}{\rho(\mu)} y(\rho(\mu); \mu_o, x^o, y^o, z^o) + \left(\frac{y^1}{\mu_1} - \frac{y^o}{\mu_o}\right)\mu,$$

(2.18)

$$z(\mu; \mu_1, x^o, y^1, z^1) = \frac{\mu}{\rho(\mu)} z(\rho(\mu); \mu_o, x^o, y^o, z^o) + \left(\frac{z^1}{\mu_1} - \frac{z^o}{\mu_o}\right) \mu$$

where $\rho(\mu) \equiv (1/\mu + 1/\mu_o - 1/\mu_1)^{-1}$.

PROOF. Let $\tilde{x}(\mu)$, $\tilde{y}(\mu)$, and $\tilde{z}(\mu)$ denote the expressions in the right side of relations (2.16), (2.17), and (2.18). It can be easily shown that $(\tilde{x}(\mu), \tilde{y}(\mu), \tilde{z}(\mu))$ is a solution of system (2.8) with $p = z^1/\mu_1 - w(x^o)^{-1}$. Since such a solution is unique, the result follows. $\square$

Finally, we would like to comment about some relationship between the trajectories defined in this section and those discussed in [3] and [8]. Let $w \in \mathbf{R}_+^n$ be given. When $p = 0$, the path of solutions $x_0(\mu)$ of problem (P_μ) is precisely the "weighted logarithmic barrier path" with respect to w as described in [8]. Note that $x^o \in S_I$ lies in the path $x_0(\mu)$ if and only if for some $\mu_o > 0$ and $y^o \in \mathbf{R}^m$, we have $(y^o, \mu_o w(x^o)^{-1}) \in T_A$. We shall refer to the path $x_0(\mu)$ as the "w-central path". In particular, when $w = e$, this path is the "affine central path" defined in [3]. We shall refer to the e-central path as the "central path".

3. Limiting behavior of the w-PAS trajectories

In this section, we examine the limiting behavior of the w-PAS trajectories based on the characterization of these trajectories presented in the previous section. Given a point $x^o \in S_I$, we show that the w-PAS trajectory through this point converges to an optimal solution of problem (P). Moreover, if the dimension of the optimal face of problem (P) is greater than zero then the limiting point is an interior point of this face that depends on the point x^o. We also show that the dual paths $(y_p(\mu), z_p(\mu))$, described in the previous section, converge to a common point, namely, the w-center of the optimal face of the dual problem, for any $p \in \mathbf{R}^n$.

We start this section by introducing the necessary terminology. Let r^* denote the common optimal value of problems (P) and (D). The optimal face of (P) (resp. (D)) is the set of points $S_O \equiv \{x; x \in S_F, c^T x = r^*\}$ (resp. $T_O \equiv \{(y, z); (y, z) \in T_F, b^T y = r^*\}$). The set S_O is a face of the polyhedron S_F and therefore can be expressed as the set of points $\{x \in S_F; x_j = 0, j \in N\}$ for some index set $N \subseteq \{1, \ldots, n\}$. We may assume that N is the maximal set (with respect to inclusion) satisfying this property; that is, $j \in N$ if and only if $x_j = 0$ for every $x \in S_O$. Let B denote the set of indices $j \in \{1, \ldots, n\}$ such that $j \notin N$. It is well known that T_O is the face of the polyhedron T_F given by $\{(y, z) \in T_F; z_j = 0, j \in B\}$ and that B is the maximal set with this property. (See for example Schrijver [12].)

The next result shows that the w-PAS trajectories and their associated dual trajectories converge in objective value to the optimal value r^*.

THEOREM 3.1. *Let* $x^o \in S_I$, $(y^o, z^o) \in T_A$, *and* $\mu_o > 0$ *be given. Consider the solution* $(x(\mu), y(\mu), z(\mu)) \equiv (x_p(\mu), y_p(\mu), z_p(\mu))$ *of system* (2.8) *with* $p = z^o/\mu_o - w(x^o)^{-1}$ *and* $\mu \in I(p)$. *Then,* $\lim_{\mu \to 0} c^T x(\mu) = \lim_{\mu \to 0} b^T y(\mu) = r^*$.

PROOF. In view of Propositions 2.4 and 2.5, we may assume without loss of generality that $z^o > 0$ and that μ_o satisfies

$$p = \frac{z^o}{\mu_o} - w(x^o)^{-1} \geq 0.$$

Since $x(\mu) > 0$ for all $\mu \in I(p)$, it then follows from (2.8.a) that $z(\mu) > 0$ for all $\mu \in I(p)$. This implies that $(x(\mu), y(\mu), z(\mu)) \in S_I \times T_I$ for all $\mu \in I(p)$, and therefore, by the weak duality theorem of linear programming, we have that $b^T y(\mu) \leq r^* \leq c^T x(\mu)$ for all $\mu \in I(p)$. On the other hand, one can easily verify that if $x \in S_A$ and $(y, z) \in T_A$ then $c^T x - b^T y = x^T z$. Therefore, $c^T x(\mu) - b^T y(\mu) = x(\mu)^T z(\mu)$ for all $\mu \in I(p)$. The theorem now follows if we show that $x(\mu)^T z(\mu)$ converges to 0 as μ tends to 0. Indeed, multiplying (2.8.a) by $x(\mu)^T$, we obtain

$$0 \leq x(\mu)^T z(\mu) = \left(\sum_{j=1}^{n} w_j \right) \mu + \left(\frac{z^o}{\mu_o} - w(x^o)^{-1} \right)^T x(\mu)\mu$$

$$\leq \left(\sum_{j=1}^{n} w_j \right) \mu + \frac{\mu}{\mu_o}(z^o)^T x(\mu)$$

$$= \left(\sum_{j=1}^{n} w_j \right) \mu + \frac{\mu}{\mu_o}(c^T x(\mu) - b^T y^o).$$

Since from Proposition 2.1 $c^T x(\mu)$ remains bounded as μ approaches 0, the last expression implies that $x(\mu)^T z(\mu)$ converges to 0 as μ approaches 0. This completes the proof of the theorem. □

We now turn our efforts towards examining the limiting behavior of the w-PAS trajectories and its associated dual trajectories. Let $x^o \in S_I$, $(y^o, z^o) \in T_A$, and $\mu_o > 0$ be given. Set $p = z^o/\mu_o - w(x^o)^{-1}$. In Section 2, we have seen that the path of solutions $x(\mu) \equiv x_p(\mu)$ of problem (P_μ), for $\mu \in I(p)$, is a w-PAS trajectory (Proposition 2.3 and Theorem 2.1). We are now interested in the behavior of $x(\mu)$ as μ approaches 0. The next result shows that the subvector of $x(\mu)$ corresponding to the index set N, namely, the vector $x_N(\mu)$, converges to 0 as μ approaches 0.

LEMMA 3.1.

(a) $\lim_{\mu \to 0} x_N(\mu) = 0$.

(b) $\lim_{\mu \to 0} c_B^T x_B(\mu) = r^*$.

PROOF. Note that by Assumption 2.1.b, Proposition 2.1, and the fact that $c^T x(\mu)$ is decreasing in μ, it follows that $x(\mu)$ lies in a compact set for all μ sufficiently small. Let $\bar{x}$ be an accumulation point of $x(\mu)$ as μ approaches 0; that is, $\bar{x} = \lim_{k \to \infty} x(\mu^k)$, where (μ^k) is a sequence of positive numbers converging to zero. By Theorem 3.1, it then follows that $c^T \bar{x} = r^*$. Obviously, $\bar{x} \in S_F$. Hence, $\bar{x}$ lies in the optimal face of problem (P) and therefore $\bar{x}_N = 0$. Since this holds for all the accumulation points of $x(\mu)$, then (a) follows. The limit in (b) is a direct consequence of (a) and Theorem 3.1. $\square$

We next analyze the limiting behavior of the subvector $x_B(\mu)$ as μ approaches 0. We first make some observations. Since $x(\mu)$ is a solution of problem (P_μ), one can easily verify that $x_B(\mu)$ is an optimal solution of the problem stated as follows.

(3.1)
$$(Q_\mu) \quad \max p_B^T x_B + \sum_{j \in B} w_j \ln x_j ,$$
$$\text{s. t. } A_B x_B = b - A_N x_N(\mu),$$
$$x_B > 0.$$

As a consequence, for $\mu \in I(p)$, $x_B(\mu)$ satisfies the following relations, which are the optimality conditions for the unique optimal solution of problem (Q_μ).

(3.2)
$$w_B x_B^{-1} + p_B \in H$$
$$A_B x_B = b - A_N x_N(\mu), \qquad x_B > 0$$

where H is the subspace spanned by the rows of the matrix A_B.

In view of Lemma 3.1, we are led to consider the following problem, which is in some sense the limit of problem (Q_μ) as μ approaches 0. Let

(3.3)
$$(Q) \quad \max p_B^T x_B + \sum_{j \in B} w_j \ln x_j ,$$
$$\text{s. t. } A_B x_B = b,$$
$$x_B > 0$$

and x_B^* be the unique optimal solution of problem (Q). Obviously, $x^* \equiv (x_B^*, 0)$ is an optimal solution for (P). The next result completes the determination of the limiting behavior of the w-PAS trajectory $x(\mu)$.

THEOREM 3.2. $\lim_{\mu \to 0} x_B(\mu) = x_B^*$.

PROOF. Let $\bar{x}$ be an accumulation point of $x(\mu)$ as μ approaches 0. Hence, $\lim_{k \to 0} x(\mu^k) = \bar{x}$ for some sequence (μ^k) of positive numbers converging to 0. Since for all k, $x_B(\mu^k)$ satisfies system (3.2) with $\mu = \mu^k$, it follows that $\bar{x}_B$ satisfies the conditions $w_B(\bar{x}_B)^{-1} + p_B \in H$ and $A_B \bar{x}_B = b$. It turns out that these conditions are exactly the optimality conditions for the unique optimal solution x_B^* of problem (Q). Hence, $\bar{x}_B = x_B^*$ and the result follows. $\square$

Note that the limit x_B^* of the w-PAS trajectory as μ tends to 0 depends on $p_B = z_B^o/\mu_o - w_B(x_B^o)^{-1}$ as can be observed by considering problem (Q). (Of course, this dependence is realized only when the dimension of the optimal face of (P) is greater than 0). In view of Proposition 2.4, it seems that x_B^* should depend only on x^o. Indeed, this is the case, since substituting for p_B in the objective function of problem (Q) leads to

$$p_B^T x_B = \left(\frac{z_B^o}{\mu_o} - w_B(x_B^o)^{-1}\right)^T x_B$$
$$= \frac{1}{\mu_o}(c_B^T - (y^o)^T A_B)x_B - (w_B(x_B^o)^{-1})^T x_B$$
$$= \frac{r^* - b^T y^o}{\mu_o} - (w_B(x_B^o)^{-1})^T x_B$$

where the last equation follows from observing the constraint of (Q) and the fact that $A_B x_B = b$ implies that $c_B^T x_B = r^*$. Since the first term in the expression for $p_B^T x_B$ is constant, it is clear that x_B^* is the solution of problem (Q) with $-w_B(x_B^o)^{-1}$ replacing p_B in the objective function.

In particular, Theorem 3.2 shows that the w-central path (resp. central path) converges to the so-called "w-center" (resp. "center") of the optimal face of (P), that is, the optimal solution of problem (Q) with $p_B = 0$. A similar result for this case was obtained in [8].

We now examine the limiting behavior of the dual trajectories associated with the w-PAS trajectories. Using Theorem 3.1 and arguments similar to the ones used in the proof of Lemma 3.1, one can show the following result.

LEMMA 3.2. *Let* $(y(\mu), z(\mu)) \equiv (y_p(\mu), z_p(\mu))$ *be the dual trajectory associated with the* w-PAS *trajectory* $x(\mu) = x_p(\mu)$. *Then* $\lim_{\mu \to 0} z_B(\mu) = 0$.

Consider the trajectory $(y(\mu), z(\mu)) \equiv (y_p(\mu), z_p(\mu))$ associated with the w-PAS trajectory $x(\mu) = x_p(\mu)$. It follows from Section 2 that these trajectories satisfy $x(\mu) \in S_I$, $(y(\mu), z(\mu)) \in T_A$, and

$$x(\mu) - \frac{\mu w}{z(\mu) - \mu p} = 0.$$

It can be verified that these conditions are the optimality conditions for the problem stated as follows.

$$(D_\mu) \quad \max b^T y + \mu \sum_{j=1}^n w_j \ln(z_j - \mu p_j),$$
(3.4)
$$\text{s.t. } A^T y + z = c,$$
$$z > \mu p.$$

The point $(y(\mu), z(\mu))$ is the unique optimal solution of problem (D_μ) whereas $x(\mu)$ is the Lagrangian multiplier associated with the equality constraint of (D_μ). One can easily verify that $(y(\mu), z_N(\mu))$ is also an optimal solution of the problem stated as follows.

$$(E_\mu) \quad \max \sum_{j \in N} w_j \ln(z_j - \mu p_j),$$

(3.5)
$$\text{s. t. } A_N^T y + z_N = c_N,$$
$$A_B^T y = c_B - z_B(\mu),$$
$$z_N > \mu p_N.$$

Now consider the following problem, which arises from problem (E_μ) as μ tends to 0.

$$(E) \quad \max \sum_{j \in N} w_j \ln z_j,$$

(3.6)
$$\text{s. t. } A_N^T y + z_N = c_{\dot{N}},$$
$$A_B^T y = c_B,$$
$$z_N > 0$$

and let y^* and z_N^* denote its unique optimal solution. Obviously, (y^*, z^*), where $z^* = (0, z_N^*)$, is an optimal solution of problem (D). Using arguments similar to the ones used in the proof of Theorem 3.2, one can show the following result.

THEOREM 3.3. *Let* $(y(\mu), z(\mu)) \equiv (y_p(\mu), z_p(\mu))$ *be the dual trajectory associated with the* w-PAS *trajectory* $x(\mu) = x_p(\mu)$. *Then*

(a) $\lim_{\mu \to 0} y(\mu) = y^*$.

(b) $\lim_{\mu \to 0} z_N(\mu) = z_N^*$.

Although the limit of a w-PAS trajectory does depend on the initial condition, Theorem 3.3 shows that all dual trajectories converge to the w-center of the optimal face of the dual problem, that is, the solution of problem (E).

4. Limiting behavior of the dual estimates

Let x^k and $(y^E(x^k), z^E(x^k))$, $k = 1, 2, \ldots$, be the sequence of feasible solutions and its associated dual estimates generated by the PAS algorithm starting with $x^o \in S_I$. (See (2.2) and (2.3).) It was shown by [2], [4], and [13] that if the linear programming problem (P) satisfies Assumption 2.1 and is primal and dual nondegenerate then x^k and $(y^E(x^k), z^E(x^k))$ converge to the unique optimal solutions x^* and (y^*, z^*) of the primal problem (P) and the dual problem (D), respectively. Actually, [4] assumes only primal nondegeneracy and shows that x^k converges to an optimal solution x^* of problem (P) satisfying $x_B^* > 0$; however, it uses smaller steps than the discrete algorithm studied in [13].

Similar results (under the same assumptions) with respect to the continuous PAS trajectories were obtained in [9].

In the previous section, we showed that a w-PAS trajectory $x(\mu)$ passing through $x^o \in S_I$ tends to an optimal solution x^* of (P), which may depend on x^o if problem (P) does not have a unique optimal solution. Let $x(\mu)$ denote an arbitrary w-PAS trajectory. In this section, we shall discuss the limiting behavior of the trajectory $(y^E(x(\mu)), z^E(x(\mu)))$ with no prior assumption of nondegeneracy.

Given $x \in S_I$, recall that the dual estimate at x (see (2.2)) is given by

(4.1)
$$y^E(x) = (AD^2A^T)^{-1}AD^2c,$$
$$z^E(x) = c - A^T(AD^2A^T)^{-1}AD^2c$$

where $D = W^{-1/2}X$. It can be verified that $(y^E(x), z^E(x))$ satisfy the optimality conditions of the following convex quadratic programming problem where the minimization is with respect to (r, s).

(4.2)
$$\min \tfrac{1}{2}\|Ds\|^2,$$
$$\text{s.t.} \, A^T r + s = c$$

Given a w-PAS trajectory $x(\mu)$, let $y^E(\mu) \equiv y^E(x(\mu))$ and $z^E(\mu) \equiv z^E(x(\mu))$, and refer to $(y^E(\mu), z^E(\mu))$ as the dual estimate trajectory associated with the w-PAS trajectory $x(\mu)$. In this section, our main purpose is to show that $(y^E(\mu), z^E(\mu))$ converges to the same limit as the dual trajectories associated with the w-PAS trajectories. Hence, as was shown in the previous section, they converge to the so-called w-center of the optimal face of problem (D).

Before stating the main result of this section, we make some observations. Note that from (2.14), (2.15), and (4.1), it follows that

(4.3) $$y^E(\mu) = y(\mu) - \mu\dot{y}(\mu),$$

(4.4) $$z^E(\mu) = z(\mu) - \mu\dot{z}(\mu).$$

Note that the quantity in the right side of (4.3) (resp. (4.4)) can be viewed as the first-order Taylor approximation of the point $y(0) \equiv \lim_{\mu \to 0} y(\mu)$ (resp. $z(0) \equiv \lim_{\mu \to 0} z(\mu)$) given that we are at the point $y(\mu)$ (resp. $z(\mu)$) of the dual trajectory $y(.)$ (resp. $z(.)$). Expressions (4.3) and (4.4) show that, although there can be many dual trajectories $(y(\mu), z(\mu))$ (depending on the choice of the initial condition (y^o, z^o)) associated with a given w-PAS trajectory $x(\mu)$, their first-order Taylor approximation as described above is the same and is equal to the dual estimate trajectory $(y^E(.), z^E(.))$.

One consequence of relations (4.3) and (4.4) is that one way to show that $y^E(\mu)$ and $y(\mu)$ (resp. $z^E(\mu)$ and $z(\mu)$) have the same limit as μ approaches 0 is to prove that $\dot{y}(\mu)$ (resp. $\dot{z}(\mu)$) is bounded for all μ sufficiently small. However, we have not been able to show the latter result directly. Rather, we show directly in the next theorem that $(y^E(\mu), z^E(\mu))$ have the same limit as $(y(\mu), z(\mu))$ when μ approaches 0. We will see in the next section that this result enables us to characterize the limiting behavior of the derivatives of the w-PAS trajectories and its associated dual trajectories.

THEOREM 4.1. *Let $(x^*, y^*, z^*) = \lim_{\mu \to 0}(x(\mu), y(\mu), z(\mu))$ where $x(\mu)$ is the w-PAS trajectory passing through some $x^o \in S_I$ and $(y(\mu), z(\mu))$ is its associated dual trajectory. Let $(y^E(\mu), z^E(\mu))$ be the dual estimate trajectory associated with the w-PAS trajectory $x(\mu)$. Then $\lim_{\mu \to 0}(y^E(\mu), z^E(\mu)) = (y^*, z^*)$.*

PROOF. Let (B, N) be the partition of the index set $\{1, \ldots, n\}$ such that $x_B^* > 0$, $x_N^* = 0$, $z_B^* = 0$, and $z_N^* > 0$. From (2.8.a) we have that

$$(4.5) \qquad x_N(\mu) = (z_N(\mu) - \mu p_N)^{-1} w_N \mu.$$

Since $(y^E(x), z^E(x))$ is the optimal solution of problem (4.2), and noting (4.5), it follows that $(y^E(\mu), z^E(\mu))$ is the optimal solution of the following problem where the minimization is with respect to (r, s).

$$(\mathrm{V}_\mu) \quad \min \frac{1}{2}\|(w_B)^{-1/2} x_B(\mu) s_B\|^2 + \frac{\mu^2}{2}\|(w_N)^{1/2}(z_N(\mu) - \mu p_N)^{-1} s_N\|^2,$$

$$\text{s.t. } A^T r + s = c$$

Consider now the following problem

$$(\mathrm{R}_\mu) \quad \min \frac{1}{2}\|(w_N)^{1/2}(z_N(\mu) - \mu p_N)^{-1} s_N\|^2,$$

$$\text{s.t. } A_B^T r = c_B,$$

$$A_N^T r + s_N = c_N$$

and let $\theta(\mu)$ and $(r(\mu), s_N(\mu))$ denote the optimal value and optimal solution of (R_μ), respectively. One can easily verify that $\theta(\mu)$ and $(r(\mu), s_N(\mu))$ converge to the optimal value and optimal solution, respectively, of the following problem, which arises from problem (R_μ) as μ tends to 0.

$$(\mathrm{R}) \quad \min \frac{1}{2}\|(w_N)^{1/2}(z_N^*)^{-1} s_N\|^2,$$

$$\text{s.t. } A_B^T r = c_B,$$

$$A_N^T r + s_N = c_N$$

Let θ^* denote the optimal value of (R). One can easily verify that (y^*, z_N^*) is the (unique) optimal solution of problem (R). Hence, $\theta^* = \|(w_N)^{1/2}\|^2/2$. On the other hand, letting $s(\mu) = (0, s_N(\mu))$, it follows that $(r(\mu), s(\mu))$ is a feasible solution for problem (V_μ). Hence, we have
(4.6)

$$\mu^2 \theta(\mu) \geq \frac{1}{2}\|(w_B)^{-1/2} x_B(\mu) z_B^E(\mu)\|^2 + \frac{\mu^2}{2}\|(w_N)^{1/2}(z_N(\mu) - \mu p_N)^{-1} z_N^E(\mu)\|^2.$$

This relation obviously implies that $\lim_{\mu \to 0}(w_B)^{-1/2} x_B(\mu) z_B^E(\mu) = 0$ and since, by Theorem 3.2, $\lim_{\mu \to 0} x_B(\mu) = x_B^* > 0$, we obtain $\lim_{\mu \to 0} z_B^E(\mu) = 0 = z_B^*$. Since, by Theorem 3.3, $\lim_{\mu \to 0}(z_N(\mu) - \mu p_N)^{-1} = (z_N^*)^{-1} > 0$, relation (4.6) also implies that $z_N^E(\mu)$ is bounded for all μ sufficiently small.

Since A has full row rank, it also follows that $y^E(\mu)$ is bounded for all μ sufficiently small. Let $(\bar{y}, \bar{z})$ be an accumulation point of $(y^E(\mu), z^E(\mu))$ as μ tends to 0. Obviously, $\bar{z}_B = 0$. It then follows from relation (4.6) that $\theta^* \geq \|(w_N)^{1/2}(z_N^*)^{-1}\bar{z}_N\|/2$. But since $(\bar{y}, \bar{z}_N)$ is feasible to (R), it follows that $(\bar{y}, \bar{z}_N)$ is an optimal solution for (R). Hence $\bar{y} = y^*$ and $\bar{z}_N = z_N^*$. Since this holds for any accumulation point of $(y^E(\mu), z^E(\mu))$, the result follows. $\square$

The following corollary is a consequence of the proof of Theorem 4.1 and shows that $z_B^E(\mu) = o(\mu)$ as μ tends to 0.

COROLLARY 4.1. $\lim_{\mu \to 0} z_B^E(\mu)/\mu = 0$.

PROOF. Dividing expression (4.6) by $\mu^2/2$ and taking the limit, we obtain

$$\theta^* \geq \frac{1}{2} \limsup_{\mu \to 0} \frac{1}{\mu^2} \|(w_B)^{-1/2} x_B(\mu) z_B^E(\mu)\|^2 + \frac{1}{2} \|(w_N)^{1/2}\|^2$$
$$= \frac{1}{2} \limsup_{\mu \to 0} \frac{1}{\mu^2} \|(w_B)^{-1/2} x_B(\mu) z_B^E(\mu)\|^2 + \theta^*,$$

which implies that

$$\lim_{\mu \to 0} \frac{1}{\mu^2} \|(w_B)^{-1/2} x_B(\mu) z_B^E(\mu)\|^2 = 0.$$

Since $w_B > 0$ and $\lim_{\mu \to 0} x_B(\mu) > 0$, the result follows. $\square$

5. Limiting behavior of the derivatives of the w-PAS trajectories

In this section, we analyze the limiting behavior of the derivatives of the w-PAS trajectories and their associated dual trajectories.

In order to simplify the development and the statements of the results of this section, we start by recalling the notation to be used throughout this section. Let $x^o \in S_I$, $(y^o, z^o) \in T_A$, and $\mu_o > 0$ be given. Let $p = z^o/\mu_o - w(x^o)^{-1}$. Let $(x(\mu), y(\mu), z(\mu))$ denote the path of solutions $(x_p(\mu), y_p(\mu), z_p(\mu))$, $\mu \in I(p)$, of system (2.8). Let x^* and (y^*, z^*) denote the limits of the paths $x(\mu)$ and $(y(\mu), z(\mu))$ as μ approaches 0, respectively. Recall from Section 3 that $x_N^* = 0$, $z_B^* = 0$, and that x_B^* and (y^*, z_N^*) are the (unique) optimal solutions of problems (Q) and (E) (see (3.3) and (3.6)), respectively. Also let $(y^E(\mu), z^E(\mu))$ denote the dual estimate associated with $x(\mu)$ for $\mu \in I(p)$. We proved in Section 4 that the limit of $(y^E(\mu), z^E(\mu))$ as μ approaches 0 is equal to (y^*, z^*). As an immediate consequence of this fact, we have the following result.

THEOREM 5.1. $\lim_{\mu \to 0} \dot{x}_N(\mu) = w_N(z_N^*)^{-1}$.

PROOF. From (2.15) and (4.1), it follows that $z^E(\mu) = z(\mu) - \mu\dot{z}(\mu)$. Using this relation and relations (2.12) and (2.8.a), we obtain

$$\dot{x}(\mu) = \frac{w(z(\mu) - \mu \dot{z}(\mu))}{(z(\mu) - p\mu)^2}$$

$$= \frac{w z^E(\mu)}{(z(\mu) - p\mu)^2}$$

from which the result easily follows. □

Theorem 5.1 was essentially first proved in [9] under nondegeneracy assumptions.

Some observations are in order at this point. When the dimension of the optimal face of problem (P) is 0, that is, when problem (P) has a unique optimal solution that is a vertex of S_F, the limiting behavior of $\dot{x}_B(\mu)$ is easily determined as follows. In this case, all the columns of the matrix A_B are linearly independent, and we have

$$A_B \dot{x}_B(\mu) + A_N \dot{x}_N(\mu) = 0,$$

which implies that

$$\dot{x}_B(\mu) = -(A_B^T A_B)^{-1} A_B^T A_N \dot{x}_N(\mu).$$

Taking the limit in the expression above, we obtain

$$\lim_{\mu \to 0} \dot{x}_B(\mu) = -(A_B^T A_B)^{-1} A_B^T A_N w_N (z_N^*)^{-1}.$$

Note that in this case the limit of $\dot{x}(\mu)$ as μ tends to 0 does not depend on the initial point $x^o \in S_I$. However, this is not necessarily the case when the dimension of the optimal face is greater than 0 as we now show.

THEOREM 5.2. *The limit of* $\dot{x}_B(\mu)$ *as* μ *tends to* 0 *exists and is equal to the* (*unique*) *optimal solution* v_B^* *of the following problem where the minimization is with respect to* v_B.

(5.1)
$$\min \frac{1}{2} \|(w_B)^{1/2} (x_B^*)^{-1} v_B\|^2,$$
$$\text{s.t. } A_B v_B = -A_N w_N (z_N^*)^{-1}.$$

PROOF. We first show that $\dot{x}_B(\mu)$ is the (unique) optimal solution of the problem stated as follows.

(5.2)
$$\min \frac{1}{2} \|(w_B)^{1/2} x_B^{-1}(\mu) v_B\|^2,$$
$$\text{s.t. } A_B v_B = -A_N \dot{x}_N(\mu).$$

Indeed, recall from Section 3 that since $x_B(\mu)$ is the optimal solution of problem (Q_μ) (see (3.1)), it must satisfy the conditions of system (3.2). Therefore, $\dot{x}_B(\mu)$ satisfies

(5.3)
$$w_B x_B^{-2}(\mu) \dot{x}_B(\mu) \in H,$$
$$A_B \dot{x}_B(\mu) = -A_N \dot{x}_N(\mu)$$

or, equivalently, $\dot{x}_B(\mu)$ satisfies the optimality condition for problem (5.2). Therefore, $\dot{x}_B(\mu)$ is the optimal solution of problem (5.2). Using this fact, one can easily verify that $\dot{x}_B(\mu)$ is bounded as μ tends to 0. If $\bar{v}_B$ is an accumulation point of $\dot{x}_B(\mu)$ as μ approaches 0, then, by (5.3), it follows that $w_B(x_B^*)^{-2}\bar{v}_B \in H$ and $A_B\bar{v}_B = -A_N w_N(z_N^*)^{-1}$. Hence, $\bar{v}_B$ satisfies the optimality condition for problem (5.1). This implies that $\bar{v}_B = v_B^*$. Since this holds for any accumulation point of $\dot{x}_B(\mu)$ as μ approaches 0, the result follows. $\square$

With the aid of the previous two theorems, we now have enough information to analyze the limiting behavior of $(\dot{y}(\mu), \dot{z}(\mu))$ as μ approaches 0.

THEOREM 5.3. $\lim_{\mu \to 0} \dot{z}_B(\mu) = p_B + w_B(x_B^*)^{-1}$.

PROOF. Observe that by relations (2.12) and (2.8.a), we have

$$(5.4) \qquad \dot{z}(\mu) = p + wx^{-2}(\mu)[x(\mu) - \mu\dot{x}(\mu)].$$

By Theorems 5.1 and 5.2, we have that $\mu\dot{x}(\mu)$ converges to 0 as μ tends to 0. From this observation, relation (5.4), and Theorem 3.2, the result follows. $\square$

Using the previous result, one can prove the following theorem.

THEOREM 5.4. *The limit of* $(\dot{y}(\mu), \dot{z}_N(\mu))$ *as* μ *tends to* 0 *exists and is equal to the (unique) optimal solution of the following problem where the minimization is with respect to* (r, s_N).

$$(5.5) \qquad \begin{aligned} &\min \frac{1}{2}\|(w_N)^{1/2}(z_N^*)^{-1}(s_N - p_N)\|^2, \\ &\text{s.t.}\ A_B^T r = -p_B - w_B(x_B^*)^{-1}, \\ &A_N^T r + s_N = 0 \end{aligned}$$

The proof of Theorem 5.4 follows by noting that $(\dot{y}(\mu), \dot{z}_N(\mu))$ is the optimal solution of the problem

$$\begin{aligned} &\min \frac{1}{2}\|(w_N)^{1/2}(z_N(\mu) - p_N\mu)^{-1}(s_N - p_N)\|^2, \\ &\text{s.t.}\ A_B^T r = -\dot{z}_B(\mu), \\ &A_N^T r + s_N = 0 \end{aligned}$$

and by using arguments similar to the ones used in the proof of Theorem 5.2.

The next result shows that the value of the primal objective function along the w-PAS trajectory converges to the optimal value more slowly than the value of the dual objective function does along the dual estimate trajectory. This fact was observed in [2] and [13] with respect to the discrete version of the PAS algorithm (under nondegeneracy assumption).

THEOREM 5.5. *Let* r^* *denote the common optimal value of problems* (P) *and* (D). *Then, we have*

(a) $c^T x(\mu) - r^* = \mu \sum_{j \in N} w_j + o(\mu)$.

(b) $r^* - b^T y^E(\mu) = o(\mu)$.

PROOF. One can easily verify that $c^T x(\mu) - r^* = (z_N^*)^T x_N(\mu)$ and that $r^* - b^T y^E(\mu) = (x_B^*)^T z_B^E(\mu)$. The result now follows from Corollary 4.1 and the fact that

$$\lim_{\mu \to 0} \frac{x_N(\mu)}{\mu} = \lim_{\mu \to 0} \dot{x}_N(\mu) = w_N (z_N^*)^{-1}$$

where the last equality follows from Theorem 5.1. $\square$

6. The dual affine scaling (DAS) trajectories

In the previous sections, we restricted our attention to linear programming problems formulated in standard form. The affine scaling algorithm can be easily applied to linear programming problems formulated differently. In particular, a substantial effort was invested in exploring the same ideas as applied to the inequality form, namely, linear programming problems in the format of problem (D) of Section 2. (See [1] and [10].) Since this form is precisely the dual of a problem in standard form, it is customary to refer to the affine scaling algorithm as applied to problem (D) as the dual affine scaling (DAS) algorithm. It is obvious that all the results presented in this paper can be applied directly to the linear programming problem (D).

Specifically, consider the linear programming problem (D) of Section 2 and assume that a vector $w \in \mathbf{R}_+^n$ is given. (Note however that, as in the PAS algorithm, the DAS algorithm is usually defined with $w = e$ (see [1]).) Given an interior point $(\bar{y}, \bar{z})$ for problem (D), that is, $(\bar{y}, \bar{z}) \in T_I$, the w-DAS algorithm computes a search direction $(\Delta y, \Delta z)$ analogous to the w-PAS algorithm (see [1]) where now the linear scaling transformation $\phi_w : \mathbf{R}^n \to \mathbf{R}^n$ defined by $\phi_w(z) = Dz$, $z \in \mathbf{R}^n$, with $D = W^{1/2} \bar{Z}^{-1}$, is applied to the slack z in order to transform the current slack vector $\bar{z}$ into the vector $w^{1/2}$. Similarly to the discussion on the w-PAS algorithm, problem (D) is transformed to

$$\max b^T y,$$
$$\text{s.t.} \, DA^T y + v = Dc,$$
$$v \ge 0$$

and the search direction d_y and d_v in the transformed space is then computed as the solution of the problem

$$\min b^T y + \frac{1}{2}\|v\|^2,$$
$$\text{s.t.} \, DA^T y + v = 0,$$

which yields the following explicit solution:

$$d_y = -(AD^2 A^T)^{-1} b,$$
$$d_v = DA^T (AD^2 A^T)^{-1} b.$$

Finally, the search directions in the original space are given by

$$\Delta y = -(AD^2A^T)^{-1}b,$$
$$\Delta z = A^T(AD^2A^T)^{-1}b.$$

Observe that the quantity

$$x^E \equiv Dd_v = D^2A^T(AD^2A^T)^{-1}b$$

satisfies $Ax^E = b$ and is the so-called "primal estimate". The next iterate $(\hat{y}, \hat{z})$ is computed as $\hat{y} = \bar{y} - \alpha\Delta y$ and $\hat{z} = \bar{z} - \alpha\Delta z$ where $\alpha \equiv \alpha(\bar{y}, \bar{z}) > 0$ is an appropriate step size that guarantees that $\hat{z} > 0$. In a manner similar to Section 2, we are interested in studying the trajectories induced by the vector field $V_w : T_I \times \mathbf{R} \to \{(u, v); A^T u + v = 0\}$ defined as

$$V_w((y, z), \mu) = \frac{1}{\mu^2}(\Delta y, \Delta z).$$

Thus, similar to the w-PAS case, the trajectories $(y(\mu), z(\mu))$ are characterized as the solutions of the logarithmic barrier family of problems

$$\max b^T y + \mu \left[p^T z + \sum_{j=1}^{n} w_j \ln x_j \right],$$
$$\text{s.t.} \ A^T y + z = c,$$
$$z > 0,$$

which leads to the following relations between the trajectories $(y(\mu), z(\mu))$ and their associated Lagrangian "primal trajectories" $x(\mu)$

$$x(\mu) - \mu w z^{-1}(\mu) = p\mu$$

where $p = (x^o/\mu^o - w/z^o)$, x^o is an arbitrary vector satisfying $Ax^o = b$ and $\mu_o > 0$. Similarly, one can show that $x^E(\mu) = x(\mu) - \mu\dot{x}(\mu)$. Then, all the results in the previous sections will apply to these constructs by simply "dualizing" in the sense that x and (y, z) are interchanged. This task is rather trivial, so we leave out the details.

7. Analysis of the primal–dual affine scaling (PDAS) trajectories

In this section, we apply all the results obtained in Sections 2–5 to the trajectories generated by the primal–dual affine scaling algorithm.

The primal–dual affine scaling algorithm has been presented and analyzed in [11]. Here, we briefly review the directions generated by the PDAS algorithm. Given a point $v = (x, y, z) \in S_I \times T_I$, the PDAS algorithm computes a search direction $\Delta v \equiv \Delta v(v) \equiv (\Delta x, \Delta y, \Delta z) \in \mathbf{R}^n \times \mathbf{R}^m \times \mathbf{R}^n$, which is a solution of the following system of linear equations.

$$Z\Delta x + X\Delta z = xz,$$

(7.1)
$$A\Delta x = 0,$$
$$A^T\Delta y + \Delta z = 0.$$

This system yields the following explicit expression for Δx, Δy, and Δz.

$$\Delta x = [Z^{-1} - Z^{-1}XZ^T(AZ^{-1}XA^T)^{-1}AZ^{-1}]xz,$$
$$\Delta y = -[(AZ^{-1}XA^T)^{-1}AZ^{-1}]xz,$$
$$\Delta z = [A^T(AZ^{-1}XA^T)^{-1}AZ^{-1}]xz$$

Letting $D = (Z^{-1}X)^{1/2}$ and using the fact that $A^Ty + z = c$ and $Ax = b$, we obtain

$$\Delta x = D[I - DA^T(AD^2A^T)^{-1}AD]Dc,$$
$$\Delta y = -(AD^2A^T)^{-1}b,$$
$$\Delta z = A^T(AD^2A^T)^{-1}b.$$

We are interested in obtaining results about the limiting behavior of the trajectories induced by the vector field $\Delta v(v)$, $v \equiv (x, y, z) \in S_I \times T_I$. With this aim, we consider the solution curves of the following reparametrized differential equation:

(7.2)
$$\dot{v}(\mu) = \frac{1}{\mu}\Delta v(v(\mu)),$$
$$v(\mu_o) = v^o$$

where $v(\mu) \equiv (x(\mu), y(\mu), z(\mu)) \in \mathbf{R}^n \times \mathbf{R}^m \times \mathbf{R}^n$ and where $\mu_o > 0$ and $v^o \equiv (x^o, y^o, z^o) \in S_I \times T_I$ are the initial conditions. Equivalently, in view of the definition of $\Delta v(v)$ given by system (7.1), the differential equation (7.2) can be rewritten as

(7.3.a) $\quad z(\mu)\dot{x}(\mu) + x(\mu)\dot{z}(\mu) = \dfrac{x(\mu)z(\mu)}{\mu}$,

(7.3.b) $\qquad\qquad\qquad A\dot{x}(\mu) = 0$,

(7.3.c) $\qquad\quad A^T\dot{y}(\mu) + \dot{z}(\mu) = 0$,

(7.3.d) $\qquad x(\mu_o) = x^o, \qquad y(\mu_o) = y^o, \qquad z(\mu_o) = z^o$.

The following result characterizes the solution of (7.3) and is an immediate consequence of the results of Section 2.

THEOREM 7.1. *Let* $v(\mu) \equiv (x(\mu), y(\mu), z(\mu))$ *be the (unique) solution of system (2.8) with* $p \equiv 0$ *and* $w \equiv x^oz^o/\mu_o$. *Then* $v(\mu)$ *is the (unique) solution of (7.3). As a consequence, it follows that* $x(\mu)$ *is the (unique) optimal solution of*

(7.4)
$$\min c^Tx - \mu\sum_{j=1}^n w_j \ln x_j,$$
$$\text{s. t. } Ax = b,$$
$$x > 0$$

while $(y(\mu), z(\mu))$ *is the (unique) optimal solution of*

(7.5)
$$\max b^T y + \mu \sum_{j=1}^{n} w_j \ln z_j,$$
$$\text{s. t. } A^T y + z = c,$$
$$z > 0.$$

PROOF. It follows from (2.10), (2.11), (2.12), (2.8.a), and the fact that $p = 0$ that $v(\mu)$ satisfies (7.3.a), (7.3.b), and (7.3.c). Also, since $v^o \in S_I \times T_I$, $p = 0$, and $w = x^o z^o / \mu_o$, it follows that v^o is a solution of system (2.8) with $\mu = \mu_o$. Hence, $v(\mu_o) = v^o$. This shows that $v(\mu)$ is a solution of (7.3). To show that $x(\mu)$ and $(y(\mu), z(\mu))$ are the optimal solutions of problems (P) and (D), respectively, just note that (2.8) with $p = 0$ is the optimality condition for both problems. □

For a detailed discussion of the relationship between problems (7.4) and (7.5), see also [8].

Theorem 7.1 shows that the family of PDAS trajectories are precisely the w-central paths as defined in Section 2. Thus, all the results obtained in Sections 2–5 are applicable to these trajectories as well. We summarize these results in the following two theorems.

THEOREM 7.2. *Let* $x^o \in S_I$ *and* $(y^o, z^o) \in T_I$ *be given. Let* $(x(\mu), y(\mu), z(\mu))$ *be the* PDAS *trajectory passing through* (x^o, y^o, z^o). *Let* $w = x^o z^o$. *Then*
$$\lim_{\mu \to 0} (x(\mu), y(\mu), z(\mu)) = (x^*, y^*, z^*)$$
where x^* *and* (y^*, z^*) *are the* w-*center points of the optimal faces of the primal* (P) *and the dual* (D) *problems, respectively.*

It should be noted that Theorem 7.2 was first proved in [8].

THEOREM 7.3. *Let* $x^o \in S_I$ *and* $(y^o, z^o) \in T_I$ *be given. Let* $(x(\mu), y(\mu), z(\mu))$ *be the* PDAS *trajectory passing through* (x^o, y^o, z^o). *Let* $w = x^o z^o$. *Let* x^* *and* (y^*, z^*) *denote the* w-*center points of the optimal faces of the primal* (P) *and the dual* (D) *problems, respectively. Then, the limit of* $(\dot{x}(\mu), \dot{y}(\mu), \dot{z}(\mu))$ *as* μ *tends to* 0 *exists, and its value is as follows.*

(a) $\lim_{\mu \to 0} \dot{x}_N(\mu) = w_N (z_N^*)^{-1}$ *and* $\lim_{\mu \to 0} \dot{x}_B(\mu)$ *is equal to the unique optimal solution of problem* (5.1).

(b) $\lim_{\mu \to 0} \dot{z}_B(\mu) = w_B (z_B^*)^{-1}$ *and* $\lim_{\mu \to 0} \dot{z}_N(\mu)$ *is equal to the unique optimal solution of problem* (5.5) *with* p_B *and* p_N *replaced by* 0.

8. Remarks

It is still an open question whether the actual discrete PAS (or DAS) algorithm converges to an optimal solution without assuming both primal and dual nondegeneracy. We believe that the results presented here are a step towards proving that all the accumulation points (and possibly the limit) of the sequence of iterates (x^k) generated by the PAS algorithm (with appropriate step size) lie in the optimal face of the primal problem.

A closely related problem to the one mentioned above is the convergence of the dual estimate in the discrete PAS (or DAS) algorithm. We again believe that the results presented in this paper provide some insight to the observed fact (see, e.g., [1] and [10]) that the "dual estimates" (or "primal estimates" in the case of the DAS algorithm) do converge to an optimal solution of the dual (primal) problem.

In contrast to the discrete PAS (or DAS) algorithm, it has been proved in [11] that the PDAS algorithm converges (in polynomial time) to an optimal and dual solution without any nondegeneracy assumption. We believe that the inherently simpler structure of the PDAS trajectories (the w-central trajectories) may provide some insight into the polynomial convergence of the PDAS algorithm.

9. Acknowledgment

This research was partially funded by the United States Navy Office of Naval Research under contract N00014-87-K-0202 and the Brazilian Postgraduate Education Federal Agency (CAPES). Their financial support is gratefully acknowledged.

References

1. I. Adler, N. Karmarkar, M. G. C. Resende, and G. Veiga, *An implementation of Karmarkar's algorithm for linear programming*, Math. Programming **44** (1989), 297–335.
2. E. R. Barnes, *A variation on Karmarkar's algorithm for solving linear programming problems*, Math. Programming **36** (1986), 174–182.
3. D. A. Bayer and J. C. Lagarias, *The nonlinear geometry of linear programming*; Part I: *Affine and projective scaling projections*; Part II: *Legendre transform coordinates and central trajectories*, Trans. Amer. Math. Soc. **314** (1989), 499–581.
4. I. I. Dikin, *Iterative solution of problems of linear and quadratic programming*, Soviet Math. Dokl. **8** (1967), 674–675.
5. A. Fiacco and G. McCormick, *Nonlinear programming: Sequential unconstrained minimization techniques*, John Wiley and Sons, New York, 1955.
6. N. Karmarkar, *A new polynomial time algorithm for linear programming*, Combinatorica **4** (1984), 373–395.
7. ____, Talk at the University of California at Berkeley (Berkeley, CA, 1984).
8. N. Megiddo, *Pathways to the optimal set in linear programming*, Progress in Mathematical Programming: Interior-point Algorithms and Related Methods (N. Megiddo, ed.), Springer-Verlag, 1989, pp. 131–158.
9. N. Megiddo and M. Shub, *Boundary behavior of interior point algorithms for linear programming*, Math. Oper. Res. **14** (1989), 97–146.
10. C. L. Monma and A. J. Morton, *Computational experience with a dual affine variant of Karmarkar's method for linear programming*, Oper. Res. Lett. **6** (1987), 261–267.
11. R. D. C. Monteiro, I. Adler, and M. G. C. Resende, *A polynomial-time primal–dual affine scaling algorithm for linear and convex quadratic programming and its power series extension*, Math. Oper. Res. **15** (1990), 191–214.
12. A. Schrijver, *Theory of linear and integer programming*, John Wiley & Sons, 1986.
13. R. J. Vanderbei, M. S. Meketon, and B. A. Freedman, *A modification of Karmarkar's linear programming algorithm*, Algorithmica **1** (1986), 395–407.

Department of Industrial Engineering and Operations Research, University of California, Berkeley, California 94720

Contemporary Mathematics
Volume **114**, 1990

Convergence and Boundary Behavior of the Projective Scaling Trajectories for Linear Programming

RENATO D. C. MONTEIRO

ABSTRACT. We analyze the convergence and limiting behavior of the continuous trajectories of the vector field induced by the projective scaling algorithm as applied to (possibly degenerate) linear programming problems in Karmarkar's standard form. We show that a projective scaling trajectory tends to an optimal solution that in general depends on the starting point. When the optimal solution is unique, we show that all projective scaling trajectories approach the optimal solution through the same asymptotic direction. Our analysis is based on the affine scaling trajectories for the homogeneous standard form linear program that arises from Karmarkar's standard form linear program by removing the unique nonhomogeneous constraint.

1. Introduction

Consider the linear programming problem in Karmarkar's standard form as follows:

$$(P) \qquad \min\{c^T w \mid Aw = 0,\, e^T w = 1,\, w \ge 0\}$$

where A is an $m \times n$ matrix ($m \le n - 1$) of full rank, c and w are n-vectors, and e denotes the n-vector of all ones. Assume that (P) has feasible solution $w > 0$. The projective scaling algorithm designed for solving problem (P), under the assumption that the optimal value of (P) is 0, was first presented by Karmarkar in his seminal paper [11]. Variants of the projective scaling algorithm have been presented by many authors, including Anstreicher [2], de Ghellinck and Vial [5], Gay [8], Gonzaga [9], Todd and Burrell [18], and Ye and Kojima [21]. Another algorithm, namely the affine scaling algorithm, designed for solving linear programming problems

1980 *Mathematics Subject Classification* (1985 *Revision*). Primary 90C05; Secondary 90C25.
Key words and phrases. Interior point methods, linear programming, Karmarkar's algorithm, logarithmic barrier function, projective and affine scaling trajectories, continuous trajectories for linear programming.
This work was done while the author was a Ph.D. student in the Department of Industrial Engineering at the University of California at Berkeley. The author is currently working at the University of Arizona, Systems and Industrial Engineering Department, Tuczon, AZ 85721.

in standard form, was presented by Dikin [6] and was later (independently) reintroduced by Barnes [3] and Vanderbei, Meketon, and Freedman [19] as a variant of the projective scaling algorithm. The continuous trajectories generated by the vector fields induced by the projective and affine scaling algorithms are commonly referred to as projective scaling trajectories and affine scaling trajectories, respectively.

Continuous trajectories in the context of nonlinear programming have been discussed long ago within the nonlinear programming community. For instance, in the book of Fiacco and McCormick [7], continuous trajectories that arise as sets of minimizers for parametrized families of weighted logarithmic barrier problems, usually called weighted barrier trajectories, are systematically studied.

With the discovery of Karmarkar's algorithm and other related interior point methods, a new effort arose towards studying their underlying continuous trajectories. Because of the special structure exhibited by linear programming problems, these continuous trajectories turn out to be more amenable to a deeper analysis than do the trajectories associated with algorithms for general nonlinear optimization problems.

The importance of analyzing the projective and affine scaling trajectories was recently pointed out by Karmarkar [12]. Bayer and Lagarias [4] presented a systematic study of some properties and mathematical structures of these trajectories. Megiddo and Shub [14] and Shub [17] analyzed the limiting behavior of these trajectories near the optimal vertex under nondegeneracy assumptions. By characterizing the affine scaling trajectories as solutions of certain parametrized logarithmic barrier families of problems, Adler and Monteiro [1] analyzed the convergence and limiting behavior of the affine scaling trajectories under no nondegeneracy assumptions. They also established convergence of the affine dual estimate along the affine scaling trajectories. Using a different approach, Witzgall, Boggs, and Domich [20] also investigated convergence of the affine scaling trajectories without nondegeneracy assumptions. The latter also contains a good historic review on the development and relations of several works that discuss continuous trajectories for linear and nonlinear programming problems.

Another effort towards analyzing continuous trajectories for linear programs is the work of Megiddo [13], which studied the weighted barrier trajectories in the framework of primal-dual complementarity relationships for the special case of linear and convex quadratic programming problems. A polynomial algorithm, namely the primal-dual affine scaling algorithm, which has the weighted barrier trajectories as its continuous trajectories, was presented and analyzed by Monteiro, Adler, and Resende [15].

The purpose of this paper is to extend the results of [14] and [17] by studying the convergence and limiting behavior of the projective scaling trajectories for the linear program (P) without imposing any nondegeneracy assumption on (P). The development in this paper is closely related to the one in [1] in the following sense: Let (HP) denote the homogeneous linear program that arises from problem (P) by removing the constraint $e^T w = 1$. As shown in Bayer and Lagarias [4], every projective scaling trajectory for (P) can be obtained from an affine scaling trajectory for problem (HP) by projecting it radially into the simplex $\{x \in \mathbf{R}^n | e^T x = 1, x \geq 0\}$. Using this fact, we will be able to translate the behavior of the affine scaling trajectories for problem (HP) to the projective scaling trajectories for problem (P). We do so under two cases: when the optimal value of (P) is zero and when it is positive. In the first and more interesting case, it turns out that the set of optimal solutions to problem (HP) is unbounded. Since [1] analyzed the behavior of the affine scaling trajectories for a linear program that has bounded optimal face, it is necessary to extend the results of [1] in order to handle this more general case.

This paper is organized as follows. In Section 2, we briefly recall the definition of the projective scaling vector field and its associated trajectories for problem (P). We then discuss the relationship between these trajectories and the affine scaling trajectories for problem (HP) by recalling the result of Bayer and Lagarias. In Section 3, we show how to extend the analysis of [1] for the affine scaling trajectories associated with linear programs that have unbounded optimal face. In Section 4, we analyze the limiting behavior of the projective scaling trajectories for the case in which the optimal value of problem (P) is 0. We show that the projective scaling trajectories converge to optimal solutions that depend on the starting points. We also analyze the limiting behavior of the dual estimate that naturally arises in the context of the projective scaling method as described in [18]. Finally, we characterize the asymptotic direction of the projective scaling trajectories near their limit. In Section 5, we consider the case when the optimal value of (P) is positive and show that every projective scaling trajectory converges to the unique point that minimizes the so-called Karmarkar potential function associated with problem (P).

2. Related results

In this section, we recall the definition of the projective scaling vector field and its associated trajectories for a linear programming problem in Karmarkar's standard form. We also motivate our approach towards analyzing the limiting behavior of the projective scaling trajectories.

Karmarkar's standard form for a linear programming problem is

$$(P) \qquad \min\{c^T w | Aw = 0, e^T w = 1, w \geq 0\}$$

where A is an $m \times n$ matrix $(m \leq n-1)$, $c, w \in \mathbf{R}^n$, and e denotes the n-vector of all ones. (In fact, Karmarkar's standard form also requires that the optimal value of (P) be equal to zero. However, we do not impose this condition beforehand.) Also associated with problem (P) is the homogeneous standard form linear programming problem as follows:

$$(HP) \qquad \min\{c^T x | Ax = 0, x \geq 0\}.$$

Problems (P) and (HP) are strongly related in the sense that information in one of them gives information in the other problem, as will be pointed out below. First, we introduce some notation to be used throughout this paper. Let

$$\mathbf{R}_+^n = \{x \in \mathbf{R}^n | x > 0\},$$
$$W = \{w \in \mathbf{R}^n | Aw = 0, e^T w = 1, w \geq 0\},$$
$$W_I = W \cap \mathbf{R}_+^n,$$
$$S = \{x \in \mathbf{R}^n | Ax = 0, x \geq 0\},$$
$$S_I = S \cap \mathbf{R}_+^n.$$

The sets W and S are the feasible sets for problems (P) and (HP), respectively. Also, W_I and S_I are, respectively, the relative interiors of W and S whenever $W_I \neq \varnothing$. The lower-case letter e denotes the vector of all ones whose dimension is dictated by the appropriate context. If x is a lower-case letter that denotes a vector $x = (x_1, \ldots, x_n)^T$, then a capital letter will denote the diagonal matrix with the components of the vector on the diagonal; i.e., $X = \operatorname{diag}(x_1, \ldots, x_n)$. If x and z are two column n-vectors, we define their product xz to be the vector $XZe = (x_1 z_1, \ldots, x_n z_n)^T$. The inverse of x under this operation is denoted by x^{-1} and is given by $x^{-1} \equiv X^{-1} e$. In this way, expressions like x^{-1} and $x^{-2} z$ are defined if all the components of x are nonzero. No confusion should arise between the expressions xz and $x^T z$ where the latter just denotes the inner product of x and z. Given an $m \times n$ matrix A and a subset B of the index set $\{1, \ldots, n\}$, we denote by A_B the submatrix of A associated with the index set B. A function $x: I \subset \mathbf{R} \to \mathbf{R}^n$, where I is an open interval, will be called a *path* or *curve*. We will say that a path $x: I \to \mathbf{R}^n$ passes through a point $x^o \in \mathbf{R}^n$ if for some $\mu^o \in I$, we have $x(\mu^o) = x^o$. The set of points traced by a path will be called a *trajectory*.

Let r^* denote the optimal value of problem (P). We impose the following assumptions on problem (P).

ASSUMPTION 2.1.

(a) $r^* \geq 0$.
(b) $W_I \neq \varnothing$.
(c) *The matrix A has rank m.*

In [11], it is assumed that the optimal value of problem (P) is 0 in order to derive a polynomially convergent algorithm for (P), namely, the so-called projective scaling algorithm. However, the analysis of the projective scaling trajectories for the case $r^* > 0$ can also be derived using the same approach as for the case $r^* = 0$. Therefore, we will consider both cases simultaneously in our analysis. It can be easily shown that $r^* \geq 0$ if and only if problem (HP) has an optimal solution, and in this case $0 \in \mathbf{R}^n$ is an optimal solution of (HP). Hence, under our assumptions, 0 is an optimal solution of (HP). The following results relating problems (P) and (HP) can be easily shown.

PROPOSITION 2.1. $r^* > 0$ if and only if 0 is the unique optimal solution of (HP).

PROPOSITION 2.2. Assume that $r^* = 0$ and let $w \in W$ be given. Then w is an optimal solution of (P) if and only if for some (every) $\lambda > 0$, λw is an optimal solution of (HP).

Observe that when $r^* = 0$, Proposition 2.2 implies that the set of optimal solutions of problem (HP) is unbounded.

We next briefly describe the projective scaling method. We refer the reader to [14] for a more detailed discussion of the projective scaling method as described below. Given a point $x \in W_I$, the projective scaling algorithm assigns to x a search direction as follows:

$$(2.1) \qquad \xi(x) = [X - xx^T][I - XA^T(AX^2A^T)^{-1}AX]Xc.$$

Note that $\xi(x)$ is a feasible direction for problem (P); that is, $\xi(x) \in \{h \in \mathbf{R}^n | Ah = 0, e^T h = 0\}$. The projective scaling algorithm determines the next iterate $\hat{x}$ according to $\hat{x} = x - \alpha(x)\xi(x)$ where $\alpha(x) > 0$ is an appropriate step size that guarantees that $\hat{x} > 0$, as well as a sufficient decrease in a potential function associated with problem (P) [11]. The transformation $\xi(x)$ then defines a vector field in W_I. In the following, we will be interested in studying the limiting behavior of the trajectories generated by $\xi(x)$, that is, the set of points traced by the solution curves of the following differential equation:

$$\dot{w}(t) = \xi(w(t)).$$

Note that, since ξ is a vector field in W_I, $w(t) \in W_I$ for every t for which $w(t)$ is defined. In our analysis, we will make use of another vector field related to problem (HP), namely, the affine scaling vector field associated with (HP) defined as follows:

$$(2.2) \qquad \Phi(x) = X[I - XA^T(AX^2A^T)^{-1}AX]Xc$$

for $x \in S_I$. The trajectories generated by $\Phi(x)$ are the sets of points traced by the solution curves of the following differential equation:

$$(2.3) \qquad \dot{x}(t) = \Phi(x(t)).$$

Note that $x(t) \in S_I$ for every t for which $x(t)$ is defined. These trajectories have been studied in [1] by considering the reparametrized differential equation

$$(2.4) \qquad \dot{x}(\mu) = \frac{1}{\mu^2}\Phi(x(\mu)), \qquad \mu > 0,$$

which yields the same family of trajectories as the one associated with (2.3) but with different parametrizations. Specifically, let $x^o \in S_I$ and $\mu^o > 0$ be given. If $v(t)$ is a solution curve of (2.3) satisfying $v(0) = x^o$ then $x(\mu) \equiv v((\mu^o)^{-1} - \mu^{-1})$ is a solution curve of (2.4) satisfying $x(\mu^o) = x^o$. In the next section, we will show that a solution curve of (2.4) is always defined over an open interval of the form $(0, a)$, where $a > 0$. ($a = \infty$ is also allowable.)

The next result is due to Bayer and Lagarias [4] and will play a crucial role in our analysis of the projective scaling trajectories.

PROPOSITION 2.3 (Bayer and Lagarias). *If* $x: I = (0, a) \to S_I$ *is a solution curve of* (2.4) *then the curve* $w: I \to W_I$ *defined by* $w(\mu) \equiv [1/e^T x(\mu)]x(\mu)$ *satisfies the relation* $\dot{w}(\mu) = [e^T x(\mu)/\mu^2]\xi(w(\mu))$ *for all* $\mu \in I$. *As a consequence,* $w: I \to W_I$ *is a parametrization of a projective scaling trajectory.*

Proposition 2.3 can be easily proved using (2.1), (2.2), and the fact that $\Phi(\lambda x) = \lambda^2 \Phi(x)$ for all $x \in S_I$ and $\lambda > 0$.

The paths $w: I \to W_I$ obtained according to Proposition 2.3 will be called *projective scaling paths*. Obviously, given a point $w^o \in W_I$, a projective scaling path $w: I \to W_I$ passing through w^o can be obtained according to Proposition 2.3 from a solution curve $x: I \to S_I$ of (2.4) passing through w^o (or any point λw^o with $\lambda > 0$).

We will analyze the projective scaling trajectories in the next sections by considering these parametrizations, namely, the projective scaling paths.

3. Affine scaling trajectories

In this section, we will be concerned with analyzing the limiting behavior of the affine scaling trajectories of a linear programming problem in standard form under weaker assumptions than the ones imposed in [1]. Specifically, [1] assumed that the set of optimal solutions of the linear program under consideration is nonempty and bounded. However, from the discussion in the previous section, we saw that the homogeneous standard form problem (HP) underlying the Karmarkar standard form problem (P) has unbounded optimal face when the optimal value of (P) is 0. In order to use the result of Proposition 2.3 to analyze the projective scaling trajectories, it is necessary to extend the results presented in [1] to ones that do not assume that the set of optimal solutions is bounded. The discussion in this section is intended to be brief. Whenever possible, we refer the reader to [1] for some omitted detail or proof. Our goal will be to point out the necessary modifications on the development in [1] in order to cover this more general case.

We start by introducing our terminology. Consider the linear programming problem in standard form

$$(\tilde{P}) \qquad \min\{c^T x | Ax = b, x \geq 0\}$$

and its dual problem

$$(\tilde{D}) \qquad \max\{b^T y \mid A^T y + z = c, \; z \geq 0\}$$

where A is an $m \times n$ matrix and b, c are vectors of length m and n, respectively. The following notation will be used throughout this section. Let

$$S_A = \{x \in \mathbf{R}^n \,;\, Ax = b\},$$
$$S_F = \{x \,;\, x \in S_A, \; x \geq 0\},$$
$$S_I = \{x \,;\, x \in S_A, \; x > 0\},$$
$$T_A = \{(y, z) \in \mathbf{R}^m \times \mathbf{R}^n \,;\, A^T y + z = c\},$$
$$T_F = \{(y, z) \,;\, (y, z) \in T_A, \; z \geq 0\},$$
$$T_I = \{(y, z) \,;\, (y, z) \in T_A, \; z > 0\}.$$

The sets S_F and T_F are, respectively, the feasible sets of problems $(\tilde{P})$ and $(\tilde{D})$, and the sets S_A and T_A are, respectively, the affine hulls of S_F and T_F.

We impose the following assumptions on problem $(\tilde{P})$:

ASSUMPTION 3.1.

  (a) *The set S_I is nonempty.*
  (b) *Problem $(\tilde{P})$ has an optimal solution.*
  (c) *Rank $(A) = m$.*

The affine scaling vector field is given by

$$\Phi(x) = X[I - XA^T(AX^2 A^T)^{-1} AX]Xc$$

for $x \in S_I$. In the following, the trajectories of this vector field will be analyzed by studying the solution curves of the following differential equation

$$(3.1) \qquad \dot{x}(\mu) = \frac{1}{\mu^2}\Phi(x(\mu))$$

where $\mu > 0$. In particular, we are interested in the behavior of a solution curve of (3.1) as μ decreases, that is, as $c^T x(\mu)$ monotonically decreases.

We now turn our efforts towards characterizing the solution curves of (3.1) as paths of solutions of a logarithmic barrier family of problems. Consider the following family of problems parametrized by the penalty parameter $\mu > 0$.

$$(P_\mu) \qquad \min c^T x - \mu \left[p^T x + \sum_{j=1}^{n} \ln x_j \right]$$

$$\text{s.t.} \;\; Ax = b$$

$$x > 0$$

where $p \in \mathbf{R}^n$. Since the objective function of problem (P_μ) is a strictly convex function, it follows that (P_μ) has at most one global solution. The global solution $x(\mu)$ of problem (P_μ), if it exists, satisfies the following Karush–Kuhn–Tucker optimality condition:

(3.2.a) $$z(\mu) - \mu x(\mu)^{-1} = \mu p$$

(3.2.b) $$Ax(\mu) = b, \qquad x(\mu) > 0$$

(3.2.c) $$A^T y(\mu) + z(\mu) = c$$

for some $y(\mu) \in \mathbf{R}^m$ and $z(\mu) \in \mathbf{R}^n$, which are uniquely determined. Let $I(p)$ denote the set of parameters $\mu > 0$ such that problem (P_μ) (and hence system (3.2)) has a solution. Since (P_μ) has a global solution if and only if the set $Y(p, \mu) \equiv \{y; A^T y < c - \mu p\} \neq \varnothing$, it follows that $I(p) = \{\mu > 0; Y(p, \mu) \neq \varnothing\}$. If the set $I(p)$ is nonempty, it has to be an open interval having 0 as one of its extreme points. This fact is stated in the following result.

PROPOSITION 3.1. *If the set* $I(p)$ *is nonempty then* $I(p) = (0, d_p)$ *for some* $d_p > 0$.

PROOF. Assume that $I(p) \neq \varnothing$ and let $\mu \in I(p)$ be given. By the definition of $I(p)$, it follows that there exists $y \in \mathbf{R}^m$ such that $A^T y < c - \mu p$. On the other hand, (b) of Assumption 3.1 implies that the dual problem $(\widetilde{D})$ is feasible; that is, there exists $y^* \in \mathbf{R}^m$ such that $A^T y^* \leq c$. Let $\lambda \in (0, 1)$ be given and consider the vector $y_\lambda \equiv (1 - \lambda)y^* + \lambda y$. Since $A^T y_\lambda < c - \lambda \mu p$, it follows that $\lambda \mu \in I(p)$. Since this holds for any $\lambda \in (0, 1)$ and $\mu \in I(p)$, the result follows. □

It has been shown in [1] that whenever the set of optimal solutions of problem $(\widetilde{P})$ is bounded, the set $I(p)$ is nonempty for all $p \in \mathbf{R}^n$. Moreover, if the set S_F is bounded then $I(p) = (0, \infty)$ for all $p \in \mathbf{R}^n$. However, when the set of optimal solutions of problem $(\widetilde{P})$ is unbounded, it may happen that $I(p) = \varnothing$. One trivial example is $p = 0$. Indeed, when the set of optimal solutions of $(\widetilde{P})$ is unbounded, $I(p) = \varnothing$ for any $p \geq 0$.

For a point $x \in S_I$, we define the affine dual estimates $y^E(x)$ and $z^E(x)$ at x by

(3.3) $$y^E(x) = (AD^2 A^T)^{-1} AD^2 c,$$

(3.4) $$z^E(x) = c - A^T (AD^2 A^T)^{-1} AD^2 c.$$

The following result was proved in [1], and its equivalent proof under our present assumptions is exactly the same.

THEOREM 3.1. *Assume that* $p \in \mathbf{R}^n$ *is given such that* $I(p) \neq \varnothing$. *Let* $(x(\mu), y(\mu), z(\mu))$, $\mu \in I(p)$ *denote the path of solutions of the parametrized system of equation* (3.2) *corresponding to the given* p. *Then* $x(\mu)$, $\mu \in I(p)$

is a solution of the differential equation (3.1), *and* $y(\mu)$ *and* $z(\mu)$ *satisfy the relations*

$$y(\mu) - \mu\dot{y}(\mu) = y^E(x(\mu)),$$
$$z(\mu) - \mu\dot{z}(\mu) = z^E(x(\mu))$$

for all $\mu \in I(p)$.

COROLLARY 3.1. *Let* $x^o \in S_I$, $(y^o, z^o) \in T_A$, *and* $\mu^o > 0$ *be given. Let* $p = (z^o/\mu^o - (x^o)^{-1})$. *Then* $(0, \mu^o] \subset I(p)$ *and the path of solutions* $x(\mu)$ *of problem* (P_μ) *is a solution curve of* (3.1) *satisfying* $x(\mu^o) = x^o$.

PROOF. By the choice of p, it follows that (x^o, y^o, z^o) is a solution of system (3.2) with $\mu = \mu^o$. Hence, $\mu^o \in I(p)$, and by Proposition 3.1, it follows that $(0, \mu^o] \subset I(p)$. Also, since the solution of system (3.2) for a given $\mu > 0$ is unique, it follows that $(x(\mu^o), y(\mu^o), z(\mu^o)) = (x^o, y^o, z^o)$. □

In the following, let θ^* denote the common optimal value of problems $(\tilde{P})$ and $(\tilde{D})$. The next result shows that the trajectory $x(\mu)$ and its associated "dual trajectory" $(y(\mu), z(\mu))$ converge in objective value to the optimal value θ^*.

THEOREM 3.2. *Let* $p \in \mathbf{R}^n$ *be given such that* $I(p) \neq \varnothing$ *and consider the solution* $(x(\mu), y(\mu), z(\mu))$, $\mu \in I(p)$ *of system* (3.2). *Then* $\lim_{\mu \to 0} c^T x(\mu) = \lim_{\mu \to 0} b^T y(\mu) = \theta^*$. *Moreover, both* $x(\mu)$ *and* $(y(\mu), z(\mu))$ *lie in a bounded set for* μ *sufficiently small.*

PROOF. Since $I(p) \neq \varnothing$, select an arbitrary $\mu^o \in I(p)$ and let $(x^o, y^o, z^o) \equiv (x(\mu^o), y(\mu^o), z(\mu^o))$. Multiplying (3.2.a) by $x(\mu)^T$, we obtain

$$x(\mu)^T z(\mu) = n\mu + \mu p^T x(\mu).$$

Using the fact that $x^T z = c^T x - b^T y$ for any $x \in S_A$ and $(y, z) \in T_A$, one can easily verify that the last expression is equivalent to the following relation:

$$(3.5) \qquad \left(1 - \frac{\mu}{\mu^o}\right)[c^T x(\mu) - b^T y(\mu)] = \frac{\mu}{\mu^o}[(c^T x^o - b^T y^o)$$
$$- (z^o - \mu^o p)^T x(\mu) - (x^o)^T z(\mu)] + n\mu.$$

Note that $x(\mu) > 0$ and $z(\mu) - \mu p > 0$ for all $\mu \in I(p)$, and, in particular, we have that $x^o > 0$ and $z^o - \mu^o p > 0$. This observation and relation (3.5) imply that

$$\left(1 - \frac{\mu}{\mu^o}\right)[c^T x(\mu) - b^T y(\mu)] \leq \frac{\mu}{\mu^o}[(c^T x^o - b^T y^o) - \mu(x^o)^T p] + n\mu.$$

Hence, as μ tends to 0, we obtain

$$(3.6) \qquad \limsup_{\mu \to 0} c^T x(\mu) - b^T y(\mu) \leq 0.$$

Obviously, since $x(\mu) \in S_I$, we have

(3.7) $$c^T x(\mu) \geq \theta^*$$

for all $\mu \in I(p)$, and therefore

(3.8) $$\liminf_{\mu \to 0} c^T x(\mu) \geq \theta^*.$$

Let x^* be an optimal solution of $(\tilde{P})$ (cf. (b) of Assumption 3.1). Since $z(\mu) - \mu p > 0$, we have $(z(\mu) - \mu p)^T x^* \geq 0$, which implies that

$$c^T x^* - b^T y(\mu) - \mu p^T x^* \geq 0$$

or equivalently,

(3.9) $$\theta^* - \mu p^T x^* \geq b^T y(\mu).$$

Hence, we have

(3.10) $$\limsup_{\mu \to 0} b^T y(\mu) \leq \theta^*.$$

Now from relations (3.6), (3.8), and (3.10), it follows that $\lim_{\mu \to 0} c^T x(\mu) = \theta^*$ and $\lim_{\mu \to 0} b^T y(\mu) = \theta^*$. To show that $x(\mu)$ and $(y(\mu), z(\mu))$ lie in a bounded set for all μ sufficiently small, first note that the components of $x(\mu)$ and $z(\mu)$ are bounded from below since $x(\mu) > 0$ and $z(\mu) > \mu p$. On the other hand, from relations (3.5), (3.7), and (3.9), it follows that

$$\mu \left(1 - \frac{\mu}{\mu^o}\right) p^T x^* \leq \frac{\mu}{\mu^o}[(c^T x^o - b^T y^o) - (z^o - \mu^o p)^T x(\mu) - (x^o)^T z(\mu)] + n\mu.$$

Dividing the last expression by μ/μ^o and rearranging yields

$$(z^o - \mu^o p)^T x(\mu) + (x^o)^T z(\mu) \leq n\mu^o + (c^T x^o - b^T y^o) - \mu^o \left(1 - \frac{\mu}{\mu^o}\right) p^T x^*,$$

which shows that the components of $x(\mu)$ and $z(\mu)$ are bounded from above and therefore bounded for all μ sufficiently small. Since A has full rank, it also follows that $y(\mu)$ is bounded for all μ sufficiently small. $\square$

At this point, we should point out that the necessary adaptations for the development in [1] to hold under our present assumptions (cf. Assumption 3.1) have already been presented in Proposition 3.1, Corollary 3.1, and Theorem 3.2. Before stating the next result, we need to introduce some terminology. The optimal face of $(\tilde{P})$ (resp. $(\tilde{D})$) is the set of points $S_O \equiv \{x; x \in S_F, c^T x = \theta^*\}$ (resp. $T_O \equiv \{(y, z); (y, z) \in T_F, b^T y = \theta^*\}$). The set S_O is a face of the polyhedron S_F and therefore can be expressed as the set of points $\{x \in S_F; x_j = 0, j \in N\}$ for some index set $N \subseteq \{1, \ldots, n\}$. We may assume that N is the maximal set (with respect to inclusion) satisfying this property; that is, $j \in N$ if and only if $x_j = 0$ for every $x \in S_O$. Let B denote the set of indices $j \in \{1, \ldots, n\}$ such that $j \notin N$. It is well known that T_O is the face of the polyhedron T_F given by $\{(y, z) \in T_F; z_j = 0, j \in B\}$ and that B is the maximal set with this property. (See, for example, Schrijver [16].) We are now ready to state the main result of this section. It generalizes the results obtained in [1] under the additional assumption that the set of optimal solutions of problem $(\tilde{P})$ is bounded.

THEOREM 3.3. Let $p \in \mathbf{R}^n$ be given such that $I(p) \neq \emptyset$ and consider the path of solutions $(x(\mu), y(\mu), z(\mu))$, $\mu \in I(p)$ of system (3.2) with respect to the given p. Then, we have

(a) The limit of $x(\mu)$ as μ tends to 0 exists and is equal to the optimal solution x^* of problem $(\tilde{P})$ (and hence $x_N^* = 0$) such that x_B^* is the (unique) optimal solution of the problem

$$\max -[(x_B^o)^{-1}]^T x_B + \sum_{j \in B} \ln x_j$$

$$\text{s. t. } A_B x_B = b, \qquad x_B > 0$$

where x^o is any point in the path $x(\mu)$; that is, $x(\mu^o) = x^o$ for some $\mu^o \in I(p)$.

(b) The limit of $(y(\mu), z(\mu))$ as μ tends to 0 exists and is equal to the optimal solution (y^*, z^*) of problem $(\tilde{D})$ (and hence $z_B^* = 0$) such that (y^*, z_N^*) is the (unique) optimal solution of the following problem

$$\max \left\{ \sum_{j \in N} \ln z_j \mid A_N^T y + z_N = c_N, A_B^T y = c_B, z_N > 0 \right\}.$$

(c) The limit of the affine dual estimates along the trajectory $x(\mu)$, $\mu \in I(p)$ as μ tends to 0 exists and is given by

$$\lim_{\mu \to 0} (y^E(x(\mu)), z^E(x(\mu))) = (y^*, z^*)$$

where (y^*, z^*) is the optimal solution of problem $(\tilde{P})$ as in statement (b) above.

(d) The limit of $\dot{x}(\mu)$ as μ tends to 0 exists, and its value is given by

$$\lim_{\mu \to 0} \dot{x}_N(\mu) = (z_N^*)^{-1}$$

and $\lim_{\mu \to 0} \dot{x}_B(\mu)$ is equal to the (unique) optimal solution of the following problem, where the minimization is with respect to v_B:

$$\min \frac{1}{2} \|(x_B^*) - 1 v_B\|^2$$

$$\text{s. t. } A_B v_B = -A_N (z_N^*)^{-1}.$$

(e) The limit of $(\dot{y}(\mu), \dot{z}(\mu))$ as μ tends to 0 exists, and its value is given by

$$\lim_{\mu \to 0} \dot{z}_B(\mu) = p_B + (x_B^*)^{-1}$$

and $\lim_{\mu \to 0}(\dot{y}(\mu), \dot{z}_N(\mu))$ *is equal to the (unique) optimal solution of the following problem, where the minimization is with respect to* (r, s_N):

$$\min \frac{1}{2}\|(z_N^*)^{-1}(s_N - p_N)\|^2$$

$$\text{s. t. } A_B^T r = -p_B - (x_B^*)^{-1}$$

$$A_N^T r + s_N = 0.$$

The proof of Theorem 3.3 follows along the same line as the results proved in [1], and therefore we do not provide the details here.

4. Limiting behavior of the projective scaling trajectories $(r^* = 0)$

In this section, we study the limiting behavior of the projective scaling trajectories for the linear programming problem in Karmarkar's standard form (P) stated in Section 2 under the assumption that the optimal value r^* of problem (P) is 0. The case $r^* > 0$ will be briefly analyzed in the next section. Since by Proposition 2.3 every projective scaling trajectory $w(\mu)$ for problem (P) can be obtained from an affine scaling trajectory by radial projection onto $\{x | e^T x = 1\}$, it is enough to use the results of the previous section in order to analyze the limiting behavior of the projective scaling trajectories.

Before going through the analysis of the limiting behavior of the projective scaling trajectories, we point out the relationship between the optimal faces of problems (P) and (HP). Let N denote the set of indices j such that $x_j = 0$ whenever x is an optimal solution of problem (HP). It follows from Proposition 2.2 that N is also the set of indices j for which $x_j = 0$ in every optimal solution of problem (P). Moreover, the set $B \equiv \{1, \ldots, n\} - N$ is nonempty.

The dual of problem (P) is the problem stated as follows:

$$(D) \qquad \max \rho$$

$$\text{s. t. } A^T y + e\rho + z = 0, \qquad z \geq 0.$$

With respect to the dual problem (D), we have that (y, ρ, z) is an optimal solution of problem (D) if and only if $A_B^T y = c_B$, $A_N^T y + z_N = c_N$, $z_N \geq 0$, $z_B = 0$, and $\rho = 0$.

We start by showing that a projective scaling trajectory always converges to an optimal solution of problem (P). In general, when the dimension of the optimal face of problem (P) is greater than 0, two different projective scaling trajectories may converge to two distinct points on the optimal face of (P).

THEOREM 4.1. *Let* $w(\mu)$ *be a projective scaling path passing through a point* $w^o \in W_I$. *Then, the limit of* $w(\mu)$ *as* μ *tends to* 0 *exists and has the value*

$$\lim_{\mu \to 0} w(\mu) = \frac{x^*}{e^T x^*}$$

where x^* *is the optimal solution of problem* (HP) *(and hence* $x_N^* = 0$*) such that* x_B^* *is the (unique) optimal solution of the problem*

$$\max -[(w_B^o)^{-1}]^T x_B + \sum_{j \in B} \ln x_j$$

$$\text{s.t. } A_B x_B = 0, \qquad x_B > 0.$$

As a consequence, the limit of $w(\mu)$ *as* μ *tends to* 0 *is an optimal solution of* (P).

The proof of Theorem 4.1 follows immediately from statement (a) of Theorem 3.3 and the definition of a projective scaling path.

The next result translates statement (c) of Theorem 3.3, namely, the convergence of the affine dual estimates along an affine scaling trajectory, to the context of the projective scaling trajectories. First, we need to define dual estimates, at a point $x \in W_I$, that naturally arises from applying the projective scaling algorithm to problem (P). The dual estimates for the projective algorithm were first introduced in [18]. They are defined as follows: Given a point $x \in W_I$, define the projective dual estimates $\hat{y}^E(x)$, $\hat{\rho}^E(x)$, and $\hat{z}^E(x)$ at x as

(4.1) $$\hat{y}^E(x) \equiv (AX^2 A^T)^{-1} X^2 c,$$

(4.2) $$\hat{\rho}^E(x) \equiv \min_j \{c_j - A_j^T \hat{y}^E(x)\},$$

(4.3) $$\hat{z}^E(x) \equiv c - A^T \hat{y}^E(x) - e\hat{\rho}^E(x)$$

where A_j denotes the jth column of the matrix A. Observe that the projective dual estimate $(\hat{y}^E(x), \hat{\rho}^E(x), \hat{z}^E(x))$ at a point $x \in W_I$ is a feasible solution to problem (D). It has been shown in [18] that when x^k, $k = 1, 2, \ldots$, is a sequence converging to a nondegenerate optimal solution of the linear program (P), the projective dual estimate $(\hat{y}^E(x^k), \hat{\rho}^E(x^k), \hat{z}^E(x^k))$ converges to the unique optimal solution $(y^*, 0, z^*)$ of the dual problem (D). The analogue of this fact with respect to the projective scaling trajectories without nondegeneracy assumptions is as follows.

THEOREM 4.2. *Let* $w: I \to W_I$ *where* $I = (0, a)$ *be a projective scaling path. Then the limit as* μ *tends to* 0 *of the projective dual estimate along the projective scaling path* $w(\mu)$ *exists and is given as follows:*

(4.4) $$\lim_{\mu \to 0} \hat{y}^E(x(\mu)) = y^*$$

(4.5) $$\lim_{\mu \to 0} \hat{\rho}^E(x(\mu)) = 0,$$

(4.6) $$\lim_{\mu \to 0} \hat{z}^E(x(\mu)) = z^*$$

where $(y^*, 0, z^*)$ is the optimal solution of problem (D) (hence $z_B^* = 0$) such that (y^*, z_N^*) is the (unique) optimal solution of the problem

$$\max \left\{ \sum_{j \in N} \ln z_j \, \middle| \, A_N^T y + z_N = c_N, \, A_B^T y = c_B, \, z_N > 0 \right\}.$$

PROOF. Let $x(\mu)$ be a solution curve of (2.4) such that $w(\mu) = [1/e^T x(\mu)] x(\mu)$ for all $\mu \in I$. From (3.3) and (4.1) it follows that $y^E(x(\mu)) = \hat{y}^E(w(\mu))$, for all $\mu \in I$. By statement (c) of Theorem 3.3, we have that $\lim_{\mu \to 0} y^E(x(\mu)) = y^*$, which implies (4.4). It then follows from (4.2) that

$$\lim_{\mu \to 0} \hat{\rho}(w(\mu)) = \min_j \{c_j - A_j^T y^*\} = \min_j z_j^* = 0$$

since $z_N^* = 0$ and $N \neq \varnothing$. This shows (4.5). Relation (4.6) now follows from (4.3), (4.4), and (4.5). □

The next result shows the behavior of the derivatives of the projective scaling paths.

THEOREM 4.3. Let $w: I \to W_I$ where $I = (0, a)$ be a projective scaling path. Let (y^*, z^*) be as in the statement of Theorem 4.1. Then, the limit of $\dot{w}(\mu)$ as μ tends to 0 exists, and the vector $\lim_{\mu \to 0} \dot{w}_N(\mu)$ is a positive multiple of the vector $(z_N^*)^{-1}$.

PROOF. Let $x(\mu)$ be a solution curve of (2.4) such that $w(\mu) = [1/e^T x(\mu)] x(\mu)$ for all $\mu \in I$. Then

$$\dot{w}(\mu) = \frac{1}{e^T x(\mu)} \dot{x}(\mu) - \frac{e^T \dot{x}(\mu)}{[e^T x(\mu)]^2} x(\mu).$$

The theorem now follows from statements (a) and (d) of Theorem 3.3 □

The next result is an immediate application of Theorem 4.3.

COROLLARY 4.1. If the optimal face of problem (P) consists of a unique vertex of W then all projective scaling trajectories approach this unique vertex with a common asymptotic direction $d \in \mathbf{R}^n$ given by

(4.7)
$$d_N = (z_N^*)^{-1},$$
$$d_B = -(\tilde{A}_B^T \tilde{A}_B)^{-1} \tilde{A}_B^T \tilde{A}_N d_N$$

where

$$\tilde{A}_B = \begin{pmatrix} A_B \\ e^T \end{pmatrix} \quad \text{and} \quad \tilde{A}_N = \begin{pmatrix} A_N \\ e^T \end{pmatrix}$$

and z_N^* is as in the statement of Theorem 4.1.

PROOF. Assume that a projective scaling trajectory approaches the unique optimal solution of problem (P) according to the asymptotic direction d. By Theorem 4.3, we may assume that $d_N = (z_N^*)^{-1}$. Obviously d satisfies

(4.8)
$$\tilde{A}_B d_B + \tilde{A}_N d_N = 0.$$

Since problem (P) has a unique optimal solution, it follows that the matrix $\tilde{A}_B$ has linearly independent columns. Hence, (4.8) implies (4.7), and the theorem follows.   □

A proof of Corollary 4.1 was given in [17] for the case when problem (P) is primal nondegenerate.

5. The limiting behavior of the projective scaling trajectories $(r^* > 0)$

In this section, we consider the limiting behavior of the projective scaling trajectories for the case when the optimal value r^* of problem (P) is positive. In this case, the projective scaling trajectories do not converge to an optimal solution of problem (P). Instead, we show that they converge to a unique point $w^* \in W_I$ that minimizes the so-called Karmarkar potential function over W_I as follows:

$$P(x) = n \ln c^T x - \sum_{j=1}^{n} \ln x_j .$$

This fact was implicitly proved in Section 9 of [14] under the assumption that problem (P) is primal nondegenerate.

OBSERVATION 5.1. Even though the function $P(x)$ is not strictly convex over W_I, the function

$$Q(x) \equiv \exp P(x) = \frac{(c^T x)^n}{\prod_{j=1}^{n} x_j}$$

is strictly convex over W_I. For a proof of this fact, see Imai [10]. This implies that $Q(x)$, and hence $P(x)$, has a unique minimizer w^* over W_I. Also, the minimizer w^* is completely characterized by Karush–Kuhn–Tucker conditions for the problem $\min\{P(x)|x \in W_I\}$.

Before stating the main result of this section, we need to introduce a lemma that is an immediate consequence of the results of Section 2.

LEMMA 5.1. *Assume that the optimal value of problem (P) is positive and let $x: I \rightarrow S_I$ where $I = (0, a)$ be a solution curve of (2.4). Then,*

$$\lim_{\mu \to 0} x(\mu) = 0$$

and

(5.1)
$$\lim_{\mu \to 0} \dot{x}(\mu) = (z^*)^{-1}$$

where z^* is the z-component of the (unique) optimal solution of the problem

(5.2)
$$\min \sum_{j=1}^{n} \ln z_j$$
$$\text{s.t. } A^T y + z = c, \qquad z > 0 .$$

PROOF. Since $r^* > 0$, it follows from Proposition 2.1 that the optimal face of problem (HP) consists only of the vector $0 \in \mathbf{R}^n$. Hence, in this case, $N = \{1, \dots, n\}$. The result now follows from statements (a) and (d) of Theorem 3.3 □

We are now ready to state the main result of this section.

THEOREM 5.1. *Assume that the optimal value of problem* (P) *is positive and let* $w: I \to W_I$ *where* $I = (0, a)$ *be a projective scaling path. Then, the limit of* $w(\mu)$ *as* μ *tends to* 0 *exists and is equal to the (unique) optimal solution of the following problem.*

(5.3)
$$\min n \ln c^T x - \sum_{j=1}^{n} \ln x_j$$
$$\text{s. t. } Ax = 0, \, e^T x = 1, \qquad x > 0$$

PROOF. Let $x(\mu)$ be a solution curve of the differential equation (3.1) such that

$$w(\mu) = \frac{1}{e^T x(\mu)} x(\mu).$$

Then, by Lemma 5.1, we have that $\lim_{\mu \to 0} x(\mu) = 0$, which implies that $\lim_{\mu \to 0} e^T x(\mu) = 0$. Hence, applying L'Hôpital's rule, we obtain

$$\lim_{\mu \to 0} w(\mu) = \lim_{\mu \to 0} \frac{\dot{x}(\mu)}{e^T \dot{x}(\mu)} = \frac{(z^*)^{-1}}{e^T (z^*)^{-1}},$$

where the last equality follows from (5.1). Using the fact that z^* satisfies the optimality conditions for problem (5.2), one can easily verify that $[1/e^T (z^*)^{-1}](z^*)^{-1}$ satisfies the optimality conditions for problem (5.3). The theorem now follows from Observation 5.1. □

6. Acknowledgment

The author wishes to thank Ilan Adler for many useful comments and discussions that have been invaluable toward the development of this work. The author is also indebted to the referee and to Jeff Lagarias for many comments and subsequent improvements of an earlier version of this paper. This research was partially funded by the Brazilian Post-Graduate Education Federal Agency (CAPES) and the United States Navy Office of Naval Research under contract N00014-87-K-0202. Their financial support is gratefully acknowledged.

REFERENCES

1. I. Adler and R. D. C. Monteiro, *Limiting behavior of the affine scaling continuous trajectories for linear programming problems*, this volume, pp. 189–211.
2. K. M. Anstreicher, *A monotonic projective algorithm for fractional linear programming*, Algorithmica 1 (1986), 483–498.

3. E. R. Barnes, *A variation on Karmarkar's algorithm for solving linear programming problems*, Math. Programming **36** (1986), 174–182.

4. D. A. Bayer and J. C. Lagarias, *The nonlinear geometry of linear programming. Part I. Affine and projective scaling trajectories, Part II. Legendre transform coordinates and central trajectories*, Trans. Amer. Math. Soc. **314** (1989), 499–581.

5. G. de Ghellinck and J. Ph. Vial, *A polynomial Newton method for linear programming*, Algorithmica **1** (1986), 425–453.

6. I. I. Dikin, *Iterative solution of problems of linear and quadratic programming*, Soviet Math. Dokl. **8** (1967), 674–675.

7. A. Fiacco and G. McCormick, *Nonlinear programming: Sequential unconstrained minimization techniques*, John Wiley and Sons, New York, 1968.

8. D. Gay, *A variant of Karmarkar's linear programming algorithm for problems in standard form*, Math. Programming **37** (1987), 81–90.

9. C. Gonzaga, *Conical projection algorithms for linear programming*, Math. Programming **43** (1989), 151–173.

10. H. Imai, *On the convexity of the multiplicative version of Karmarkar's potential function*, Math. Programming **40** (1988), 29–32.

11. N. Karmarkar, *A new polynomial-time algorithm for linear programming*, Combinatorica **4** (1984), 373–395.

12. ____, Talk at the University of California at Berkeley (Berkeley, CA, 1984).

13. N. Megiddo, *Pathways to the optimal set in linear programming*, in Progress in mathematical programming: Interior-point algorithms and related methods (N. Megiddo, ed.), Springer-Verlag, Berlin and New York, 1989, pp. 131–158.

14. N. Megiddo and M. Shub, *Boundary behavior of interior point algorithms for linear programming*, Math. Oper. Res. **14** (1989), 97–146.

15. R. D. C. Monteiro, I. Adler, and M. G. C. Resende, *A polynomial-time primal-dual affine scaling algorithm for linear and convex quadratic programming and its power series extension*, Math. Oper. Res. **15** (1990), 191–214.

16. A. Schrijver, *Theory of linear and integer programming*, John Wiley & Sons, (1986).

17. M. Shub, *On the asymptotic behavior of the projective rescaling algorithm for linear programming*, J. Complexity **3** (1987), 258–269.

18. M. J. Todd and B. P. Burrell, *An extension of Karmarkar's algorithm for linear programming using dual variables*, Algorithmica **1** (1986), 409–424.

19. R. J. Vanderbei, M. S. Meketon, and B. A. Freedman, *A modification of Karmarkar's linear programming algorithm*, Algorithmica **1** (1986), 395–407.

20. C. Witzgall, P. Boggs, and P. Domich, *On the convergence behavior of trajectories for linear programming*, this volume, pp. 161–187.

21. Y. Ye and M. Kojima, *Recovering optimal dual solutions in Karmarkar's polynomial algorithm for linear programming*, Math. Programming **39** (1987), 305–317.

DEPARTMENT OF INDUSTRIAL ENGINEERING AND OPERATIONS RESEARCH UNIVERSITY OF CALIFORNIA, BERKELEY, CALIFORNIA 94720

Current address: SYSTEMS AND INDUSTRIAL ENGINEERING DEPARTMENT, THE UNIVERSITY OF ARIZONA, TUCSON, ARIZONA 85721

E-mail address: tucson.sie.arizona.edu!renu

- 4 -
Nonlinear Optimization

Contemporary Mathematics
Volume **114**, 1990

On the Complexity of a Numerical Algorithm for Solving Generalized Convex Quadratic Programs by Following a Central Path

F. JARRE, G. SONNEVEND, AND J. STOER

ABSTRACT. This paper is divided into two parts. In the first part we give an outline of a complexity analysis concerning the global convergence of a zero-order extrapolation algorithm along a curve of analytic centers for solving generalized convex quadratic programs. It can be shown (see [**Ja**] and [**SoSt**]) that the analytic center of the feasible set provides a two-sided ellipsoidal approximation of this set, whose tightness—as well as the global rate of convergence of the algorithm—only depends on the number of constraints but not on the data of the constraint functions. So far, similar results were known only for programs with linear constraints. The second part presents some preliminary numerical experiments, which were obtained with a modified algorithm. The modification includes that higher-order extrapolation schemes are used to significantly accelerate this path-following method, so that the number of iterations seems to be a very slowly growing function of the number of variables.

Introduction

In 1984 [**Ka**] published a new method for solving linear programs by using interior points. Since then there have been several papers applying (modified) interior point methods to more general classes of problems. [**KaVai**] and [**YeTs**] allowed a convex quadratic objective function to be minimized over a set bounded by linear constraints and obtained a sufficiently accurate solution after $O(NL)$ iterations. In [**YeTo**], the same problem was solved using ellipsoids, which contain all optimal primal and dual slack vectors and whose volumes shrink at a ratio of $2^{-O(\sqrt{n})}$. [**KoMiNo**] developed a continuation method for uniform P-functions, and recently [**MeSu**] developed a method that is based on ideas similar to the one presented here.

1980 *Mathematics Subject Classification* (1985 *Revision*). Primary 90C25, 90C20, 65B99.

This work was supported by the Deutsche Forschungsgemeinschaft, Schwerpunktprogramm für anwendungsbezogene Optimierung und Steuerung.

This paper is a summary of the presentation given at the AMS–IMS–SIAM joint summer research conference in Bowdoin, Maine, June 1988.

In 1985 [So1] suggested a continuation method using "analytic centers" for solving general convex programs. In the underlying paper, the latter is applied to a generalized convex quadratic program, where also the constraints are given by convex quadratic functions. It is shown that for the latter problems, a simple zero-order extrapolation method for following the path of analytic centers has a convergence rate of $(1 - (\text{const}/\sqrt{m}))$ per iteration. We point out that the proof outlined here can be extended to a class of more general convex analytic programs. The exact proofs can be found in [Ja]; see also [SoSt] for other results on ellipsoidal approximations and higher-order extrapolation for generalized quadratic programs.

1. Definitions

The problem under study is to find

$$\lambda^* := \min\{f_0(x) | f_i(x) \leq 0 \text{ for } 1 \leq i \leq m\},$$

where $x \in \mathbf{R}^n$ and the f_i are convex quadratic functions. Let

$$P := \{x | f_i(x) \leq 0 \text{ for } 1 \leq i \leq m\}.$$

We assume that the interior P^o of P is nonempty and bounded. For $\lambda > \lambda^*$, let $P(\lambda)$ denote the polyhedron P constrained by the additional inequality $f_0(x) \leq \lambda$,

$$P(\lambda) := P \cap \{x | f_0(x) \leq \lambda\}.$$

The method follows a homotopy path $\lambda_0 \gg \lambda^*$, $\lambda_k \downarrow \lambda^*$ of some interior point in $P(\lambda)$ that is easily computable and depends smoothly on all constraints: A very convenient point ([So1] 1985) is the *analytic center* $x(\lambda)$ of $P(\lambda)$, which is the unique point x in $P(\lambda)^o$ minimizing the strictly convex logarithmic barrier function

$$\varphi(x, \lambda) := -\ln(\lambda - f_0(x)) - \sum_{i=1}^{m} \ln(-f_i(x)).$$

The analytic center $P(\lambda)$ also maximizes $(\lambda - f_0(x)) \cdot \prod_{i=1}^{m}(-f_i(x))$, i.e., the product of the 'distances' to the constraint functions. It is invariant under affine transformations of P and invariant under scaling of the functions f_i.

Given $\lambda_0 > \lambda^*$ and a first center $x(\lambda_0)$ we outline how one can find a sequence $\lambda_k > \lambda_{k+1} \to \lambda^*$ (λ^* needs not to be known in advance) and approximations $y_k \in P(\lambda_k)^o$ for $x(\lambda_k)$, so that $0 < \lambda_{k+1} - \lambda^* < (1 - c/\sqrt{m})(\lambda_k - \lambda^*)$ with some fixed positive constant c. (Note that $y_k \in P(\lambda_k)^o$ implies that y_k is feasible and that $0 < f_0(y_k) - \lambda^* < \lambda_k - \lambda^*$.) Here, y_{k+1} is obtained from y_k by one Newton step for finding the zero $x(\lambda_{k+1})$ of $D_x\varphi(x, \lambda_{k+1}) = 0$ using y_k as the starting point.

2. Newton's method

The analysis of the algorithm is based on an examination of Newton's method applied to minimizing the logarithmic barrier function. For this examination we restrict ourselves to the consideration of the set P that has the same structure as the set $P(\lambda)$; i.e., we disregard λ and the objective function $f_0(x)$. For $x \in P^o$ (i.e., $f_i(x) < 0$) we therefore consider the following functions:

$$\varphi(x) := -\sum_{i=1}^{m} \ln(-f_i(x)), \quad g(x)^T := D\varphi(x), \quad \text{and} \quad H(x) := D^2\varphi(x).$$

Here, the first derivative $D\varphi(x)$ is represented by a row vector and the second derivative by a square matrix. We note that $H(x)$ is positive definite.

Essential for all examinations is the use of a suitable norm:

DEFINITION. For a point $y \in P$ the norm $\|\cdot\|_{H(y)}$ is given by the Hessian matrix of φ in y, i.e., $\|x\|_{H(y)} := (x^T H(y)x)^{1/2}$.

The quality of Newton's iteration depends on the closeness of a quadratic approximation to the log function. Defining the quadratic approximation of φ at the point y by

$$q_y(x) := \tfrac{1}{2}(x-y)^T H(y)(x-y) + g(y)^T(x-y) + \varphi(y)$$

one can show

LEMMA 1 (see [Ja], Lemma 2.10). *Let a point* $y \in P^o$ *and a point* x *with* $\|x - y\|_{H(y)} \leq 0.1$ *be given, then*

$$|q_y(x) - \varphi(x)| < \|x - y\|_{H(y)}^3.$$

This lemma is the basis for a global bound on the domain of convergence and the speed of convergence of Newton's method and does not depend on the data of the functions f_i. The following corollary is a direct consequence of Lemma 1 and explains the behavior of the norms given by $H(y)$.

COROLLARY (ellipsoidal approximation of P). *For* $y \in P^o$ *define* $E(y) :=$ $\{x|\ \|x\|_{H(y)} \leq 1\}$. *The following holds:*

 (1) $y + 0.8E(y) \subset P^o$.

Let $\bar{x}$ *be the center of* P, *then also*

 (2) $P \subset \bar{x} + 20mE(\bar{x})$.

Part one of this corollary implies that the norms $\|\cdot\|_{H(y)}$ become singular as y approaches the boundary of P. Simple examples show that the order $O(m)$ of the ratio $20m/0.8$ is optimal. An application of the corollary is the possibility to identify inactive constraints long before reaching the optimal solution. Such an ellipsoidal approximation was first given by [SoSt] using duality considerations. For the "algebraic" proof given there, it is essential that all the functions are quadratic, while the somewhat less elegant proof using estimations like the one given in Lemma 1 has the advantage in that it can be generalized to a larger class of functions $f_1, \ldots, f_m$ satisfying a "relative Lipschitz condition" as specified below.

The following lemma shows the continuity of the H-norms.

LEMMA 2 (equivalence of the H-norms, [Ja], Lemma 2.20). *If $y \in P^o$, $z \in \mathbf{R}^n$, and $\|x - y\|_{H(y)} \leq 0.1$ then there exists an ε with $|\varepsilon| \leq 3\|x - y\|_{H(y)}$, such that*

$$\|z\|_{H(x)} = \|z\|_{H(y)}(1 + \varepsilon).$$

From the above lemmas one can already deduce that the domain of convergence for Newton's method is a fixed portion of the feasible set P independent of the data of the functions f_i. This is the key part for the proof of the main theorem. For completeness we shortly mention that in the analysis here (but not necessarily in an implementation of the algorithm that is based on extrapolation, see [JaSoSt]), it is advantageous to count the restriction $f_0(x) \leq \lambda$ m times, as first suggested by [Re].

3. The algorithm

The algorithm that we analyze is an iterative algorithm generating pairs

$$(y_0, \lambda_0) \to (y_1, \lambda_1) \to \cdots \to (y_k, \lambda_k) \to (y_{k+1}, \lambda_{k+1}) \to \cdots,$$

where $\lambda_k > \lambda_{k+1} \to \lambda^*$ is a monotone sequence of numbers and each y_k is to be regarded as an approximation to $x(\lambda_k)$. For finding a first pair (y_0, λ_0) we refer to [Ja]. Here we describe the basic step:

$$(y_k, \lambda_k) \to (y_{k+1}, \lambda_{k+1}).$$

Given (y_k, λ_k) one first reduces λ_k to

(1) $$\lambda_{k+1} := (1 - \rho)\lambda_k + \rho f_0(y_k),$$

where $0 < \rho < 1$ is a suitable fixed number. ($\rho = 0.005$, e.g.) Then, y_{k+1} is obtained from y_k by one step of Newton's method for finding the solution $x(\lambda_{k+1})$ of $D_x\varphi(x, \lambda_{k+1}) = 0$:

(2) $$y_{k+1} := y_k - (D_x^2\varphi(y_k, \lambda_{k+1}))^{-1}D_x\varphi(y_k, \lambda_{k+1})^T.$$

It can be shown that if $y_k = y_{k-1} + h_{k-1}$ is close to $x(\lambda_k)$ in the sense that

$$\|h_{k-1}\|_{\breve{H}} \leq \sigma, \qquad \sigma \text{ sufficiently small},$$

($\sigma = 0.01$, e.g.), where h_{k-1} is the previous Newton step starting from y_{k-1} for finding the solution $x(\lambda_k)$ of $D_x\varphi(x, \lambda_k) = 0$ measured in terms of the norm

$$\|z\|_{\breve{H}}^2 := z^T \breve{H} z \quad \text{with } \breve{H} := D_x^2\varphi(x, \lambda_k)|_{x=y_{k-1}},$$

then y_{k+1} is also close to $x(\lambda_{k+1})$ in the analogous sense, namely

$$\|h_k\|_H \leq \sigma \quad \text{with } H := D_x^2\varphi(x, \lambda_{k+1})|_{x=y_k}.$$

It is further shown in [**Ja**], that the error in the objective function is bounded by $c^T y_k - \lambda^* \leq \frac{5}{4}(\lambda_k - c^T y_k)$ for all k. Step 3 therefore applies the following stopping test:

(3) $\qquad\qquad$ if $\lambda_k - c^T y_k \leq 0.8\varepsilon$ then stop, else goto 1.)

From the previous results one can derive the

MAIN THEOREM. *If we start the algorithm with a point* y_0 *close to a first center* $x(\lambda_0)$ *of* $P(\lambda_0)$ ($\|y_0 - x(\lambda_0)\|_{H(y)} \leq 0.01$, *e.g.) the upper bound for the error* $\lambda_k - \lambda^* > c^T y_k - \lambda^*$ *is reduced by a factor of* $(1 - (1/450\sqrt{m}))$ *with each Newton step, or by a factor of* ε *within* $450\sqrt{m}|\ln \varepsilon|$ *iterations.*

As far as we know, this is the first theorem that proves the (polynomial) complexity of $O(\sqrt{m}\ln(1/\varepsilon))$ Newton iterations to reduce the error in the objective function value by a factor of ε for a class of problems with non-linear constraints independently of the data of the functions f_i.

The algorithm as examined in the main theorem is *not* intended for a practical implementation (see below), and no attempt has been made to improve the constant. We note that the complexity of $O(\sqrt{m}|\ln \varepsilon|)$ Newton steps for reducing an initial error of the quadratic program by a factor of ε was first obtained by [**Re**] and by [**Go**] for the case of a linear program, but so far it could not be improved, not even in the linear case.

The implementation described in Chapter 4 uses a modified algorithm that shows a remarkable speedup due to the following extrapolation procedure. Based on the values $(\lambda_{k-j}, y_{k-j}), (\lambda_{k-j+1}, y_{k-j+1}), \ldots, (\lambda_k, y_k)$ compute an interpolating function $\hat{x}$ with $\hat{x}(\lambda_i) = y_i$ for $k - j \leq i \leq k$ and evaluate $\hat{x}(\lambda_{k+1})$ as an approximation for the next center $x(\lambda_{k+1})$. Then perform a small number of Newton steps starting from $\hat{x}(\lambda_{k+1})$. (In our implementation we have determined the value of λ_{k+1} adaptively as large as possible so that Newton iteration still converges rapidly.)

The results of Chapters 2 and 3 can also be generalized with qualitatively similar outcomes to nonquadratic functions f_i whose Hessian matrices exist and fulfill the following "relative Lipschitz condition":

$$\exists M > 0 : \forall x, y \in P^o, \forall z \in \mathbf{R}^n :$$
$$|z^T (D^2 f_i(x) - D^2 f_i(y))z| \leq M\|x - y\|_{H(y)} z^T D^2 f_i(x)z.$$

In this case the speed of convergence can be shown to be of order $O(\sqrt{m}(1 + M^2)|\ln \varepsilon|)$ and the two-sided ellipsoidal approximation of the set P can be maintained with a similarity ratio of $40m(1 + M)^{4/3}$. For further details we refer to [**Ja2**].

When following the path by a first-order extrapolation method from $\lambda = \infty$ to $\lambda = \lambda^* + \delta$, the number of steps is of order $O(l^\delta(x(\cdot)))$, where

$$l^\delta(x(\cdot)) = \int_{\lambda^* + \delta}^{\infty} \|x''(\lambda)\|_{H(x(\lambda))}^{1/2} d\lambda.$$

Concerning this estimation and related techniques of step size regulation, see
[So3].

Finally we mention a variant that has already been considered by [KoMiNo]
and [MoAd]: If the optimal solution is nondegenerate following the trajectory
$(x(\lambda), \mu(\lambda)) \in \mathbf{R}^{n+m}$, where

$$\nabla f_0(x) + \sum_{i=1}^{m} \mu_i \nabla f_i(x) = 0, \qquad \mu_i f_i(x) = f_0(x) - \lambda, \qquad i = 1, \ldots, m$$

may yield a much faster method, mainly because by this formulation, the
singularity of the matrix $H(x(\lambda))$ at $\lambda = \lambda^*$ is eliminated, so that in the end
phase we have quadratic convergence.

4. Preliminary experiments

To gain insight into the practical behavior of the method using some extrapolation, we have implemented the method in Fortran 77. Small-size problems (with up to 300 unknowns and dense matrices) have been tested, and
further experiments with larger problems are in preparation. Our implementation is intended to illustrate the dependence of the number of iterations
on the dimension of the problem and to test different types and orders of
extrapolation. For a detailed description of different extrapolation schemes
and for further experiments we refer to [JaSoSt].

When interpreting the numbers of the following tables one should note
that the recomputation and Cholesky factorization of Hessian matrices needs
$O(mn^2)$ operations and is most expensive. An evaluation of $g(x, \lambda)$ (all
gradients) and any feasibility check for a vector x each needs $O(mn)$ operations, which is in general more than the number of $O(n^2)$ operations to
solve one Newton system if the Cholesky factors of the Hessian matrices are
known. The computation of an interpolating polynomial of degree j in $\mathbf{R}^n$
requires only $O(jn)$ operations and is thus relatively cheap.

EXAMPLE 1. The first set of test problems consists of random problems of
the form

$$\min_{x \in P} -\sum_{j=1}^{n} x_j$$

$$P = \{x \geq 0 : Ax \leq 10^4\}$$

where A is an $n \times n$ matrix whose entries are random numbers uniformly
distributed between 1 and 1000 [IrIm]. The connection between problem size
and running time is examined. For each dimension n the method was tested
5 times with different random numbers. Table 1 shows the average number of
iterations (#it), Hessian matrix computations (#Hess, $O(m \cdot n^2)$ operations),
and evaluations of the gradient (#grad) used for the test of the extrapolated
vector and the Newton steps ($O(m \cdot n)$ operations). These calculations dominate all computations of the algorithm. The numbers are totals for Phase 1

and Phase 2. The average convergence factor $(f_0(x(\lambda_{k+1})) - \lambda^*)/(f_0(x(\lambda_k)) - \lambda^*)$ per iteration in Phase 2 is named ω. The extrapolation scheme used here is based on fourth-order polynomials $p(r/(r+\rho))$ with $r := \lambda - f_0(x(\lambda))$.

TABLE 1. Random problems (Phase 1 and Phase 2)

n	#it	#Hess	#grad	ω
10	25	24	256	0.513
30	29	29	351	0.596
60	29	33	429	0.621
100	32	37	427	0.632
300	32	42	597	0.680

EXAMPLE 2. Within these examples we show the influence of the order j of interpolation on the performance of the method. $j + 1$ is the number of iterates y_i, $k - j \leq i \leq k$ used for extrapolation. The results refer to a randomly chosen problem of Example 1 with $n = 10$. Also, the objective function was weighted 10 times. The results in Table 2 show that it is advantageous to use higher than first-order interpolation, thereby exploiting the smoothness of the curve $x(\lambda)$ of analytic centers. Using the order $j = 0$ for extrapolation roughly corresponds to a variant of a method used by [Re], and $j = 1$ seems to be comparable to the ones of [AdReVei] and [Go]. This table, however, cannot serve as a basis for a comparison of the methods, as the underlying algorithm uses extrapolation in order to be able to make large steps, but efficient updates of the Hessian matrix are easier to devise when the step lengths are short.

TABLE 2. The influence of the order j of extrapolation

j	#it	#Hess	$\lambda^* - f_0(x_{end})$
0	556	308	3.7×10^{-6}
1	34	34	3.0×10^{-6}
2	28	29	2.7×10^{-6}
3	27	27	4.7×10^{-6}
4	26	26	2.1×10^{-6}
5	26	27	1.5×10^{-6}
5^*	22	23	$5.2 \cdot 10^{-6}$

All examples were based on extrapolation by polynomials $\hat{x}(\lambda)$ in the parameter λ, with the exception of the last example in row 5^*, where polynomials $\hat{x}(t)$ in the parameter $t = r(\lambda)/(r(\lambda)+\rho)$ were used, which of course corresponds to certain rational functions in r. Here the proper choice of ρ is essential for fast convergence. From other experiments we obtained an acceptable value of $\rho \approx c^T x(\lambda_0) - \lambda^*$, which can be obtained from an ellipsoidal approximation of P using the first center. The stopping test always was $c^T x - \lambda^* \leq 10^{-5}$. The results reported here for $n = 10$ are typical: Similar results were obtained for other dimensions n as well. This example is a first attempt to use rational extrapolation, a further speedup can perhaps be obtained when using other suitable rational extrapolation schemes. An extrapolation method—for which the use of increasingly higher-order rational (multipoint Pade) extrapolation can be justified theoretically—is described in [SoSt]. It is exact for the case $m = 1$ with f_0, f_1 quadratic and can be viewed as a generalization of the method of conjugate gradients. Since the set P can be approximated by one quadratic function ($\tilde{f}_1(x) \leq 0$, see corollary to Lemma 1) the latter case is used as a "semilocal" approximation of the general case.

EXAMPLE 3. Another set of test problems was given by

$$f_i(x) := x^T Q_i x + b_i^T x + \gamma_i$$

where the Q_i are diagonal dominant tridiagonal matrices with random entries, the b_i are uniformly distributed vectors, and the γ_i are negative real numbers. Each problem was run three times with different random numbers; the average is listed in Table 3. The number of constraints was chosen as $m = n/3$.

TABLE 3. Quadratic constraints (Phase 2)

n	#it	#Hess	#grad	ω
10	10	12	161	0.202
20	12	12	159	0.329
30	12	12	151	0.305
60	13	15	157	0.346
100	13	14	153	0.328

We cannot report all numerical experiments here, but we mention briefly that the weak dependence of the number of iterations from the dimension was typical for all examples computed so far. In our method, similar matrix structures appear as in the methods of [Ka] and [AdReVei], and this offers further possibilities of improvement by using the same sparse matrix techniques, sophisticated update procedures, etc., as well. In the preceding examples the reader could notice a relatively high number of Newton steps (with the same inexact Hessian matrix) per local iteration. This behavior of the algorithm is due to the fact that inexact Hessian matrices have been used for as long as possible before computing a complete new Hessian matrix. Updates of the Hessian matrix like the one suggested by [Ka] are straightforward to apply, but we did not use them here.

Acknowledgments

Thanks to the unknown referee for helping to clarify parts of this paper.

REFERENCES

[AdReVei] I. Adler, M. G. C. Resende, G. Veiga, *An implementation of Karmarkar's algorithm for linear programming*, ORC Report, May 1986, Dept. of Indust. Engineering and OR, University of California, Berkeley, CA 94720.

[BaLa] D. A. Bayer and J. C. Lagarias, *Karmarkar's linear programming algorithm and Newton's method*, preprint, AT&T Bell Laboratories, Murray Hill, NJ, 1987.

[Go] C. C. Gonzaga, *An algorithm for solving linear programming problems in* $O(n^3 L)$ *operations*, Memo no. UCB/ERL M87/10, 5 March 1987, College of Engineering, University of California, Berkeley, CA 94720.

[IrIm] M. Iri and H. Imai, *A multiplicative barrier function method for linear programming*, Algorithmica **1** (1986), 455–482.

[Ja] F. Jarre, *On the convergence of the method of analytic centers when applied to convex quadratic programs*, Report No. 35, Dec. 1987, Schwerpunktprogramm der DFG für anwendungsbezogene Optimierung und Steuerung, to appear in Math. Programming in revised form.

[Ja2] F. Jarre, *Convergence of the method of analytic centers for generalized convex programs*, DFG, Report No. 67, May 1988, revised July 1988, submitted to Lecture Notes in Mathematics, proceedings of the 5th French–German conference in Optimization Varetz, Oct. 1988.

[JaSoSt] F. Jarre, G. Sonnevend, and J. Stoer, *An implementation of the method of analytic centers*, Lecture Notes in Control and Inform. Sci., vol. 111, Springer-Verlag, Berlin and New York, 1988, pp. 297–307.

[Ka] N. Karmarkar, *A new polynomial-time algorithm for linear programming*, Combinatorica **4** (1984), 373–395.

[KaVai] S. Kapoor and P. M. Vaidya, *Fast algorithms for convex quadratic programming and multicommodity flows*, J. ACM, (1986), 147–159.

[KoMiNo] M. Kojima, S. Mizuno, and T. Noma, *A new continuation method for complementarity problems with uniform P-functions*, 1987, Report B-194, Dept. of Information Sciences, Tokyo Institute of Technology, Oh-okayama, Meguro-ku, Tokyo 152, Japan.

[MeSu] S. Mehrotra and Y. Sun, *A method of analytic centers for quadratically constrained convex quadratic programs*, Technical report, Jan. 1988, revised April 1988, Dept. of Industrial Engineering and Management Sciences, Northwestern University, Evaston, IL 60208.

[MoAd] R. C. Monteiro and I. Adler, *An* $O(n^3)$ *primal-dual interior-point algorithm for linear programming*, technical report, Operations Research Center, University of California, Berkeley, CA, 1987.

[Re] J. Renegar, *A polynomial-time algorithm based on Newton's method for linear programming*, Math. Programming **40** (1988), 59–94.

[So1] G. Sonnevend, *An "analytic centre" for polyhedrons and new classes of global algorithms for linear (smooth convex) programming*, Proc. 12th IFIP Conference on System Modelling and Optimization, Budapest 1985, Lecture Notes in Control and Inform. Sci., vol. 84, Springer-Verlag, Berlin and New York, 1986, pp. 866–876.

[So2] G. Sonnevend, *New algorithms in convex programming based on a notion of "centre" (for systems of analytic inequalities) and on rational extrapolation*, Trends in Mathematical Optimization (K.-H. Hoffmann et al., eds.), ISNM **84** 311–327, Birkhäuser-Verlag, 1988.

[So3] G. Sonnevend, *Sequential algorithms of optimal order, global error for the uniform recovery of functions with monotone rth derivatives*, Anal. Math. **10** (1984), 311–355.

[SoSt] G. Sonnevend and J. Stoer, *Global ellipsoidal approximations and homotopy methods for solving convex analytic programs*, Appl. Math. Optim. **21** (1990) 139–165.

[YeTo] Y. Ye and M. Todd, *Containing and shrinking ellipsoids in the path-following algorithm*, Math. Programming **47** (1990), 1–9.

[YeTs] Y. Ye and E. Tse, *A polynomial-time algorithm for convex quadratic programming*, 1986, report, Dept. of Engineering–Economic Systems, Stanford University, Stanford CA 94305.

INSTITUT FÜR ANGEWANDTE MATHEMATIK UND STATISTIK, UNIVERSITÄT WÜRZBURG, AM HUBLAND, WEST GERMANY

E-mail address: Angm026@vax.tz.uni-wuerzburg.dbp.de

Contemporary Mathematics
Volume **114**, 1990

Canonical Problems for Quadratic Programming and Projective Methods for Their Solution

BAHMAN KALANTARI

ABSTRACT. Define (P) to be the decision problem that asks if a given quadratic form ϕ attains the value of zero over a simplex. We prove that (P) is NP-complete. We associate a generalized Karmarkar potential function f to (P) having the property that if its infimum over the interior of the simplex is nonpositive, then ϕ has a zero over the simplex. We describe two procedures, W (weak) and S (strong), for optimization or f using projective transformations. Procedure $W(S)$ either solves (P) of obtains an interior unconstrained stationary (unconstrained local minimum) point of f. Under convexity of ϕ, either procedure is capable of solving (P). In particular, linear programming can be formulated as (P) with a convex ϕ. We describe a third procedure that for convex ϕ reduces to our previous extension of Karmarkar's algorithm. Define (GP) to be the more general problem that asks if ϕ has a nontrivial nonnegative zero in an arbitrary subspace containing the vector of ones. Procedure W extends to (GP), and in the special case where ϕ is the square of a linear function, it reduces to Karmarkar's algorithm using optimal stepsize in each iteration. A general quadratic programming problem can be solved by solving a sequence of (GP)'s, each of which can be approximated by solving two associated (P)'s.

1. Introduction

Karmarkar's ingenious projective algorithm [13] has motivated the development of several other interior algorithms for linear and convex quadratic programming. For the NP-hard problem of computing the global minimum of a nonconvex quadratic over linear constraints, except for Karmarkar's interest [14], [15] and the work of Ye [24], we are not aware of any interior algorithms. While there exist methods for solving this problem (e.g., see [10], [11]), it is natural to investigate extensions of the interior linear or convex

1980 *Mathematics Subject Classification* (1985 *Revision*). Primary 90C20.
Key words and phrases. Karmarkar's algorithm, projective transformation, linear programming, quadratic programming.
This research was supported in part by the National Science Foundation under Grant No. CCR 88–07518.

programming algorithms. Ye [24] gives an interior procedure that generates a point satisfying the first- and second-order optimality conditions for a given quadratic program. In this paper, we present a different approach, which views such an optimization problem in a specific canonical form motivated by that of Karmarkar's [13] (see also [8]). This canonical problem asks if a given quadratic form has a nontrivial nonnegative zero. In order to solve this decision problem and as in Karmarkar's approach, we replace the direct optimization of the quadratic form with that of an associated potential function. We show that either the quadratic form has a nontrivial nonnegative zero, or the potential function has an unconstrained positive stationary point. This relationship is exclusive under the assumption of the convexity of the quadratic form. We give projective algorithms that either find a zero of the quadratic form, or find an unconstrained stationary or local minimum point of the potential function.

More specifically, given a quadratic form $\phi(x) = x^T Q x$, where Q is an $n \times n$ symmetric matrix, define the *quadratic feasibility* (P) to be the decision problem that asks if $\phi(d) = 0$ for some d in $S = \{x \in \mathbf{R}^n : e^T x = n, x \geq 0\}$, where $e^T = (1, 1, \ldots, 1)$. Assume that $\phi(e) > 0$. The feasibility problem in linear programming, for which de Ghellinck and Vial [7] gave a variant of Karmarkar's algorithm, and hence linear programming itself, can be formulated as (P) with a convex quadratic form (see [9]).

We prove that in general (P) is NP-complete. We associate a generalized Karmarkar potential function f with (P) having the following property: If its infimum over S_+, the interior of S, is nonpositive, $\phi(d) = 0$ for some $d \in S$. If ϕ is convex, we prove that ϕ does not have a desired zero if and only if there exist $d \in S_+$, $f(d) > 0$, which is an unconstrained stationary point of f. It turn out that such a point d is an unconstrained local minimum of f. We describe two procedures, W (*weak*) and S (*strong*), for solving (P) based on the optimization of f using projective transformations. Procedure $W(S)$ either solves (P) or obtains a point $d \in S_+$, which is an unconstrained stationary (unconstrained local minimum) point of f. Thus, these procedures are capable of solving (P) in the case where the quadratic form is convex.

The convergence result of Procedure W with respect to solving (P) with a convex quadratic form is comparable to the convergence of the affine-scaling variants of Karmarkar's algorithm. For affine scaling, see Barnes [2], Vanderbei et al. [22], Dikin [3], and Vanderbei and Lagarias [23].

The study of this canonical decision problem as opposed to the general minimization problem is more appealing in the sense that we obtain simple characterizations in the case of the nonexistence of a desired zero. The relationship between (P) and a general quadratic programming problem is as follows: To test if a given quadratic form attains a specific value subject to linear constraints is equivalent to solving (GP), which asks if a given quadratic form ϕ has a zero in a subspace of $\mathbf{R}^n$ containing the vector of ones. This more restrictive decision problem can be approximated by solving two decision problems as (P). Thus, we view (P) as the more fundamental problem in both linear and quadratic programming. Procedure W extends to (GP) and in the special case where ϕ is the square of a linear function, it reduces to Karmarkar's algorithm using optimal stepsize in each iteration.

The above procedures make use of projective transformations. For each $d \in S_+$ there exist a projective transformation T_d that maps S one-to-one and onto itself, satisfying $T_d(d) = e$. T_d induces a corresponding quadratic form ϕ_d and a potential f_d. The functions ϕ_d and f_d preserve all the relationships between ϕ and f. The iterative step of Procedure W computes the infimum of f_d in the direction of the projected steepest descent at e (which coincides with the steepest descent direction), while Procedure S computes the infimum of f_d over the largest inscribed sphere in S, centered at e. As in Karmarkar's algorithm [13], the iterative step of each procedure is repeated, replacing d by the image of T_d^{-1} at the computed point. We describe a third procedure, Procedure A, for solving (P), which under convexity of ϕ reduces to our previous extension of Karmarkar's algorithm [9]. The iterative step of A computes the infimum of f_d in the descent direction connecting e to the point that globally minimizes ϕ_d over the smallest sphere circumscribing S.

1.1. Notations and summary of the results. Let $S = \{x \in \mathbf{R}^n : e^T x = n, x \geq 0\}$, where $e^T = (1, 1, \ldots, 1)$. Given an $n \times n$ symmetric matrix Q, let

$$\phi(x) = x^T Q x.$$

Assume that $\phi(e) > 0$. Otherwise, replace Q with $-Q$. Consider the set

$$V = \{x \in S : \phi(x) = 0\}.$$

Define (P) to be the decision problem that asks if V is nonempty. We prove that when Q is arbitrary, (P) is NP-complete. Define

$$\phi^* = \min\{\phi(x) : x \in S\}.$$

The word *min* above, as well as throughout the paper, means the global minimum. With the assumption that $\phi(e) > 0$, $\phi^* \leq 0$ if and only if $V \neq \varnothing$. Let

$$S_+ = \{x : e^T x = n, x > 0\} \quad \text{and} \quad \partial S = S - S_+.$$

One S_+ define

$$f(x) = \frac{\phi(x)}{\pi^2(x)},$$

where $\pi(x) = (\prod_{i=1}^n x_i)^{1/n}$, the geometric mean. The function f is a generalized Karmarkar potential function. Let

$$f^* = \inf\{f(x) : x \in S_+\}.$$

If $f^* \leq 0$, then $\phi^* \leq 0$ and hence $V \neq \emptyset$. Thus, if we compute f^*, we may be able to solve (P). If ϕ is convex, we prove that $d \in S_+$ with $f(d) > 0$ is an unconstrained stationary point of f if and only if d is an unconstrained local minimum of f if and only if $V = \emptyset$.

We describe two procedures, W (*weak*) and S (*strong*), for solving (P) based on the optimization of f using projective transformations. Procedure $W(S)$ either solves (P) or obtains $d \in S_+$, which is an unconstrained stationary (unconstrained local minimum) point of f.

We now describe the iterative step of the procedures. Given $d \in S_+$, define

$$D = \mathrm{diag}(d_1, \ldots, d_n),$$

$$\phi_d(y) = y^T DQDy,$$

$$V_d = \{y \in S : \phi_d(y) = 0\}, \quad \text{and} \quad \phi_d^* = \min\{\phi_d(y) : y \in S\}.$$

When $d = e$, $f_d = f$, $\phi_d = \phi$, and $V_d = V$. We prove $V_d \neq \emptyset$ if and only if $\phi_d^* \leq 0$. Moreover, $V \neq \emptyset$ if and only if $V_d \neq \emptyset$. Also, define

$$f_d(y) = \frac{\phi_d(y)}{\pi^2(y)} \quad \text{and} \quad f_d^* = \inf\{f_d(y) : y \in S_+\}.$$

We show that $f_d^* \leq 0$ implies that $V_d \neq \emptyset$.

We prove that $d \in S_+$ is an unconstrained stationary (unconstrained local minimum) point of f if and only if e is an unconstrained stationary (unconstrained local minimum) point of f_d. Furthermore, e is a constrained stationary (constrained local minimum) point of f_d over S if and only if e is an unconstrained stationary (unconstrained local minimum) point of f_d.

Procedure W either determines that $V \neq \emptyset$ or it obtains a point $\hat{d} \in S_+$ where e is a constrained stationary point of $f_{\hat{d}}$ over S_+. (Equivalently, $\hat{d}$ will be an unconstrained stationary point of f.) Its iterative step takes as input a point $d \in S_+$ with $f(d) > 0$ and computes the infimum of f_d in the direction of the projected steepest descent at e. This direction coincides with the steepest descent direction at e. This results in either the conclusion that V is nonempty, or gives a point y_d necessarily in S_+. In the latter case, d is replaced with $x_d = nDy_d/e^T Dy_d \in S_+$.

Procedure S either solves (P) or it obtains a point $\hat{d} \in S_+$ where e is a local minimum of $f_{\hat{d}}$ over S. (Equivalently, $\hat{d}$ will be an unconstrained local minimum of f.) To describe the iterative step of Procedure S on input $d \in S_+$ satisfying $f(d) > 0$, define

$$L(d) = \min\{\phi_d(y) : e^T y = n, \|y - e\| \leq R\},$$

$$U(d) = \min\{\phi_d(y) : e^T y = n, \|y - e\| \leq r\},$$

where $r = \sqrt{n/(n-1)}$ and $R = \sqrt{n(n-1)}$ are the radii of the largest inscribed and smallest circumscribed spheres to S centered at e, respectively.

Thus,

$$L(d) \le \phi_d^* \le U(d).$$

The iterative step of procedure S first computes $L(d)$. If $L(d) > 0$, $V = \varnothing$. Otherwise, it computes $U(d)$. If $U(d) \le 0$, $V \ne \varnothing$. If $L(d) \le 0$ and $U(d) > 0$, the iterative step computes

$$\gamma(d) = \inf \left\{ \frac{f_d(y)}{f_d(e)} : e^T y = n, \ \|y - e\| < r \right\},$$

a corresponding optimal solution y_d, which will necessarily belong to S_+, and replaces d by $x_d = nDy_d/e^T Dy_d$. We will first prove that $\gamma(d)$ is continuous at any $d \in S_+$ such that $U(d) > 0$. Using this result, we will then prove convergence either to a zero of ϕ over S, or to an interior local minimum of f.

The iterative step of our third procedure, Procedure A, for a given $d \in S_+$ satisfying $f(d) > 0$, computes $L(d)$. If $L(d) > 0$, $V = \varnothing$. Otherwise, it computes $\delta(d)$, the infimum of $f_d(y)/f_d(e)$ along the direction defined by the center and a solution of $L(d)$. If $\delta(d) \le 0$, $V \ne \varnothing$. Otherwise, $\delta(d)$ is attained at a point $y_d \in S_+$, in which case d is replaced by $x_d = nDy_d/e^T Dy_d$. In the special case where ϕ is convex, Procedure A is a variant of our earlier extension of Karmarkar's algorithm [9].

Define the *general quadratic feasibility problem* (GP) to be the decision problem that asks if the following set is nonempty:

$$V = \{x \in S \cap W : \phi(x) = 0\},$$

where

$$W = \{x : Ax = 0\},$$

A an $m \times m$ matrix satisfying $Ae = 0$. We prove that (GP) is NP-complete. Procedure W extends to (GP) and either solves it or obtains a point $\hat{d} \in S_+ \cap W$, which is a constrained stationary point of f with respect to W. If ϕ is convex, this procedure solves (GP). In the special case of (GP) where $Q = cc^T$ for some $c \in \mathbf{R}^n$, Procedure W reduces to Karmarkar's algorithm using optimal stepsize in each iteration.

A general quadratic programming problem can be solved by solving a sequence of (GP)'s. Given a (GP) and an $\varepsilon > 0$, we prove that by solving two associated quadratic feasibility problems we can either solve (GP), or obtain a point $x \in S$ satisfying $\phi(x) = 0$ and $\|Ax\| \le \sqrt{\varepsilon}$.

In §2, we prove the NP-completeness of (P) and (GP). In §3, we investigate invariant properties under projective transformations. In §§4, 5, and 6, we describe Procedures W, S, and A; respectively. In §7, we analyze the relationship between (P) and (GP). Furthermore, we show how a general quadratic programming optimization over a set of linear constraints can be solved by the solving of a sequence of (GP)'s. In §8, we discuss per iteration computational complexities of the three procedures.

2. NP-completeness of (P) and (GP)

In this section we prove the NP-completeness of (P) and (GP). Define

$$V(\alpha, A) = \{x : \phi(x) = 0, \, Ax = 0, \, e^T x = \alpha, \, x \geq 0\},$$

where $\alpha > 0$,

$$\widehat{V}(A) = \{x : \phi(x) = 0, \, Ax = 0, \, x \geq 0, \, x \neq 0\},$$

and

$$\phi^* = \min\{\phi(x) : x \in S, \, Ax = 0\},$$

where it is assumed that $\phi(e) > 0$ and $Ae = 0$. Note that to ask if $V(n, A) \neq \varnothing$ is (GP). If $A \equiv 0$, to ask if $V(n, 0) \neq \varnothing$ is (P).

LEMMA 2.1. $\phi^* \leq 0 \Leftrightarrow V(n, A) \neq \varnothing \Leftrightarrow V(\alpha, A) \neq \varnothing$, for all $\alpha > 0 \Leftrightarrow \widehat{V}(A) \neq \varnothing$.

PROOF. Assume $\phi^* \leq 0$. Since $\phi(e) > 0$, continuity of ϕ, and convexity of $\{x \in S : Ax = 0\}$ implies $V(n, A) \neq \varnothing$. $V(n, A) \neq \varnothing$ trivially implies $\phi^* \leq 0$. Let $x \in V(n, A)$, then $(\alpha/n)x \in V(\alpha, A)$. Clearly, $x \in \widehat{V}(A)$. Suppose $x \in \widehat{V}(A)$, then $\hat{x} = n/(e^T x)x$ is well-defined and belongs to $V(n, A)$. $\square$

LEMMA 2.2. *The decision problem that asks "is $\widehat{V}(0)$ nonempty?" is NP-hard.*

PROOF. The subset sum problem is an NP-complete problem [5] defined as follows: Given positive integers d_i, $i = 0, \ldots, n$, does there exist y_i, $i = 1, \ldots, n$ such that $\sum_{i=1}^{n} y_i d_i = d_0$, where $y_i = 0$, or 1, $i = 1, \ldots, n$. Define

$$\hat{\phi}(y, s) = \hat{\phi}_1(y, s) + \hat{\phi}_2(y, s) + \hat{\phi}_3(y, s),$$

where

$$\hat{\phi}_1(y, s) = \left(\sum_{i=1}^{n} y_i d_i - \frac{d_0}{n} \sum_{i=1}^{n}(y_i + s_i)\right)^2,$$

$$\hat{\phi}_2(y, s) = \sum_{i=1}^{n} \left((y_i + s_i) - \frac{1}{n} \sum_{i=1}^{n}(y_i + s_i)\right)^2,$$

and

$$\hat{\phi}_3(y, s) = \sum_{i=1}^{n} y_i s_i.$$

Define

$$\widehat{V}(0) = \{(y, s) : \hat{\phi}(y, s) = 0, \, (y, s) \geq 0, \, (y, s) \neq 0\}.$$

We claim that the previously stated subset sum problem has a solution if and only if $\widehat{V}(0)$ is nonempty. Let y be a solution to that subset sum problem.

For $1 \leq i \leq n$, let $s_i = 1 - y_i$. Clearly, $(y, s) \in \widehat{V}(0)$. Conversely, let $x = (y, s) \in \widehat{V}(0)$. Thus, we have $\hat{\phi}_1(x) = \hat{\phi}_2(x) = \hat{\phi}_3(x) = 0$. From $\hat{\phi}_2(x) = 0$, we get $(y_i + s_i) = \sigma > 0$, for some σ and for all $1 \leq i \leq n$. This, together with the fact that $\hat{\phi}_3(x) = 0$, implies that for all $1 \leq i \leq 2n$ either $x_i = 0$ or $x_i = \sigma$. Substituting for x_i in the equation $\hat{\phi}_1(x) = 0$, and dividing by σ, we get a solution to the subset sum problem. $\quad\square$

To show that the decision problem that asks "is $\phi^* \leq 0$?" is in NP, we sketch the argument of Murty and Kabadi [18], who prove NP-completeness of some quadratic programming decision problems (see also Murty [19]). Consider a general quadratic programming $\psi^* = \min\{\psi(x) = \frac{1}{2}x^T Q x + c^T x : Bx \geq b, x \geq 0\}$. From Karush-Kuhn-Tucker optimality conditions, it can be shown that if $-\infty < \psi^* \leq 0$, then there exists a basic feasible solution $(\hat{x}, \hat{y}, \hat{u}, \hat{v})$ of the linear complementarity problem (LCP) defined by

$$\begin{pmatrix} u \\ v \end{pmatrix} - \begin{pmatrix} Q & -B^T \\ B & 0 \end{pmatrix} \begin{pmatrix} x \\ y \end{pmatrix} = \begin{pmatrix} c \\ -b \end{pmatrix},$$

$$\begin{pmatrix} u \\ v \end{pmatrix} \geq 0, \qquad \begin{pmatrix} x \\ y \end{pmatrix} \geq 0, \qquad \begin{pmatrix} u \\ v \end{pmatrix}^T \begin{pmatrix} x \\ y \end{pmatrix} = 0,$$

for which $\psi(\hat{x}) \leq 0$. But the basic feasible solutions of the above LCP have polynomial encoding in terms of m, n, B, and b. From this and the above lemmas, we have

THEOREM 2.1. (P) and (GP) are NP-complete.

3. Invariant properties under projective transformations

In this section we prove some properties of (P) and (GP) in terms of stationary and local minimum points of f, their characterization, and invariance under projective transformation. We will use these properties for proving convergence results on the procedures to be described in subsequent sections.

Let $d \in S_+ = \{y : e^T y = n, y > 0\}$ where $\phi(d) > 0$. We recall that $\phi_d(y) = y^T DQDy$, where $D = \text{diag}(d_1, \ldots, d_n)$ and $f_d(y) = \phi_d(y)/\pi^2(y)$. Define

$$W_d = \{y : ADy = 0\}, \qquad V_d = \{y \in S \cap W_d : \phi_d(y) = 0\},$$

where A is an $m \times n$ matrix satisfying $ADe = 0$ and $Ae = 0$. Also, define

$$\phi_d^* = \min\{\phi_d(y) : y \in S \cap W_d\} \quad \text{and} \quad f_d^* = \inf\{f_d(y) : y \in S_+ \cap W_d\}.$$

If $d = e$, we drop the subscript on W_e, V_e, ϕ_e, and f_e. We will assume that either A has full rank or $A \equiv 0$. Note that to test if $V \neq \varnothing$ is (GP) and, if $A \equiv 0$, it is (P).

LEMMA 3.1. Assume $\phi(d) \geq 0$. $V_d \neq \varnothing \Leftrightarrow \phi_d^* \leq 0$, and $f_d^* \leq 0 \Rightarrow \phi_d^* \leq 0$.

PROOF. That $V_d \neq \varnothing \Leftrightarrow \phi_d^* \leq 0$ has already been proved in Lemma 2.1. Suppose $f_d^* \leq 0$. If $f_d^* < 0$, since $\phi_d(e) = \phi(d) \geq 0$, clearly $\phi_d^* < 0$.

Assume $f_d^* = 0$. If f_d^* is attained at some point of $S_+ \cap W_d$, then $\phi_d^* = 0$. Assume f_d^* is not attained. Given $y \in S_+ \cap W_d$, from the well-known inequality relating geometric and arithmetic means of nonnegative numbers, for $y \in S$ we have

$$\pi(y) \le \frac{e^T y}{n} = 1.$$

From this and that $f_d^* = 0$, it follows that

$$0 \le \phi_d(y) \le f_d(y).$$

From the above inequality, the continuity of ϕ_d, and the compactness of $S \cap W_d$, we get $\phi_d^* = 0$. □

The set V_d is the image of V under the projective transformation

$$T_d(x) = \frac{nD^{-1}x}{e^T D^{-1}x},$$

which maps S one-to-one and onto itself. Its inverse is given by

$$T_d^{-1}(y) = \frac{nDy}{e^T Dy}.$$

To see that these functions are well-defined at an arbitrary point z of S, let d_M and d_m be the maximum and minimum component of d, respectively. Then, $e^T D^{-1}z \ge n/d_M > 0$, and $e^T Dz \ge nd_m > 0$. Under T_d, d gets mapped to the center of the simplex. We have

LEMMA 3.2. $V \ne \varnothing \Leftrightarrow V_d \ne \varnothing$.

PROOF. Let $x^* \in S \cap W$ satisfying $\phi(x^*) = 0$. It can easily be checked that $y^* = T_d(x^*) \in S \cap W_d$ and $\phi_d(y^*) = 0$. Conversely, given $y^{**} \in S \cap W_d$ satisfying $\phi_d(y^{**}) = 0$, $x^{**} = T_d^{-1}(y^{**}) \in S \cap W$ and $\phi(x^{**}) = 0$. □

COROLLARY 3.1. $\phi^* \le 0 \Leftrightarrow \phi_d^* \le 0$. □

LEMMA 3.3. d is a local minimum of f over $S_+ \cap W \Leftrightarrow d$ is a local minimum of f over W. In particular, $P_e \nabla f(d) = 0$, where P_e is the orthogonal projection matrix with respect to W.

PROOF. Assume d is a local minimum over $S_+ \cap W$ but not a local minimum over W. Let $N_\varepsilon(d) = \{x \in S_+ \cap W : \|x - d\| < \varepsilon\}$ be a neighborhood of d such that d is a minimum in this neighborhood. Consider the function

$$g(x) = \left\| \frac{nx}{e^T x} - d \right\|.$$

Note that $g(d) = 0$ and that $g(x)$ is continuous at d. Thus, there exists $\delta > 0$ such that

$$\|x - d\| < \delta \Rightarrow |g(x) - g(d)| = g(x) < \varepsilon.$$

Since d is not a local minimum of f over W, there exists $x_0 > 0$, $x_0 \in W$ satisfying $\|x_0 - d\| < \delta$, and $f(x_0) < f(d)$. If $\alpha > 0$ is a scalar and $z > 0$ an n-vector, then $f(\alpha z) = f(z)$. Thus,

$$f\left(\frac{n}{e^T x_0} x_0\right) = f(x_0) < f(d).$$

But $nx_0/e^T x_0$ is in S_+, and since W is a subspace, it is also in W. Thus, $nx_0/e^T x_0$ is in $N_\varepsilon(d)$, a contradiction. That $P_e \nabla f(d) = 0$ follows from first-order optimality condition. $\square$

LEMMA 3.4. Let $x = T_d^{-1}(y) = nDy/e^T Dy \in S_+$ (equivalently, $y = T_d(x)$). Then,

$$\frac{f(x)}{f(d)} = \frac{f_d(y)}{f_d(e)}.$$

PROOF. Recall that $f_d(e) = \phi_d(e) = \phi(d)$, and $f(\alpha z) = f(z)$, where $\alpha > 0$ is a scalar and $z > 0$ an n-vector. We have

$$\frac{f(x)}{f(d)} = \frac{f\left(\frac{n}{e^T Dy} Dy\right)}{f(d)} = \frac{f(Dy)}{f(d)} = \frac{y^T DQDy}{\pi^2(Dy)f(d)}$$

$$= \frac{y^T DQDy}{\pi^2(d)\pi^2(y)f(d)} = \frac{f_d(y)}{\pi^2(d)f(d)} = \frac{f_d(y)}{\phi(d)} = \frac{f_d(y)}{f_d(e)}. \quad \square$$

THEOREM 3.1. d is a local minimum of f over $S_+ \cap W \Leftrightarrow e$ is a local minimum of f_d over $S_+ \cap W_d$.

PROOF. T_d and T_d^{-1} are continuous at d and e, respectively. Also, $T_d(d) = e$. Assume d is a local minimum of f over $S_+ \cap W$ but e is not a local minimum of f_d over $S_+ \cap W_d$. Let $N_\varepsilon(d) = \{x \in S_+ \cap W : \|x - d\| < \varepsilon\}$ be a neighborhood of d such that d is a minimum of f in this neighborhood. Since e is not a local minimum of f_d, and from the continuity of T_d^{-1} at e, there exists $y \in S_+ \cap W_d$ such that $f_d(y) < f_d(e)$ and $x = T_d^{-1}(y)$ is in $N_\varepsilon(d)$. From Lemma 3.4, $f(x) < f(d)$, a contradiction. The converse follows in a similar fashion. $\square$

THEOREM 3.2. Assume $A \equiv 0$. $d \in S_+$ is an unconstrained local minimum of $f \Leftrightarrow e$ is an unconstrained local minimum of f_d.

PROOF. If $A \equiv 0$, $W = W_d = \mathbf{R}^n$. From Theorem 3.1 and by applying Lemma 3.3 both to f and f_d, the result is immediate. $\square$

Let $P = I - (1/n)ee^T$, i.e., the projection matrix of e^T.

LEMMA 3.5. Let $d \in S_+$. Then, $\nabla f_d(e) = P \nabla \phi_d(e)$, and $\nabla f(d) = 1/(\pi^2(d))D^{-1}\nabla f_d(e)$.

PROOF. For $y \in S_+$, we have

$$\nabla f_d(y) = \frac{\nabla \phi_d(y)\pi^2(y) - 2\pi(y)\nabla\pi(y)\phi_d(y)}{\pi^4(y)}.$$

Substituting

$$\nabla \pi(y) = \frac{1}{n}\pi(y)Y^{-1}e,$$

where $Y = \mathrm{diag}(y_1, \ldots, y_n)$, we get

$$\nabla f_d(y) = \frac{1}{\pi^2(y)}\left(\nabla\phi_d(y) - \frac{2}{n}Y^{-1}e\phi_d(y)\right).$$

Letting $y = e$, we get

$$\nabla f_d(e) = \nabla\phi_d(e) - \frac{2}{n}e\phi_d(e) = 2DQDe - \frac{2}{n}ee^T DQDe$$

$$= \left(I - \frac{1}{n}ee^T\right)2DQDe = P\nabla\phi_d(e).$$

Taking $d = e$ and $y = d$ in the equation of $\nabla f_d(y)$, we get

$$\nabla f(d) = \frac{1}{\pi^2(d)}\left(\nabla\phi(d) - \frac{2}{n}D^{-1}e\phi(d)\right).$$

Multiplying by D, using $\phi(d) = \phi_d(e)$, and noting that $\nabla\phi(d) = 2QDe$, we get

$$D\nabla f(d) = \frac{1}{\pi^2(d)}\left(2DQDe - \frac{2}{n}ee^T DQDe\right) = \frac{1}{\pi^2(d)}P\nabla\phi_d(e).\quad \square$$

From Lemma 3.5 and since $P^2 = P$, we have

THEOREM 3.3. *Let $d \in S_+$. Then,* $\nabla f(d) = 0 \Leftrightarrow \nabla f_d(e) = 0 \Leftrightarrow P\nabla f_d(e) = 0$ $\Leftrightarrow P\nabla\phi_d(e) = 0 \Leftrightarrow DQDe = (e^T DQDe/n)e.\quad \square$

According to Theorem 3.3, the following statements are equivalent: $d \in S_+$ is an unconstrained stationary point of f, e is an unconstrained stationary point of f_d, e is a constrained stationary point of f_d, e is a constrained stationary point of ϕ_d, e is an eigenvector of DQD. If A has full rank, we let $P_d = I - DA^T(AD^2A^T)^{-1}AD$, the projection matrix with respect to AD, and $\widehat{P}_d = I - B^T(BB^T)^{-1}B$, where

$$B = \begin{pmatrix} AD \\ e^T \end{pmatrix}.$$

$(BB^T)^{-1}$ exists since given that $ADe = 0$, we have

$$BB^T = \begin{pmatrix} AD^2A^T & 0 \\ 0 & n \end{pmatrix}.$$

Furthermore, it can be shown that $\widehat{P}_d = P_d P$. This was noted by Todd and Burrell [**21**]. If $A \equiv 0$, we can define $P_d = I$ and $\widehat{P}_d = P$. From these, and since $P_d e = e$, we get the more general version of Theorem 3.3

THEOREM 3.4. *Let* $d \in S_+ \cap W$. *Then,* $P_d \nabla f_d(e) = 0 \Leftrightarrow \widehat{P}_d \nabla f_d(e) = 0 \Leftrightarrow$ $\widehat{P}_d \nabla \phi_d(e) = 0 \Leftrightarrow P_d DQDe = (e^T DQDe/n)e$. $\square$

According to Theorem 3.4, the following statements are equivalent: e is a stationary point of f_d over W_d, e is a stationary point of f_d over $S \cap W_d$, e is a stationary point of ϕ_d over $S \cap W_d$, e is an eigenvector of $P_d DQD$. For (GP), under convexity of ϕ (equivalently, Q positive semidefinite), the following result characterizes the case where V is empty.

THEOREM 3.5. *Assume* ϕ *is convex. Then,* $\phi^* > 0 \Leftrightarrow f^* > 0 \Leftrightarrow$ *there exists* d, $f(d) > 0$, *which is a local minimum of* f *over* $S_+ \cap W \Leftrightarrow e$ *is a local minimum of* f_d *over* $S_+ \cap W_d \Leftrightarrow P_d \nabla f_d(e) = 0 \Leftrightarrow \widehat{P}_d \nabla \phi_d(e) = 0 \Leftrightarrow$ $P_d DQDe = \lambda e$, $\lambda > 0 \Leftrightarrow e$ *is a local minimum of* ϕ_d *over* $S \cap W_d$.

PROOF. Assume $\phi^* > 0$. We have already seen that this implies $f^* > 0$. If $f^* > 0$, it must be attained at some point $d \in S_+ \cap W$. From Theorem 3.1, e is a local minimum of f_d over $S_+ \cap W_d$. In particular, e is a constrained stationary point of f_d. From Theorem 3.4, all that remains to be proved is that $\widehat{P}_d \nabla \phi_d(e) = 0$ implies $\phi^* > 0$. Since ϕ is convex, ϕ_d is convex, which in turn implies that e is a global minimum of ϕ_d over $S \cap W_d$. Since $f(d) > 0$, this implies $\phi_d(e) = \phi_d^* > 0$. From Corollary 3.1, this implies $\phi^* > 0$. $\square$

From Theorems 3.1, 3.2, and 3.5, for (P) we get the following characterization of the case where V is empty.

THEOREM 3.6. *Assume* ϕ *is convex and* $A \equiv 0$. *Then,* $\phi^* > 0 \Leftrightarrow f^* > 0 \Leftrightarrow$ *there exists* $d \in S_+$, $f(d) > 0$, *which is a local minimum of* $f \Leftrightarrow \nabla f(d) = 0 \Leftrightarrow e$ *is an eigenvector of* DQD *with positive eigenvalue* $\lambda \Leftrightarrow Qd = \lambda d^{-1}$, *where* $d^{-1} = (1/d_1, \ldots, 1/d_n)^T = D^{-1}e$. $\square$

For the special case where Q is positive semidefinite, the first if-and-only-if-part of the above theorem was proved in [9]. The algorithm in [9] can be viewed as a constructive proof of the only-if part. We are now ready to describe the procedures.

4. Procedure W

In this section we describe Procedure W for (P) and (GP) and prove its convergence properties. For a given $d \in S_+$, $f(d) > 0$, the iterative step of this procedure computes the infimum of f_d in the direction of the projected steepest descent at e. In the case of (P), this direction coincides with the steepest descent direction at e. Indeed, we will describe the procedure under a more general setting; that is, we allow using arbitrary feasible descent directions. We assume that $\phi(e) > 0$ and $Ae = 0$.

Step 0. Let $k = 0$ and choose $d_k = e$.

Step 1. If $P_{d_k} \nabla f_{d_k}(e) = 0$, stop. Otherwise, let u_k, $\|u_k\| = 1$ be a feasible descent direction of $f_{d_k}(y)$ at e. If $\phi_{d_k}(e + \alpha u_k)$ has a root α in $[0, r]$, stop.

Step 2. Compute $f_{d_k}(y_k) = g_k(\alpha_k) = \inf\{g_k(\alpha) = f_{d_k}(e + \alpha u_k) : \alpha \in (0, r)\}$.
Let $d_{k+1} = nD_k y_k / e^T D_k y_k$, where $D_k = \text{diag}(d_k)$. Set $k = k + 1$ and go to Step 1.

Note that in Step 2, the infimum is attained. Thus, $d_{k+1} \in S_+ \cap W$. Clearly, if the procedure halts, either it has determined that V is nonempty or it has found e to be a constrained stationary point of f_{d_k} for some $d_k \in S_+ \cap W$.

THEOREM 4.1. *Assume the procedure does not halt. Let $\hat{d}$ be any accumulation point of the sequence of d_k's. Then, $\hat{d} \in S \cap W$. If $\hat{d} \in \partial S \cap W$, then $V \neq \varnothing$. Otherwise, if K_1 is the index set of a convergent subsequence and if u is any accumulation point of u_k's, $k \in K_1$, then $u^T \nabla f_{\hat{d}}(e) = 0$.*

PROOF. It is clear that $\hat{d} \in S \cap W$. Suppose $\hat{d} \in \partial S \cap W$. From Lemma 3.4, and the way in which y_k's are obtained, the sequence of $f(d_k)$'s is nonincreasing. Since $\pi(d_k)$ approaches 0, $\phi(d_k)$ must converge to 0. Hence, $\phi(\hat{d}) = 0$ and $V \neq \varnothing$. Suppose $\hat{d} \in S_+$. Let u_k for $k \in K_2 \subset K_1$ converge to u. Let y_k for $k \in K_3 \subset K_2$ converge to $\hat{y}$. We claim $\hat{y} = e$. Taking the limit of $d_{k+1} = nD_k y_k / e^T D_k y_k$ as $k \in K_3$ approaches infinity and noting that $e^T D_k y_k$ converges to $e^T \widehat{D} \hat{y} > 0$, we get

$$\hat{d} = \frac{n\widehat{D}\hat{y}}{e^T \widehat{D}\hat{y}}.$$

Multiplying both sides of the above by $\widehat{D}^{-1}$, we get

$$e = \frac{n\hat{y}}{e^T \widehat{D}\hat{y}}.$$

Multiplying the above by e^T and noting that $\hat{y} \in S$, we get $e^T \widehat{D}\hat{y} = n$. Hence the proof of the claim. Since α_k is a stationary point of $g_k(\alpha)$, we have

$$u_k^T \nabla f_{d_k}(y_k) = 0.$$

Taking the limit of the above equation as $k \in K_3$ approaches infinity and using continuity of the function $\psi(d, y) = \nabla f_d(y)$ over $S_+ \times S_+$, we get

$$u^T \nabla f_{\hat{d}}(e) = 0. \quad \square$$

COROLLARY 4.1. *Suppose in Step 1 we choose*

$$u_k = -P_{d_k} \nabla f_{d_k}(e) / \|P_{d_k} \nabla f_{d_k}(e)\|.$$

If $\hat{d} \in S_+$ is an accumulation point of d_k's, then $P_{\hat{d}} \nabla f_{\hat{d}}(e) = 0$. In particular, if $A \equiv 0$, i.e., the underlying problem is (P), then $\nabla f(\hat{d}) = 0$.

PROOF. In this case we get

$$\nabla f_{d_k}(e)^T P_{d_k} \nabla f_{d_k}(y_k) = 0.$$

For $d \in S_+$, P_d is continuous in d. Using this and taking the limit of the above as $k \in K_3$ goes to infinity, together with the fact that $P_{\hat{d}}^2 = P_{\hat{d}}$, we get $P_{\hat{d}} \nabla f_{\hat{d}}(e) = 0$. For (P), $P_{\hat{d}} = I$, thus $\nabla f_{\hat{d}}(e) = 0$. From Lemma 3.5, this implies $\nabla f(\hat{d}) = 0$. □

REMARKS. Let us make some comparisons to Karmarkar's algorithm [13]. Karmarkar's canonical linear programming problem is to determine if $v = \min\{c^T x : e^T x = n, Ax = 0, x \geq 0\} = 0$, where it is assumed that $Ae = 0$, and $c^T x > 0$ for all $x \in \{x : e^T x = n, Ax = 0, x > 0\}$. This problem can be written as (GP) by replacing $c^T x$ with $x^T Q x$, where $Q = cc^T$. For this special Q, Procedure W is Karmarkar's algorithm with optimal stepsize. This is because for a given $d \in S_+$, we have

$$P_d \nabla f_d(e) = \hat{P}_d \nabla \phi_d(e) = \hat{P}_d DQDe = \hat{P}_d Dcc^T De = (c^T d)\hat{P}_d Dc.$$

For a positive semidefinite matrix Q, it is easy to show that $x^T Q x = 0$ if and only if $Qx = 0$. Alternatively, Karmarkar's canonical problem (hence linear programming) can be formulated as (P): Let

$$B = \begin{pmatrix} A \\ c^T \end{pmatrix} \quad \text{and} \quad Q = B^T B.$$

Then, $v = 0$ if and only if $\{x : e^T x = n, Bx = 0, x \geq 0\} \neq \varnothing$ if and only if $\{x : e^T x = n, x^T Q x = 0, x \geq 0\} \neq \varnothing$. de Ghellinck and Vial [6], [7] also gave a variant of Karmarkar's algorithm that determines if $\{x : e^T x = n, Bx = 0, x \geq 0\} \neq \varnothing$. As we have seen, if Q is positive semidefinite, Procedure W when using steepest descent is capable of solving (P). Its convergence in this case is comparable to the convergence of the affine-scaling algorithm. Although we have not been able to prove polynomial convergence, it is interesting to note that since the projection matrix is fixed, the main complexity within each iteration is that of function evaluations. Each function evaluation requires only $O(n^2)$ arithmetic operations.

5. Procedure S

In this section we describe Procedure S. Recall that for $d \in S_+$, $L(d) = \min\{\phi_d(y) : e^T y = n, \|y - e\| \leq R\}$, $U(d) = \min\{\phi_d(y) : e^T y = n, \|y - e\| \leq r\}$, and $\gamma(d) = \inf\{f_d(y)/f_d(e) : e^T y = n, \|y - e\| < r\}$. Clearly, $L(d) \leq \phi_d^* \leq U(d)$. We first describe Procedure S and then prove its convergence properties.

Step 0. Let $k = 0$, $d_k \in S_+$, where $f(d_k) > 0$.

Step 1. Compute $L(d_k)$. If $L(d_k) > 0$, $V = \varnothing$, stop.

Step 2. Compute $U(d_k)$. If $U(d_k) \leq 0$, $V \neq \varnothing$, stop.

Step 3. Compute $\gamma(d_k)$. Let y_k be the corresponding solution, $d_{k+1} = nD_k y_k / e^T D_k y_k$, where $D_k = \operatorname{diag}(d_k)$. Let $k = k + 1$ and go to Step 1.

Clearly, if Procedure S halts, it has solved (P). To consider the case where it does not halt, we first prove a continuity result for the functions γ, which in turn requires the next lemma.

LEMMA 5.1. *Let* $\hat{d} \in S_+$. U *is continuous at* $\hat{d}$.

PROOF. Let $\hat{y}$ be a solution corresponding to $U(\hat{d})$. Let $d_k \in S_+$, $k = 1, \ldots$, be a sequence of points converging to $\hat{d}$. For each k, let y_k be a solution corresponding to $U(d_k)$. Let $\overline{y}$ be an accumulation point of y_k's. Let K be the index set of any subsequence converging to $\overline{y}$. On the one hand, from the definition of U, we have

$$U(d_k) = y_k^T D_k Q D_k y_k \leq \hat{y}^T D_k Q D_k \hat{y}.$$

Taking the limit of the above as $k \in K$ approaches infinity and using the continuity of $\phi_d(y) = y^T DQDy$ in (d, y), we get

$$\overline{y}^T \hat{D} Q \hat{D} \overline{y} \leq U(\hat{d}).$$

On the other hand, we have

$$U(\hat{d}) \leq \overline{y}^T \hat{D} Q \hat{D} \overline{y}. \quad \square$$

THEOREM 5.1. *If* $\hat{d} \in S_+$ *and* $U(\hat{d}) > 0$, *then* γ *is continuous at* $\hat{d}$.

PROOF. If $U(d) > 0$ for $d \in S_+$, then we can replace infimum with minimum in the definition of $\gamma(d)$. Let $\hat{y} \in S_+$ be an optimal solution corresponding to $\gamma(\hat{d})$. Since $U(\hat{d}) = \varepsilon > 0$, continuity of U implies the existence of a neighborhood N of $\hat{d}$ so that

$$d \in N \cap S_+ \Rightarrow U(d) > \frac{\varepsilon}{2}.$$

Let d_k, $k = 1, \ldots$, converge to $\hat{d}$ and without loss of generality assume $d_k \in N \cap S_+$, for all k. Let y_k be a solution corresponding to $\gamma(d_k)$. Thus, $y_k \in S_+$, for all k. Let $\overline{y}$ be an accumulation point of y_k's. Let K be the index set of any convergent subsequence. For $(d, y) \in S_+ \times S_+$, let $h(d, y) = f_d(y)/f_d(e)$. Note that $h(d, y)$ is continuous on $S_+ \times S_+$. Since $y_k \in S_+$, for all k,

$$0 < \gamma(d_k) = h(d_k, y_k) \leq 1.$$

We claim $\overline{y} \in S_+$. Otherwise, $\overline{y} \in \partial S$, which together with the fact that $U(d_k) > \varepsilon/2$ would imply that $h(d_k, y_k)$ is unbounded, a contradiction. On the one hand, from the definition of γ, we have

$$\gamma(d_k) = h(d_k, y_k) \leq h(d_k, \hat{y}).$$

Taking the limit of the above as $k \in K$ approaches infinity and using the continuity of h, we get

$$h(\hat{d}, \overline{y}) \leq h(\hat{d}, \hat{y}) = \gamma(\hat{d}).$$

On the other hand, we have

$$\gamma(\hat{d}) \le h(\hat{d}, \overline{y}).$$

Hence $h(\hat{d}, \overline{y}) = h(\hat{d}, \hat{y})$. $\square$

Now, assume Procedure S does not halt generating a sequence of points d_k, $k = 0, 1, \dots$. Let $\hat{d}$ be any accumulation point of the sequence.

THEOREM 5.2. *If $\hat{d} \in \partial S$, then $V \ne \varnothing$. Suppose $\hat{d} \in S_+$. If $\gamma(\hat{d}) < 1$, then $V \ne \varnothing$. Otherwise, $\hat{d}$ is an unconstrained local minimum of f.*

PROOF. From the definition of γ and Lemma 3.4, we have $f(d_{k+1}) \le f(d_k)\gamma(d_k)$ for all k. This implies that if $\hat{d} \in \partial S$, then $\phi(\hat{d}) = 0$. Furthermore,

$$f(d_k) \le f(d_0) \prod_{i=0}^{k-1} \gamma(d_i).$$

If $\gamma(\hat{d}) < 1$, continuity of γ implies $\gamma(d_k) \le \varepsilon < 1$, for infinitely many k's. Thus, in this case $f(d_k)$ converges to zero. From the definition of γ, $\gamma(\hat{d}) = 1 \Leftrightarrow e$ is a local minimum of $f_{\hat{d}}$ over S_+. From Theorem 3.1, $\hat{d}$ is a local minimum of f over S_+. From Lemma 3.3, $\hat{d}$ is an unconstrained local minimum of f. $\square$

6. Procedure A

In this section we consider an alternate procedure for solving (P) that is the generalization of the earlier algorithm in [9]. It is based on the reduction of $f_d(y)$ along the direction defined by the solution of $L(d)$. Let $d \in S_+$, $f(d) > 0$, and $L(d) \le 0$. Let $\hat{y}_d$ be the solution corresponding to $L(d)$. Consider

$$\delta(d) = \inf\left\{\frac{f_d(y)}{f_d(e)} : y \in S_+ \cap \{y : y = \alpha e + (1-\alpha)\hat{y}_d, \ \alpha \in (0, 1)\}\right\}.$$

Note that if $\delta(d) \le 0$, $V \ne \varnothing$. Otherwise, let $y_d \in S_+$ be a solution corresponding to $\delta(d)$ and $x_d = nDy_d/e^T Dy_d$. Procedure A operates as follows:

Given $d \in S_+$ satisfying $f(d) > 0$, compute $L(d)$. If $L(d) > 0$, $V = \varnothing$, stop. Otherwise, compute $\delta(d)$. If $\delta(d) \le 0$, $V \ne \varnothing$, stop. Otherwise, replace d by x_d and repeat. Procedure A is essentially a special case of Procedure W with the property that the descent direction in each iteration is based on the computation of the solution to $L(d)$. Under convexity assumption, we have

THEOREM 6.1 (see [9]). *Assume Q is positive semidefinite and $n \ge 12$. If $d \in S_+$, $f(d) > 0$, and $L(d) = 0$, then $\delta(d) \le (\delta_0)^{1/n}$, for some constant $0 < \delta_0 < 1$.*

We now define some classes of matrices. Let $\aleph_0$ be the set of all $n \times n$ symmetric matrices. Let $\widehat{S}(Q) = \{d \in S_+ : f(d) > 0, L(d) \leq 0\}$. Define

$$\aleph_1 = \{Q \in \aleph_0 : d \in \widehat{S}(Q) \Rightarrow \delta(d) < 1\},$$
$$\aleph_2 = \{Q \in \aleph_1 : d \in \widehat{S}(Q) \Rightarrow \delta(d) \leq \delta_n < 1\},$$

where δ_n depends only on n. Let

$$\aleph_3 = \{Q \in \aleph_2 : d \in \widehat{S}(Q) \Rightarrow \delta_n^{p(n)} \leq \delta_0 < 1\},$$

where δ_0 is a constant, and $p(n)$ a polynomial in n. Let

$$\aleph_4 = \{Q : Q \text{ is positive semidefinite}\}.$$

From these definitions and Theorem 6.1, we have $\aleph_4 \subset \aleph_3 \subset \aleph_2 \subset \aleph_1$. It is easy to prove that if $Q \in \aleph_2$ Procedure A solves (P). Furthermore, given $\varepsilon > 0$, in a finite number of steps we can determine if $\phi^* \leq \varepsilon$. But, whenever $Q \in \aleph_3$, the number of steps is $p(n)\lfloor \ln \varepsilon / \ln \delta_0 \rfloor$, a polynomial in n and $\ln \varepsilon$. It is natural to investigate if any or all of the above containments are proper.

7. Relationship between (P), (GP), and quadratic programming

In this section we consider the application of (P) and (GP) in solving quadratic programming. We will show later in this section that to test if a quadratic program attains a certain value is equivalent to solving the (GP) that asks if $V = \{x : \phi(x) = x^T Q x = 0, Ax = 0, e^T x = n, x \geq 0\}$ is nonempty. If Q is positive semidefinite, $V \neq \varnothing \Leftrightarrow \{x : x^T(Q + A^T A)x = 0, e^T x = n, x \geq 0\} \neq \varnothing$. Thus, in this case, solving (GP) is equivalent to solving a single (P). In general, this is not the case. In the next theorem we show that by solving two quadratic feasibility problems, we can either determine that V is empty or obtain a *near* feasible point. While computing approximate solutions in the context of a decision problem may appear to be meaningless, it is appropriate for approximating the minimum of a quadratic program.

Given $x \in S$ we have

$$|\phi(x)| \leq |x^T Q x| \leq \|x\|^2 \|Q\| \leq n^2 \|Q\| = M.$$

Let

$$g(x) = x^T A^T A x = \|Ax\|^2.$$

Given $\varepsilon > 0$, define

$$V_1(\varepsilon) = \left\{x : \phi(x) + \frac{M}{\varepsilon} g(x) = 0, e^T x = n, x \geq 0\right\},$$
$$V_2(\varepsilon) = \left\{x : \phi(x) - \frac{M}{\varepsilon} g(x) = 0, e^T x = n, x \geq 0\right\}.$$

We observe that $V \subset V_i(\varepsilon)$ for $i = 1, 2$.

THEOREM 7.1. *If* $V_i(\varepsilon) \neq \varnothing$ *for* $i = 1, 2$, *then there exists* $\hat{x}(\varepsilon) \in S$ *satisfying* $\phi(\hat{x}(\varepsilon)) = 0$ *and* $g(\hat{x}(\varepsilon)) \leq \varepsilon$.

PROOF. Let $x^i(\varepsilon) \in V_i(\varepsilon)$ for $i = 1, 2$. Thus,

$$0 \leq g(x^1(\varepsilon)) = -\frac{\phi(x^1(\varepsilon))\varepsilon}{M} \leq \varepsilon,$$

and

$$0 \leq g(x^2(\varepsilon)) = +\frac{\phi(x^2(\varepsilon))\varepsilon}{M} \leq \varepsilon.$$

We must have $\phi(x^1(\varepsilon)) \leq 0$ and $\phi(x^2(\varepsilon)) \geq 0$. From the continuity of ϕ, there exists $\hat{x}(\varepsilon) = \lambda x^1(\varepsilon) + (1 - \lambda)x^2(\varepsilon) \in S$ where $\phi(\hat{x}(\varepsilon)) = 0$. Convexity of g implies that $g(\hat{x}(\varepsilon)) \leq \varepsilon$. $\square$

COROLLARY 7.1. $V = \varnothing \Leftrightarrow$ *there exists* $\varepsilon > 0$ *such that either* $V_1(\varepsilon) = \varnothing$ *or* $V_2(\varepsilon) = \varnothing$.

PROOF. Suppose $V = \varnothing$ but for all $\varepsilon > 0$, $V_i(\varepsilon) \neq \varnothing$ for $i = 1, 2$. Let $\varepsilon_k > 0$, $k = 1, \ldots$, converge to 0. From Theorem 7.1, for each k there exists $\hat{x}(\varepsilon_k)$ satisfying $\phi(\hat{x}(\varepsilon_k)) = 0$ and $0 \leq g(\hat{x}(\varepsilon_k)) \leq \varepsilon_k$. Let x_* be any accumulation point of $\hat{x}(\varepsilon_k)$'s. Clearly, $x_* \in V$. $\square$

If $V = \varnothing$, it is possible that for some ε one of the $V_i(\varepsilon)$ sets is nonempty. For instance, consider the case where $\phi(x) = x_1^2 + x_2^2$, $Ax = x_1 - x_2$. Then, $V_2(M) = \{x : 2x_1x_2 = 0, x_1 + x_2 = 2, x \geq 0\} \neq \varnothing$.

Consider the quadratic programming problem

$$(QP) : q^* = \min\{q(x) : x \in \Omega_0\},$$

where $q(x)$ is a quadratic function and

$$\Omega_0 = \{x \in \mathbf{R}^n : Ax = b, x \geq 0\},$$

where A is an $m \times n$ matrix. We first assume that Ω_0 is bounded and that we are given a bound μ, on the 1-norm of its feasible solution. Thus, Ω_0 can be replaced by

$$\Omega_1 = \{x \in \mathbf{R}^n : Ax = b, e^T x \leq \mu, x \geq 0\}.$$

Using a single slack variable x_{n+1}, we can replace Ω_1 by

$$\Omega_2 = \{x \in \mathbf{R}^{n+1} : A_1 x = b, e_1^T x = \mu, x \geq 0\},$$

where $A_1 = [A, 0]$ and $e_1^T = (e^T, 1)$. By replacing b with $(e_1^T x/\mu)b$, we can rewrite Ω_2 as

$$\Omega_2 = \{x \in \mathbf{R}^{n+1} : A_2 x = 0, e_1^T x = \mu, x \geq 0\},$$

where $A_2 = \mu A_1 - B_b$, and B_b is the $m \times n$ matrix whose ith row is $(b_i, \ldots, b_i)$. Finally, scaling the variables, we can replace Ω_2 by

$$\Omega_3 = \{x \in \mathbf{R}^{n+1} : A_2 x = 0, e_1^T x = n + 1, x \geq 0\}.$$

The above modifications will also have to be applied to the objective function $q(x)$. Thus, without loss of generality when the feasible region is bounded, we may assume that (QP) is in the form

$$\min\{q(x) = c^T x + x^T H x : Ax = 0, e^T x = n, x \geq 0\},$$

where H is symmetric and $n \times n$, A is $m \times n$, c is an n-vector, and e the n-vector of ones. For $x \in S$, we have

$$|q(x)| \leq |c^T x| + |x^T H x| \leq \|c\|\|x\| + \|x\|^2 \|H\| \leq n\|c\| + n^2 \|H\| = \widehat{M}.$$

Suppose we wish to test if there is a feasible point where $q(x) = \rho$, where $\rho \in [-\widehat{M}, \widehat{M}]$. Equivalently, we test if

$$V = \{x : c^T x + x^T H x = \rho, Ax = 0, e^T x = n, x \geq 0\},$$

is nonempty. Using the constraint $e^T x = n$, V can be written as

$$V = \{x : (e^T x/n)c^T x + x^T H x - (e^T x/n)^2 \rho = 0, Ax = 0, e^T x = n, x \geq 0\}.$$

Equivalently,

$$V = \{x : x^T Q' x = 0, Ax = 0, e^T x = n, x \geq 0\},$$

where

$$Q' = \frac{1}{n} e c^T + H - \frac{\rho}{n^2} e e^T.$$

Finally, we let

$$V = \{x : x^T Q x = 0, Ax = 0, e^T x = n, x \geq 0\},$$

where $Q = (Q' + Q'^T)$ if $e^T Q' e > 0$, or $Q = -(Q' + Q'^T)$ otherwise. The question of feasibility of V is (GP).

We now consider the more general case of (QP), where Ω_0 is either unbounded or that no bound for it is known. As was stated earlier, if q^* is finite, its value is attained at a basic feasible solution of the associated LCP (see [19]). We can write a trivial bound μ on the 1-norm of the basic feasible solutions of the LCP. Thus, if q^* is finite, solving (QP) is equivalent to solving

$$\min\{q(x) : x \in \Omega_1\},$$

where $\Omega_1 = \{x \in \mathbf{R}^n : Ax = b, e^T x \leq \mu, x \geq 0\}$. The new feasible set since bounded allows the computation of a trivial bound M on $|q(x)|$ for $x \in \Omega_1$, and hence for q^*. It is easy to prove that in (QP), $q^* = -\infty$ if and only if there exists $\hat{\mu} < \infty$ such that

$$\{x : q(x) = -M - 1, Ax = b, e^T x \leq \hat{\mu}, x \geq 0\} \neq \emptyset.$$

The above feasibility problem can be formulated as (GP).

8. Computational considerations

In this section we consider the complexity of the three procedures described earlier. The cost per iteration in Procedure W is essentially that of function evaluations, which is $O(n^2)$ for (P) and $O(n^3)$ for (GP) (since a projection matrix need be computed). As we have seen under convexity assumption, Procedure W is capable of solving (P). This is interesting, since it implies that linear programming can be solved with inexpensive iterations. It is also worth mentioning that Procedure W in the case of (P) admits parallelization quite naturally, since each function evaluation requires the multiplication of a matrix by a vector. However, for a given $d \in S_+$, the reduction of f in Procedure W is likely to be less than $\delta(d)$, the corresponding reduction of Procedure A, which for convex ϕ is uniformly bounded above $(\delta_0)^{1/n}$ for some constant $0 < \delta_0 < 1$, whenever $L(d) = 0$ (see [9]). For arbitrary Q, the corresponding reduction $\gamma(d)$ of Procedure S is at least as good as the other two procedures.

In terms of solving (P) and in contrast to Procedure W, Procedures A and S are more expensive. For the case where Q is positive semidefinite, the calculation of $L(d)$ can be carried out in $O(n^3)$ arithmetic operations, since in this case $y^T DQDy = 0 \Leftrightarrow QDy = 0$. We can reduce the $n \times n$ system of linear equations $QDy = 0$ to an independent $m \times n$ system, $ADy = 0$. If $Q' = DA^T AD$, we can compute $\min\{\|y - e\| : L(d) = 0, e^T y = n\}$, i.e., the nearest zero of $\phi_d(y) = y^T DQ'Dy$ (equivalently, $y^T DQDy$) to the center e, in $O(m^2 n)$ arithmetic operations (see [9]). For general Q, the calculation of $L(d)$ and $U(d)$, ignoring the constraint $e^T y = n$, is minimization of a quadratic over a sphere (see [4]), which arises in trust region methods; e.g., see [17], [20]. Recently, Ye [24] has analyzed the complexity of computing the minimum of an arbitrary quadratic over the intersection of a sphere and the hyperplane $e^T y = n$. The calculation of $\gamma(d)$ is a harder problem because of the division by $\pi^2(y)$. When attempting to compute $\gamma(d)$ in practice, we can use a quadratic approximation to $f_d(y)$ at e. In a forthcoming paper [1], we will consider other approximation techniques for computing $L(d)$, $U(d)$, and $\gamma(d)$. These techniques make use of conjugate directions.

With respect to general Q, a major drawback in using the three procedures of this paper is that they may not solve the canonical decision problems considered here; i.e., the sequence of point may converge to an interior point d that is either a stationary point (in case of W) or a local minimum point (in case of S) of f. Even if $f^* > 0$ is computed, in general it does not follow that $\phi^* > 0$. Needless to say, that the direct optimization of ϕ via any algorithm that computes a point satisfying only first- and second-order optimality conditions, e.g., Ye's interior algorithm [24], also may fail to solve these decision problems. In contrast, other methods for global optimization of quadratic problems exist that will decisively solve the underlying problem; e.g., see [10] for concave quadratic minimization, and [11] for indefinite quadratic minimization. Furthermore, while for convex quadratic minimization, polynomial-time algorithms exist (e.g., see [16] and [12]), it is not clear whether its formulation as either (P) or (GP) is solvable by the procedures of this paper in polynomial time. However, we belive that the polynomial convergence of some of the given procedures extends beyond the positive semidefinite cases. Moreover, given the success of Karmarkar's algorithm for linear programming, one would expect that in practice these procedures will at least obtain a stationary or a local minimum point of f in reasonable time.

Acknowledgments

I am grateful to an anonymous referee for a careful reading of the paper and for corrections and recommendations that resulted in the improvement of the presentation.

REFERENCES

1. A. Bagchi and B. Kalantari, *A method for computing approximate solution of the trust region problem with application to projective methods for quadratic programming*, working paper, 1988.
2. E. R. Barnes, *A variation on Karmarkar's algorithm for solving linear programming problems*, Math. Programming **36** (1986), 174–182.
3. I. I. Dikin, *Iterative solution of linear and quadratic programming*, Soviet Math. Dokl. **8** (1967), 674–675.
4. G. E. Forsythe and G. H. Golub, *On the stationary values of a second-degree polynomial on the unit sphere*, J. Soc. Indus. Appl. Math. **13** (1965), 1050–1068.
5. M. R. Garey and D. J. Johnson, *Computers and intractability, a guide to the theory of NP-completeness*, Freeman, New York, 1980.
6. G. de Ghellinck and J. P. Vial, *A polynomial Newton method for linear programming*, Algorithmica **1** (1986), 425–453.
7. G. de Ghellinck and J. P. Vial, *An extension of Karmarkar's algorithm for solving a system of linear homogeneous equations on the complex*, Math. Programming **39** (1987), 79–92.
8. B. Kalantari, *Karmarkar's algorithm with improved steps*, Math. Programming **46** (1990), 73–78.
9. B. Kalantari, *Solving linear programming by bisection and a projective feasibility algorithm*, Technical Report LCSR-TR-91, Department of Computer Science, Rutgers University, New Brunswick, New Jersey, 1987.
10. B. Kalantari and J. B. Rosen, *An algorithm for global minimization of linearly constrained concave quadratic functions*, Math. Oper. Res. **12** (1987), 544–561.
11. B. Kalantari and J. B. Rosen, *Convex envelopes for indefinite quadratic programs*, Technical Report LCSR-TR-78, Department of Computer Science, Rutgers University, New Brunswick, New Jersey, 1986.
12. S. Kapoor and P. M. Vaidya, *Fast algorithms for convex quadratic programming and multicommodity flows*, Proceedings of 18th ACM Symposium on Theory of Computing, 1986, pp. 147–159.
13. N. Karmarkar, *A new polynomial-time algorithm for linear programming*, Combinatorica **4** (1984), 373–395.
14. N. Karmarkar, private communication, 1987.
15. N. Karmarkar, *An interior-point approach to NP-complete problems*, Part 1, this volume, pp. 297–308.
16. M. K. Kozlov, S. P. Tarasov, and L. G. Khachiyan, *Polynomial solvability of quadratic programming*, Soviet Math. Dokl. **20** (1979), 1081–1111.
17. J. J. More and D. C. Sorensen, *Computing a trust region step*, SIAM J. Sci. Statist. Comput. **4** (1983), 553–572.
18. K. G. Murty and N. Kabadi, *Some NP-complete problems in quadratic and nonlinear programming*, Math. Programming **39** (1987), 117–129.

19. K. G. Murty, *Linear complementarity*, Linear and Nonlinear Programming, Heldermann-Verlag, West Berlin, 1988.

20. G. A. Shultz, R. B. Schnabel and R. H. Byrd, *A family of trust region-based algorithms for unconstrained minimization with strong global convergence properties*, SIAM J. Numer. Anal. **22** (1985), 47–67.

21. M. J. Todd and B. P. Burrell, *An extension of Karmarkar's algorithm for linear programming using dual variables*, Algorithmica **1** (1986), 409–424.

22. R. J. Vanderbei, M. S. Meketon, and B. A. Freedman, *A modification of Karmarkar's linear programming algorithm*, Algorithmica **1** (1986), 395–407.

23. R. J. Vanderbei and J. C. Lagarias, *I. I. Dikin's convergence result for the affine-scaling algorithm*, this volume, pp. 109–119.

24. Y. Ye, *On the interior algorithms for nonconvex quadratic programming*, manuscript, Integrated Systems Inc., Santa Clara, California, 1988.

DEPARTMENT OF COMPUTER SCIENCE, RUTGERS UNIVERSITY, NEW BRUNSWICK, NEW JERSEY 08903

E-mail address: Kalantar@cs.rutgers.edu

Contemporary Mathematics
Volume **114**, 1990

An Interior Point Algorithm for Solving Smooth Convex Programs Based on Newton's Method

SANJAY MEHROTRA AND JIE SUN

ABSTRACT. An interior point algorithm for smooth convex programming is developed. The seminorms defined by Hessians of the functions in the problem are assumed to be equivalent on the feasible set. The algorithm constructs a sequence of nested convex sets within the feasible region and finds approximate centers of these sets by using a partial Newton step. We show that starting from a point near the center of the first set the difference of the optimal value and its iterative bound is reduced by a constant rate at each iteration. This rate depends on the number of constraints and a curvature constant.

1. Introduction

We consider the convex programming problem

$$(\mathrm{CP}) \begin{cases} \text{maximize} & c(x) \\ \text{subject to} & c_i(x) \geq 0, \qquad i = 1, 2, \ldots, m, \end{cases}$$

where $x \in \mathfrak{R}^n$, $c(x)$, $c_i(x)$, $i = 1, \ldots, m$, are concave and twice continuously differentiable. The optimal objective value of (CP) is denoted by z^*. Let $\underline{z}$ be a lower bound on the objective function value of (CP) and let

$$(1.1.1) \qquad P_z \equiv \{x \in \mathfrak{R}^n | c(x) \geq z, c_i(x) \geq 0, i = 1, 2, \ldots, m\}$$

be the convex set obtained by restricting the feasible region of (CP) with an additional constraint $c(x) \geq z$, where z is a parameter satisfying $\underline{z} \leq z < z^*$. We assume that the set $\text{int} \, P_z = \{x \in \mathfrak{R}^n | c(x) > z, c_i(x) > 0, i = 1, 2, \ldots, m\}$ is nonempty and bounded for some $z < z^*$. By convex analysis [16], the assumption is then true for all $z < z^*$. Since P_z is bounded,

1980 *Mathematics Subject Classification* (1985 *Revision*). Primary 90C20.
Key words and phrases. Convex programming, interior point methods, path-following methods.
Research of the first author was supported in part by the ONR under grant N00014-87-K-0214 and by the NSF under grant CCR-8810107.
Research of the second author was supported in part by the NSF under grant ECS-8721709.

without loss of generality we assume that $c_1(x)$ is of the form $r^2 - x^T x \geq 0$. This assumption ensures uniqueness of the center of P_z discussed below. It can be replaced by other conditions on $c(x)$ or $c_i(x)$, $i = 1, \ldots, m$. For instance, we may assume that one of the $\nabla^2 c_i(x)$ is negative definite on P_z instead. For simplicity of notation, for $i = m + 1, \ldots, 2m$, let us define $c_i(x) \equiv c(x) - z$, $\nabla c_i(x) \equiv \nabla_x c(x)$, and $\nabla^2 c_i(x) \equiv \nabla_x^2 c(x)$, respectively. We also assume that there exists a "curvature constant" $\kappa \geq 1$ such that the following condition is valid:

$$(1.1.2) \qquad \kappa^{-1} \|p\|_{(i,y)}^2 \leq \|p\|_{(i,x)}^2 \leq \kappa \|p\|_{(i,y)}^2$$

for $i = 1, \ldots, 2m$ and for $x, y \in \operatorname{int} P_z$, where $\|p\|_{(i,x)} \equiv \sqrt{-p^T \nabla^2 c_i(x) p}$. Functions satisfying this assumption include linear, concave quadratic, strongly concave functions, and functions whose Hessian is negative definite on P_z, for example.

A point $\omega(z) \in \mathfrak{R}^n$ is called a *center* of P_z if it maximizes

$$(1.1.3) \qquad F(x, z) \equiv \sum_{i=1}^{2m} \ln c_i(x)$$

subject to $x \in \operatorname{int} P_z$. The function $F(x, z)$ defined in (1.1.3) is called the *potential function*. Because of the strict concavity of $\ln c_1(x)$, $F(x, z)$ is strictly concave in x on $\operatorname{int} P_z$, and hence $\omega(z)$ is unique. Let $\{\omega(z) | \underline{z} \leq z < z^*\}$ be the *trajectory of centers* obtained by continuously improving $\underline{z}$ toward z^*. By a theorem of Fiacco (see [8], Section 11.3), $\omega(z)$ is a continuously differentiable function of z in $[\underline{z}, z^*)$.

In this paper we show that the trajectory $\{\omega(z) | \underline{z} \leq z < z^*\}$ can be followed by using Newton's method. This trajectory has been previously considered by Renegar [15] and Vaidya [18] to develop algorithms for linear programming. The results of Renegar [15] and Vaidya [18] have been extended to linearly constrained convex quadratic programs in Mehrotra and Sun [10] and to quadratically constrained convex quadratic programs in Jarre [3] and Mehrotra and Sun [11, 12]. These studies were motivated by the work of Bayer and Lagarias [1], Huard [2], Karmarkar [7], Meggido [9], and Sonnevend [17].

In our analysis we extend the results in Jarre [3] and Mehrotra and Sun [10, 11, 12] to problem (CP). It should be noted that recently Jarre [4]; Jarre, Sonnevend, and Stoer [5]; Kojima, Mizuno, and Noma [6]; and Nesterov and Nemirovsky [13] have also considered interior methods for convex programming problems. Our work differs from theirs in assumptions on functions, algorithms, ways of analysis, and rates of convergence.

In the next section we state our basic algorithm (Algorithm 2.1). Several preliminary results about the potential function and an ellipsoid inscribed in the feasible region are provided in Section 3. The results of Section 3 are frequently used in Section 4 to analyze Algorithm 2.1. This analysis establishes three key results. First we relate the value of potential functions with some ellipsoidal norms. The second result proves a lower bound on the amount of improvement in the potential function obtained by taking a partial Newton step, and the last result establishes an upper bound on the change in the potential function as z increases. Using these results we establish a global convergence theorem for our algorithm. This theorem shows that the total number of arithmetic operations required by this algorithm to converge to an approximate solution x^k satisfying $z^* - c(x^k) \leq \nu$ is polynomial in $|\ln[\nu/(z^* - \underline{z})]|$, m, n, and the curvature constant κ.

2. The basic algorithm

Let $F(x, z)$ be defined as in (1.1.3) and $\omega(z)$ be the corresponding center. Let $f(x, z) \equiv F(\omega(z), z) - F(x, z)$. The function $f(x, z)$ is called the *normalized potential function*. Note that $\omega(z)$ is the center of P_z if and only if it minimizes $f(x, z)$ over $\operatorname{int} P_z$. We write $\omega(z)$ as ω, whenever it is clear from the context. Let $\nabla f(x, z)$ and $\nabla^2 f(x, z)$ denote the gradient and the Hessian of $f(x, z)$ with respect to x. It is easy to see that

$$\nabla f(x, z) = -\sum_{i=1}^{2m} \frac{\nabla c_i(x)}{c_i(x)}$$

and

$$\nabla^2 f(x, z) = \sum_{i=1}^{2m} \frac{1}{c_i(x)^2} \nabla c_i(x) \nabla c_i(x)^T - \sum_{i=1}^{2m} \frac{1}{c_i(x)} \nabla^2 c_i(x).$$

In this paper we are concerned with the following algorithm for solving (CP).

ALGORITHM 2.1.
Initialization: *Let* $x^0 \in \operatorname{int} P_{z^0}$ *and* $z^0 = \underline{z}$ *be such that* $f(x^0, z^0) \leq \varepsilon$.
For $k = 0, 1, \dots$ **until** $c(x^k) - z^k \leq \nu/4$ *is satisfied* **do:**
Determine a step direction p *by solving*

$$\nabla^2 f(x^k, z^k)p = -\nabla f(x^k, z^k).$$

Let

$$x^{k+1} \leftarrow x^k + \frac{\beta}{\sqrt{p^T \nabla^2 f(x^k, z^k)p}} p,$$

$$z^{k+1} \leftarrow z^k + \frac{\alpha}{\sqrt{m}}(c(x^{k+1}) - z^k).$$

Here β, α, *and* ε *are suitable positive numbers.*
End

Algorithm 2.1 can be understood as follows. Assume that at the beginning of iteration k the current x^k is close to $\omega(z^k)$; i.e., $f(x^k, z^k) \leq \varepsilon$. We take a partial Newton step to get x^{k+1}, a point even closer to $\omega(z^k)$. Then we increase z^k by an amount $\alpha(c(x^{k+1}) - z^k)/\sqrt{m}$ to obtain z^{k+1} so that x^{k+1} remains close to $\omega(z^{k+1})$; i.e., $f(x^{k+1}, z^{k+1}) \leq \varepsilon$. The process is repeated until x^k becomes a good approximate optimal solution to problem (CP).

Note that the stopping criterion in Algorithm 2.1 is in terms of quantities that are computable at each iteration. In the remark on Lemma 3.4 we will show that this criterion ensures $z^* - z^k \leq \nu$, where ν is a given tolerance.

We make the choices of parameters ε, β, and α more precise in terms of κ in our analysis of Algorithm 2.1 (Theorem 4.7). Because κ is seldom known in advance, in practice the parameters ε and α may have to be determined experimentally and β may be replaced by performing a (possibly approximate) line search that maximizes the potential function $F(x, z^k)$ in P_{z^k} along the Newton direction.

For finding the initial point that satisfies the assumptions of Algorithm 2.1 we refer the reader to Mehrotra and Sun [11, 12]. The procedures given in those papers and their analysis generalize for (CP).

The main computational work in the implementation of Algorithm 2.1 involves solving a system of linear equations

$$(2.1.1) \qquad \nabla^2 f(x, z)p = -\nabla f(x, z).$$

The matrix $\nabla^2 f(x, z)$ defining (2.1.1) is symmetric and positive semidefinite. Direct (e.g., symmetric Gaussian elimination) and iterative methods (e.g., preconditioned conjugate gradient methods) may be used to solve (2.1.1).

3. Building blocks for convergence analysis

3.1. Bounds on function $c_i(x)$ $(1 \leq i \leq 2m)$.

LEMMA 3.1. *Let* $c_i(x)$, $i = 1, \ldots, 2m$ *satisfy* (1.1.2). *Let*

$$(3.1.1) \qquad Q_i = \int_0^1 \nabla^2 c_i(x + \tau(y - x)) \, d\tau$$

$$(3.1.2) \, |c_i(y) - c_i(x)| \leq |\nabla c_i(y)^T(x - y)| + \frac{\kappa}{2}|(x - y)^T \nabla^2 c_i(y)(x - y)|,$$

and

$$(3.1.3) \qquad \begin{aligned} \kappa^{-1}|(x - y)^T \nabla^2 c_i(y)(x - y)| &\leq |(x - y)^T Q_i(x - y)| \\ &\leq \kappa|(x - y)^T \nabla^2 c_i(y)(x - y)|, \end{aligned}$$

where

$$Q_i = \int_0^1 \nabla^2 c_i(x + \tau(y - x)) \, d\tau.$$

PROOF. The proof of (3.1.2) follows by using Taylor's expansions of $c_i(x)$,

(3.1.4)
$$c_i(x) = c_i(y) + \nabla c_i(y)^T (x - y)$$
$$+ \frac{1}{2}(x - y)^T \nabla^2 c_i(\hat{y})(x - y), \qquad \hat{y} \in [x, y],$$

and assumption (1.1.2). (3.1.3) is implied by (1.1.2) and the definition (3.1.1). Q.E.D.

3.2. Inscribed ellipsoids. Let $E(y, \delta) \equiv \{x | (x-y)^T \nabla^2 f(y, z)(x-y) \leq \delta^2\}$ be the ellipsoid of radius δ around $y \in \text{int } P_z$. The next lemma bounds the relative variation of function values in $E(y, \delta)$.

LEMMA 3.2. *If* $x \in E(y, \delta)$, $y \in \text{int } P_z$, *and* $\delta > 0$, *then*

(3.2.1)
$$\frac{|c_i(y) - c_i(x)|}{c_i(y)} \leq \delta + \frac{\kappa}{2}\delta^2, \qquad i = 1, \ldots, m,$$

and

(3.2.2)
$$\frac{|c_i(y) - c_i(x)|}{c_i(y)} \leq \frac{\delta}{\sqrt{m}} + \frac{\kappa}{2m}\delta^2, \qquad i = m+1, \ldots, 2m.$$

PROOF. Since $c_i(y) > 0$, from (3.1.2) for $i = 1, \ldots, 2m$

(3.2.3)
$$\frac{|c_i(y) - c_i(x)|}{c_i(y)} \leq \frac{|\nabla c_i(y)^T (x - y)|}{c_i(y)} + \frac{\kappa}{2} \frac{|(x - y)^T \nabla^2 c_i(y)(x - y)|}{c_i(y)}.$$

The definition of $E(y, \delta)$ now gives us
(3.2.4)
$$\sum_{i=1}^{2m} \left| \frac{\nabla c_i(y)^T (x - y)}{c_i(y)} \right|^2 \leq \delta^2 \quad \text{and} \quad \sum_{i=1}^{2m} \frac{|(x - y)^T \nabla^2 c_i(y)(x - y)|}{c_i(y)} \leq \delta^2,$$

which together with (3.2.3) implies (3.2.1) and (3.2.2). Q.E.D.

COROLLARY 3.3. *Let* $y \in \text{int } P_z$. *Then for all* $x \in E(y, 0.5/\sqrt{\kappa})$, $|1 - c_i(x)/c_i(y)| < 1$ *for all* i. *Furthermore,* $E(y, 0.5/\sqrt{\kappa}) \subset \text{int } P_z$.

PROOF. If $y \in \text{int } P_z$ and $\delta = 0.5/\sqrt{\kappa}$, then the inequality (3.2.1) implies $|1 - c_i(x)/c_i(y)| < 1$ for all i and for all $x \in E(y, 0.5/\sqrt{\kappa})$. Thus $c_i(x) > 0$; i.e., $E(y, 0.5/\sqrt{\kappa}) \subset \text{int } P_z$. Q.E.D.

LEMMA 3.4. *Let* ω *be the center of the convex set* P_z. *Then*

$$\frac{c(x) - z}{c(\omega) - z} \leq 2 \quad \text{for any } x \in P_z.$$

PROOF. Since $c_i(x)$ are concave functions on P_z and $c_i(\omega) > 0$,

$$\frac{c_i(x)}{c_i(\omega)} \leq 1 + \frac{1}{c_i(\omega)}\nabla c_i(\omega)^T (x - \omega) \quad \text{for } i = 1, \ldots, 2m.$$

Hence

$$m\frac{c(x) - z}{c(\omega) - z} \le \sum_{i=1}^{2m} \frac{c_i(x)}{c_i(\omega)} \le \sum_{i=1}^{2m} \left[1 + \frac{\nabla c_i(\omega)^T (x - \omega)}{c_i(\omega)}\right]$$

$$= 2m - \nabla f(\omega, z)^T (x - \omega) = 2m \quad (\text{since } \nabla f(\omega, z) = 0). \quad \text{Q.E.D.}$$

REMARK. As a consequence of Lemma 3.4 and (3.2.2) we have

$$z^* - z^k \le 2(c(\omega(z^k)) - z^k) \le 2\left(1 - \frac{\delta}{\sqrt{m}} - \frac{\delta^2}{2m}\right)^{-1}(c(x^k) - z^k)$$

for $z^k < z^*$ and $x^k \in E(\omega(z^k), \delta)$. In Theorem 4.7 it is shown that $x^k \in E(\omega(z^k), 10^{-1}\kappa^{-2})$. Hence Algorithm 2.1 stops with $z^* - z^k \le 4(c(x^k) - z^k) \le \nu$.

3.3. Quadratic approximation of the normalized potential function. Let

$$q(x, y, z) = f(y, z) + \nabla f(y, z)^T (x - y) + \frac{1}{2}(x - y)^T \nabla^2 f(y, z)(x - y)$$

be the quadratic approximation of the normalized potential function $f(x, z)$ at y. In the following lemma we show that the normalized potential function is well behaved on points in an ellipsoid $E(y, \delta)$.

LEMMA 3.5. *For all* $y \in \text{int } P_z$, $x \in E(y, \delta)$ *with* $0 \le \delta < 0.5/\sqrt{\kappa}$, *we have*

$$f(x, z) - q(x, y, z) \le \rho\delta^3 + \frac{\kappa - 1}{2}\delta^2,$$

where

$$\rho = \frac{\kappa}{2} + \frac{\kappa^2\delta}{8} + \frac{(1 + \frac{\kappa\delta}{2})^3}{3(1 - \delta - \frac{\kappa\delta^2}{2})}.$$

PROOF. Since from Corollary 3.3, for $x \in E(y, \delta)$,

$$\left|1 - \frac{c_i(x)}{c_i(y)}\right| < 1 \quad \text{for } i = 1, 2, \ldots, 2m,$$

one has

$$(3.5.1) \qquad f(x, z) - f(y, z) = -\sum_{i=1}^{2m} \ln\frac{c_i(x)}{c_i(y)} = \sum_{i=1}^{2m}\sum_{k=1}^{\infty} \frac{1}{k}\left[1 - \frac{c_i(x)}{c_i(y)}\right]^k.$$

In order to bound $f(x, z) - q(x, y, z)$ we bound the third and higher powered terms in (3.5.1) and the difference of the first two terms in (3.5.1) with $q(x, y, z)$ separately.

Let

$$\alpha_i = \left|\frac{\nabla c_i(y)^T (x - y)}{c_i(y)}\right| \quad \text{and} \quad \beta_i = -\frac{(x - y)^T \nabla^2 c_i(y)(x - y)}{c_i(y)}$$

$$\text{for } i = 1, \ldots, 2m$$

and let $\| \cdot \|_r$ stand for the r-norm of vectors. Note that

$$(x - y)^T \nabla^2 f(y, z)(x - y) = \sum_{i=1}^{2m} (\alpha_i^2 + \beta_i),$$

and hence for $x \in E(y, \delta)$, we have

(3.5.2)
$$\sum_{i=1}^{2m} (\alpha_i^2 + \beta_i) \le \delta^2.$$

Let

$$\bar{\alpha} = (\alpha_1, \alpha_2, \dots, \alpha_{2m}) \in \mathfrak{R}^{2m} \quad \text{and} \quad \bar{\beta} = (\beta_1, \beta_2, \dots, \beta_{2m}) \in \mathfrak{R}^{2m}.$$

Then from (3.5.2) we have

(3.5.3) $\|\bar{\alpha}\|_2 \le \delta \le 0.5/\sqrt{\kappa}, \quad \|\bar{\beta}\|_1 \le \delta^2 \le 0.25/\kappa, \quad \text{and} \quad \alpha_i + \dfrac{\kappa}{2}\beta_i < 1.$

Now,

(3.5.4)

$$\sum_{i=1}^{2m} \sum_{k=3}^{\infty} \frac{1}{k} \left[1 - \frac{c_i(x)}{c_i(y)} \right]^k \le \sum_{i=1}^{2m} \sum_{k=3}^{\infty} \frac{1}{k} \left| 1 - \frac{c_i(x)}{c_i(y)} \right|^k$$

$$\le \sum_{i=1}^{2m} \sum_{k=3}^{\infty} \frac{1}{k} \left(\alpha_i + \frac{\kappa}{2}\beta_i \right)^k \quad \text{(by (3.1.2))}$$

$$\le \sum_{i=1}^{2m} \frac{(\alpha_i + \frac{\kappa}{2}\beta_i)^3}{3} \sum_{k=0}^{\infty} \left(\alpha_i + \frac{\kappa}{2}\beta_i \right)^k$$

$$= \sum_{i=1}^{2m} \frac{(\alpha_i + \frac{\kappa}{2}\beta_i)^3}{3} \frac{1}{1 - \alpha_i - \frac{\kappa}{2}\beta_i}$$

$$\le \frac{1}{3(1 - \delta - \frac{\kappa}{2}\delta^2)} \sum_{i=1}^{2m} \left(\alpha_i + \frac{\kappa}{2}\beta_i \right)^3 \quad \text{(from (3.5.3), } \alpha_i \le \delta, \beta_i \le \delta^2, \forall i)$$

$$= \frac{1}{3(1 - \delta - \frac{\kappa}{2}\delta^2)} \left\| \bar{\alpha} + \frac{\kappa}{2}\bar{\beta} \right\|_3^3 \qquad \text{(continued)}$$

$$\leq \frac{1}{3(1-\delta-\frac{\kappa}{2}\delta^2)}\left(\|\bar{\alpha}\|_3 + \frac{\kappa}{2}\|\bar{\beta}\|_3\right)^3 \quad \text{(using the triangle inequality)}$$

$$\leq \frac{1}{3(1-\delta-\frac{\kappa}{2}\delta^2)}\left(\|\bar{\alpha}\|_2 + \frac{\kappa}{2}\|\bar{\beta}\|_1\right)^3 \quad \text{(using norm inequalities)}$$

$$\leq \frac{1}{3(1-\delta-\frac{\kappa}{2}\delta^2)}\left(\delta + \frac{\kappa}{2}\delta^2\right)^3.$$

Consider
(3.5.5)
$$\sum_{i=1}^{2m}\sum_{k=1}^{2}\frac{1}{k}\left[1-\frac{c_i(x)}{c_i(y)}\right]^k + f(y,z) - q(x,y,z)$$

$$\leq \sum_{i=1}^{2m}\left[1 - \frac{c_i(x)}{c_i(y)} + \frac{\nabla c_i(y)^T(x-y)}{c_i(y)} + \frac{(x-y)^T\nabla^2 c_i(y)(x-y)}{2c_i(y)}\right]$$

$$+ \frac{1}{2}\sum_{i=1}^{2m}\left[\left(1 - \frac{c_i(x)}{c_i(y)}\right)^2 - \left(\frac{\nabla c_i(y)^T(x-y)}{c_i(y)}\right)^2\right]$$

$$\text{(by using the definition of } q(x,y,z) \text{ and}$$
$$\text{negative semidefiniteness of } \nabla^2 c_i(x))$$

$$\leq \frac{\kappa-1}{2}\sum_{i=1}^{m}2\beta_i + \frac{1}{2}\sum_{i=1}^{2m}\left[\left(\frac{|\nabla c_i(y)^T(x-y)|}{c_i(y)} + \frac{\kappa}{2}\beta_i\right)^2 - \left(\frac{\nabla c_i(y)^T(x-y)}{c_i(y)}\right)^2\right]$$

$$\text{(using (3.1.1), (3.1.2), and definitions of } \beta_i)$$

$$\leq \frac{\kappa-1}{2}\|\bar{\beta}\|_1 + \frac{\kappa}{2}\sum_{i=1}^{2m}\alpha_i\beta_i + \frac{\kappa^2}{8}\|\bar{\beta}\|_2^2$$

$$\leq \frac{\kappa-1}{2}\delta^2 + \frac{\kappa}{2}\delta\|\bar{\beta}\|_1 + \frac{\kappa^2}{8}\|\bar{\beta}\|_2^2 \quad \text{(from (3.5.3))}$$

$$\leq \frac{\kappa-1}{2}\delta^2 + \frac{\kappa}{2}\delta\|\bar{\beta}\|_1 + \frac{\kappa^2}{8}\|\bar{\beta}\|_1^2 \leq \frac{\kappa}{2}\delta^3 + \frac{\kappa^2}{8}\delta^4 \quad \text{(from (3.5.3))}.$$

The proof of this lemma now follows by combining (3.5.4) with (3.5.5).
Q.E.D.

4. Analysis of the basic algorithm

Let ω be the center of P_z. In this section we first show (Lemma 4.1) that the distance measured by $f(x,z)$ is related to the distance measured in an ellipsoidal norm defined by using the Hessian of $f(x,z)$ at ω. Lemma 4.3 provides an estimate for the progress made by the partial Newton step in Algorithm 2.1. Lemma 4.2 is preparatory for Lemma 4.3.

Lemma 4.5 is used in the proof of Lemma 4.6 and shows that when z is increased to z^+ the change in the objective function values at the corresponding centers should be nonnegative and cannot exceed $z^+ - z$. Lemma 4.6 provides an upper bound on the increase in the new normalized potential function value from the current normalized potential value, when z is increased to z^+.

LEMMA 4.1. *Let* $0 \le \delta < 0.5/\sqrt{\kappa}$, $\bar{\delta} \equiv 1/(1 + \delta + \kappa\delta^2/2)$, *and*

$$f(x, z) < \frac{(\delta\bar{\delta})^2}{2} \left(\frac{1}{\kappa} - 2\kappa\delta \right).$$

Then $x \in E(\omega, \delta)$.

PROOF. Since $f(x, z)$ is a strictly convex function and $f(\omega, z) = 0$, its minimum value over the region $\{x \in R^n | x \in P_z,\ x \notin \text{int } E(\omega, \delta)\}$ occurs on the boundary of $E(\omega, \delta)$. It is therefore sufficient to show that

$$f(x, z) \ge \frac{(\delta\bar{\delta})^2}{2} \left(\frac{1}{\kappa} - 2\kappa\delta \right)$$

for all points on the boundary of $E(\omega, \delta)$. The Taylor expansion of $f(x, z)$ at ω gives

(4.1.1)
$$
\begin{aligned}
f(x, z) &= f(\omega, z) + \nabla f(\omega, z)^T (x - \omega) \\
&\quad + \frac{1}{2}(x - \omega)^T \nabla^2 f(\hat{\omega}, z)(x - \omega) \\
&\qquad \text{(where } \hat{\omega} = \omega + \lambda(x - \omega) \text{ for some } \lambda \in [0, 1]) \\
&= \frac{1}{2}(x - \omega)^T \nabla^2 f(\hat{\omega}, z)(x - \omega) \\
&\qquad \text{(since } f(\omega, z) = 0 \text{ and } \nabla f(\omega, z) = 0) \\
&= \frac{1}{2} \sum_{i=1}^{2m} \left\{ \left[\frac{\nabla c_i(\hat{\omega})^T (x - \omega)}{c_i(\hat{\omega})} \right]^2 - \frac{(x - \omega)^T \nabla^2 c_i(\hat{\omega})(x - \omega)}{c_i(\hat{\omega})} \right\}.
\end{aligned}
$$

For $i = 1, \ldots, 2m$,

$$
\left[\frac{\nabla c_i(\hat{\omega})^T (x - \omega)}{c_i(\hat{\omega})} \right]^2 - \frac{(x - \omega)^T \nabla^2 c_i(\hat{\omega})(x - \omega)}{c_i(\hat{\omega})}
$$

$$
= \left[\frac{\nabla c_i(\omega)^T (x - \omega) + (x - \omega)^T \hat{Q}_i(\hat{\omega} - \omega)}{c_i(\hat{\omega})} \right]^2 - \frac{(x - \omega)^T \nabla^2 c_i(\hat{\omega})(x - \omega)}{c_i(\hat{\omega})}
$$

(where $\hat{Q}_i = \int_0^1 \nabla^2 c_i(\omega + \tau(\hat{\omega} - \omega))\, d\tau$, $i = 1, \ldots, 2m$) (continued)

$$\geq \bar{\delta}^2 \left\{ \left[\frac{\nabla c_i(\omega)^T(x - \omega) + (x - \omega)^T \hat{Q}_i(\hat{\omega} - \omega)}{c_i(\omega)} \right]^2 \right.$$

$$\left. - \frac{(x - \omega)^T \nabla^2 c_i(\hat{\omega})(x - \omega)}{c_i(\omega)} \right\}$$

(since $\hat{\omega} \in E(\omega, \delta)$, using (3.2.1)

and (3.2.2) we have $c_i(\omega)/c_i(\hat{\omega}) \geq \bar{\delta}$)

$$\geq \bar{\delta}^2 \left\{ \left[\frac{\nabla c_i(\omega)^T(x - \omega)}{c_i(\omega)} \right]^2 - \frac{(x - \omega)^T \nabla^2 c_i(\hat{\omega})(x - \omega)}{c_i(\omega)} \right.$$

$$\left. + 2 \frac{\nabla c_i(\omega)^T(x - \omega)}{c_i(\omega)} \frac{(x - \omega)^T \hat{Q}_i(\hat{\omega} - \omega)}{c_i(\omega)} \right\}$$

$$\geq \bar{\delta}^2 \left\{ \left[\frac{\nabla c_i(\omega)^T(x - \omega)}{c_i(\omega)} \right]^2 - \frac{(x - \omega)^T \nabla^2 c_i(\omega)(x - \omega)}{\kappa c_i(\omega)} \right.$$

$$\left. - 2\kappa\delta \left| \frac{(x - \omega)^T \nabla^2 c_i(\omega)(x - \omega)}{c_i(\omega)} \right| \right\}$$

(using (1.1.2), (3.1.3), and noting that

$\hat{\omega} = \omega + \lambda(x - \omega)$ for some $\lambda \in [0, 1]$).

From (4.1.1) we how have

$$f(x, z) \geq \frac{\bar{\delta}^2}{2} \left[\frac{(x - \omega)^T \nabla^2 f(\omega, z)(x - \omega)}{\kappa} \right.$$

$$\left. - 2\kappa\delta \sum_{i=1}^{2m} \left| \frac{(x - \omega)^T \nabla^2 c_i(\omega)(x - \omega)}{c_i(\omega)} \right| \right]$$

(using the definition of $\nabla^2 f(\omega, z)$ and $k \geq 1$)

$$\geq \frac{\bar{\delta}^2}{2} \left(\frac{\delta^2}{\kappa} - 2\kappa\delta^3 \right)$$

(since x is on the boundary of $E(\omega, \delta)$ and using (3.2.4)).   Q.E.D.

LEMMA 4.2. *Let* $0 < \delta < 0.125/\kappa$ *and* $x \in E(\omega, \delta)$. *Then*
(4.2.1)
$$\nabla f(x, z)^T (x - \omega)$$
$$\geq \left[\frac{1}{\kappa} (1 - 4\kappa\delta - 2\kappa\delta^2) f(x, z)(x - \omega)^T \nabla^2 f(x, z)(x - \omega) \right]^{1/2}.$$

PROOF. Since $f(\omega, z) = 0$ and $f(x, z)$ is convex, we have

$$(4.2.2) \qquad 0 \leq f(x, z) = f(x, z) - f(\omega, z) \leq \nabla f(x, z)^T (x - \omega).$$

Hence (4.2.1) can be implied by showing that
(4.2.3)
$$\nabla f(x, z)^T (x - \omega) \geq \frac{1}{\kappa} (1 - 4\kappa\delta - 2\kappa\delta^2)(x - \omega)^T \nabla^2 f(x, z)(x - \omega) \geq 0.$$

Since

$$(4.2.4) \qquad \nabla c_i(x)^T (x - \omega) = \nabla c_i(\omega)^T (x - \omega) + (x - \omega)^T Q_i(x - \omega),$$

where Q_i is the same as in (3.1.3) with $y = \omega$, we have
(4.2.5)
$$\nabla f(x, z)^T (x - \omega) = -\sum_{i=1}^{2m} \frac{\nabla c_i(x)^T (x - \omega)}{c_i(x)}$$
$$= -\sum_{i=1}^{2m} \left[\frac{\nabla c_i(\omega)^T (x - \omega)}{c_i(x)} + \frac{(x - \omega)^T Q_i(x - \omega)}{c_i(x)} \right].$$

Since $\nabla f(\omega, z)^T (x - \omega) = 0$, from (4.2.5) we have

$$
\begin{aligned}
\nabla f(x, z)^T (x - \omega) = &-\sum_{i=1}^{2m} \frac{(x - \omega)^T Q_i(x - \omega)}{c_i(x)} \\
&+ \sum_{i=1}^{2m} \nabla c_i(\omega)^T (x - \omega) \left[\frac{1}{c_i(\omega)} - \frac{1}{c_i(x)} \right] \\
= &-\sum_{i=1}^{2m} \frac{(x - \omega)^T Q_i(x - \omega)}{c_i(x)} \\
&+ \sum_{i=1}^{2m} \frac{[c_i(x) - c_i(\omega)][\nabla c_i(\omega)^T (x - \omega)]}{[c_i(x)][c_i(\omega)]}.
\end{aligned}
$$

(4.2.6)

Now for $i = 1, \dots, 2m$ we have

(4.2.7)

$$\frac{[c_i(x) - c_i(\omega)][\nabla c_i(\omega)^T (x - \omega)]}{[c_i(x)][c_i(\omega)]}$$

$$= \frac{c_i(x)}{c_i(\omega)} \frac{c_i(x) - c_i(\omega)}{\nabla c_i(\omega)^T (x - \omega)} \left[\frac{\nabla c_i(\omega)^T (x - \omega)}{c_i(x)} \right]^2$$

$$\geq \frac{c_i(x)}{c_i(\omega)} \left[1 + \kappa \frac{(x - \omega)^T Q_i (x - \omega)}{|\nabla c_i(\omega)^T (x - \omega)|} \right] \left[\frac{\nabla c_i(\omega)^T (x - \omega)}{c_i(x)} \right]^2$$

(using Taylor's expansion, (1.1.2), and the definition of Q_i)

$$= \frac{c_i(x)}{c_i(\omega)} \left[\frac{\nabla c_i(\omega)^T (x - \omega)}{c_i(x)} \right]^2 + \kappa \frac{|\nabla c_i(\omega)^T (x - \omega)|}{c_i(\omega)} \frac{(x - \omega)^T Q_i (x - \omega)}{c_i(x)}$$

$$\geq \frac{c_i(x)}{c_i(\omega)} \left[\frac{\nabla c_i(x)^T (x - \omega) - (x - \omega)^T Q_i (x - \omega)}{c_i(x)} \right]^2$$

$$+ \kappa \delta \frac{(x - \omega)^T Q_i (x - \omega)}{c_i(x)} \quad \text{(using (4.2.4) and (3.2.4))}$$

$$\geq \frac{c_i(x)}{c_i(\omega)} \left\{ \left[\frac{\nabla c_i(x)^T (x - \omega)}{c_i(x)} \right]^2 - 2 \frac{\nabla c_i(x)^T (x - \omega)}{c_i(x)} \frac{(x - \omega)^T Q_i (x - \omega)}{c_i(x)} \right\}$$

$$+ \kappa \delta \frac{(x - \omega)^T Q_i (x - \omega)}{c_i(x)}$$

$$= \frac{c_i(x)}{c_i(\omega)} \left[\frac{\nabla c_i(x)^T (x - \omega)}{c_i(x)} \right]^2$$

$$- 2 \left[\frac{\nabla c_i(\omega)^T (x - \omega)}{c_i(\omega)} + \frac{(x - \omega)^T Q_i (x - \omega)}{c_i(\omega)} \right]$$

$$\times \frac{(x - \omega)^T Q_i (x - \omega)}{c_i(x)} + \kappa \delta \frac{(x - \omega)^T Q_i (x - \omega)}{c_i(x)} \qquad \text{(continued)}$$

$$\geq \left(1 - \delta - \frac{\kappa \delta^2}{2}\right) \left[\frac{\nabla c_i(x)^T(x - \omega)}{c_i(x)}\right]^2$$

$$+ 2(\delta + \kappa \delta^2 + \kappa \delta) \frac{(x - \omega)^T Q_i(x - \omega)}{c_i(x)} \quad \text{(using (3.2.3) and (3.2.4))}$$

$$\geq \left(1 - \delta - \frac{\kappa \delta^2}{2}\right) \left[\frac{\nabla c_i(x)^T(x - \omega)}{c_i(x)}\right]^2 + 2\kappa(2\delta + \delta^2)\frac{(x - \omega)^T Q_i(x - \omega)}{c_i(x)}.$$

From (4.2.6) and (4.2.7) we have

$$\nabla f(x, z)^T(x - \omega) = -\sum_{i=1}^{2m} \frac{(x - \omega)^T Q_i(x - \omega)}{c_i(x)}$$

$$+ \sum_{i=1}^{2m} \frac{[c_i(x) - c_i(\omega)][\nabla c_i(\omega)^T(x - \omega)]}{[c_i(x)][c_i(\omega)]}$$

$$\geq -\sum_{i=1}^{2m} \frac{(x - \omega)^T Q_i(x - \omega)}{c_i(x)}$$

$$+ \sum_{i=1}^{2m} \left[\left(1 - \delta - \frac{\kappa \delta^2}{2}\right) \left[\frac{\nabla c_i(x)^T(x - \omega)}{c_i(x)}\right]^2\right.$$

$$\left. + 2\kappa(2\delta + \delta^2)\frac{(x - \omega)^T Q_i(x - \omega)}{c_i(x)}\right]$$

$$\geq (1 - 4\kappa\delta - 2\kappa\delta^2) \sum_{i=1}^{2m} \left\{-\frac{(x - \omega)^T Q_i(x - \omega)}{c_i(x)} + \left[\frac{\nabla c_i(x)^T(x - \omega)}{c_i(x)}\right]^2\right\}$$

$$\geq (1 - 4\kappa\delta - 2\kappa\delta^2) \sum_{i=1}^{2m} \left\{-\frac{1}{\kappa} \frac{(x - \omega)^T \nabla^2 c_i(x)(x - \omega)}{c_i(x)}\right.$$

$$\left. + \left[\frac{\nabla c_i(x)^T(x - \omega)}{c_i(x)}\right]^2\right\}$$

$$\text{(using (3.1.3))}$$

$$\geq \frac{1}{\kappa}(1 - 4\kappa\delta - 2\kappa\delta^2)(x - \omega)^T \nabla^2 f(x, z)(x - \omega)$$

$$(k \geq 1 \text{ and using the definition of } \nabla^2 f(x, z)). \quad \text{Q.E.D.}$$

LEMMA 4.3. *Let* $x \in E(\omega, \delta)$, $0 \leq \delta < 0.125/\kappa$, *and* β *be a parameter such that* $0 \leq \beta < 0.5/\sqrt{\kappa}$. *The point* x^+ *that minimizes* $\nabla f(x, z)^T y$ *over*

$E(x, \beta) \equiv \{y|(y - x)^T \nabla^2 f(x, z)(y - x) \leq \beta^2\}$ *satisfies*

$$f(x^+, z) \leq f(x, z) - \beta \sqrt{\frac{1}{\kappa} f(x, z)(1 - 4\kappa\delta - 2\kappa\delta^2)} + \frac{\kappa\beta^2}{2} + \rho\beta^3,$$

where

$$\rho = \frac{\kappa}{2} + \frac{\beta\kappa^2}{8} + \frac{(1 + \frac{\kappa\beta}{2})^3}{3(1 - \beta - \frac{\kappa\beta^2}{2})}.$$

PROOF. Lemma 3.5 gives

$$f(x^+, z) \leq f(x, z) + \nabla f(x, z)^T(x^+ - x)$$

(4.3.1)
$$+ \frac{1}{2}(x^+ - x)^T \nabla^2 f(x, z)(x^+ - x) + \rho\beta^3 + \frac{\kappa - 1}{2}\beta^2$$

$$\leq f(x, z) + \nabla f(x, z)^T(x^+ - x) + \frac{\kappa\beta^2}{2} + \rho\beta^3.$$

Let $\bar{x}$ be the point where the straight line joining x to the center ω intersects with the boundary of the ellipsoid $E(x, \beta)$. Since $\nabla f(x, z)^T x^+ \leq \nabla f(x, z)^T \bar{x}$, from (4.3.1) and (4.2.1) we have

$$f(x^+, z) \leq f(x, z) + \nabla f(x, z)^T(\bar{x} - x) + \frac{\kappa\beta^2}{2} + \rho\beta^3$$

$$= f(x, z) - \frac{\beta\nabla f(x, z)^T(x - \omega)}{\sqrt{(\omega - x)^T \nabla^2 f(x, z)(\omega - x)}} + \frac{\kappa\beta^2}{2} + \rho\beta^3$$

$$\leq f(x, z) - \beta\sqrt{\frac{1}{\kappa} f(x, z)(1 - 4\kappa\delta - 2\kappa\delta^2)} + \frac{\kappa\beta^2}{2} + \rho\beta^3.$$

The last inequality follows by using Lemma 4.2.   Q.E.D.

REMARK. We note that the condition that $\nabla f(x, z)^T y$ is maximized over $E(x, \beta)$ in Lemma 4.3 is somewhat stronger than what is needed to prove it. A point x^+ satisfying

$$\nabla f(x, z)^T x^+ \leq \nabla f(x, z)^T \bar{x}$$

ensures enough reduction in the normalized potential function. This observation indicates that obtaining the exact Newton search direction is not necessary to achieve sufficient reduction in the potential function. Thus we can use inexact solutions of (2.1.1) in implementing Algorithm 2.1. This is particularly important when iterative methods are used to solve (2.1.1).

COROLLARY 4.4. *If* $\delta = 10^{-1}\kappa^{-2}$, $f(x, z) < 10^{-3}\kappa^{-5}$, *and* $\beta = 10^{-2}\kappa^{-4}$, *then* $f(x^+, z) \leq 10^{-3}\kappa^{-5} - 10^{-4}\kappa^{-7}$.

PROOF. Plugging $\delta = 10^{-1}\kappa^{-2}$, $\kappa \geq 1$, and

$$f(x, z) < 10^{-3}\kappa^{-5} = 10^{-1}\kappa^{-1}\delta^2$$

$$\leq 0.4 \left(1 + \delta + \frac{\kappa\delta^2}{2}\right)^{-2} \kappa^{-1}\delta^2 = \frac{(\delta\bar{\delta})^2}{2}\left(\frac{1}{\kappa} - 2\kappa\delta\right)$$

in Lemma 4.1, one gets $x \in E(\omega, \delta)$. Then setting $\delta = 10^{-1}\kappa^{-2}$ and $\beta = 10^{-2}\kappa^{-3}$ in Lemma 4.3, one gets

$$f(x^+, z) \leq f(x, z) - 0.7 * 10^{-2}\kappa^{-4.5}\sqrt{f(x, z)} + 0.5 * 10^{-4}\kappa^{-7} + 10^{-6}\kappa^{-8},$$

which gives

$$f(x^+, z) \leq 10^{-3}\kappa^{-5} - 10^{-4}\kappa^{-7}$$

for all $f(x, z) \in [0, 10^{-3}\kappa^{-5}]$.   Q.E.D.

LEMMA 4.5. *Let* $z^* > z^+ \geq z$ *and let* $f(x, z^+) = F(\omega^+, z^+) - F(x, z^+)$, *where*

$$F(x, z^+) = \sum_{i=1}^{m} \ln c_i(x) + m \ln[c(x) - z^+],$$

and ω^+ *is the unique maximum of* $F(x, z^+)$ *over the convex set*

$$P_{z^+} = \{x \in R^n | c_i(x) \geq 0, \ i = 1, \ldots, m \text{ and } c(x) \geq z^+\}.$$

Then

(4.5.1) $$0 \leq c(\omega^+) - c(\omega) \leq z^+ - z.$$

PROOF. Let $z(t) = z + t(z^+ - z)$ and let $\omega(t)$ be the point that maximizes the potential function

$$F(x, z(t)) = \sum_{i=1}^{m} \ln c_i(x) + m \ln[c(x) - z(t)].$$

Since the gradient of $F(x, z(t))$ vanishes at $\omega(t)$, we have

(4.5.2)
$$0 = \frac{d}{dt}\left[\sum_{i=1}^{m} \frac{\nabla c_i(\omega(t))}{c_i(\omega(t))} + m\frac{\nabla c(\omega(t))}{c(\omega(t)) - z(t)}\right]$$

$$= \sum_{i=1}^{m} \frac{1}{[c_i(\omega(t))]^2}\left[\frac{d\nabla c_i(\omega(t))}{d\omega(t)}\frac{d\omega(t)}{dt}c_i(\omega(t)) - \nabla c_i(\omega(t))\frac{dc_i(\omega(t))}{dt}\right]$$

$$+ \frac{m}{[c(\omega(t)) - z(t)]^2}\left\{\frac{d\nabla c(\omega(t))}{d\omega(t)}\frac{d\omega(t)}{dt}[c(\omega(t)) - z(t)]\right.$$

$$\left. - \nabla c(\omega(t))\left[\frac{dc(\omega(t))}{dt} - \frac{dz(t)}{dt}\right]\right\}$$

$$= \sum_{i=1}^{m} \frac{1}{[c_i(\omega(t))]^2}\left[c_i(\omega(t))\nabla^2 c_i(\omega(t))\frac{d\omega(t)}{dt} - \nabla c_i(\omega(t))\frac{dc_i(\omega(t))}{dt}\right]$$

$$+ \frac{m}{[c(\omega(t)) - z(t)]^2}\left\{[c(\omega(t)) - z(t)]\nabla^2 c(\omega(t))\frac{d\omega(t)}{dt}\right.$$

$$\left. - \nabla c(\omega(t))\left[\frac{dc(\omega(t))}{dt} - \frac{dz(t)}{dt}\right]\right\}.$$

Since

$$\nabla c(\omega(t))^T \frac{d\omega(t)}{dt} = \frac{dc(\omega(t))}{dt}, \qquad \nabla c_i(\omega(t))^T \frac{d\omega(t)}{dt} = \frac{dc_i(\omega(t))}{dt},$$
$$i = 1, \dots, m,$$

and $\nabla^2 c(\omega(t))$, $\nabla^2 c_i(\omega(t))$, $i = 1, \dots, m$ are negative semidefinite, upon taking the inner product of (4.5.2) with $d\omega(t)/dt$ we have,

$$0 \geq \frac{dc(\omega(t))}{dt} \left[\frac{dc(\omega(t))}{dt} - \frac{dz(t)}{dt} \right],$$

which implies

$$\left[\frac{dc(\omega(t))}{dt} \right]^2 \leq \frac{dc(\omega(t))}{dt} \frac{dz(t)}{dt} = \frac{dc(\omega(t))}{dt}(z^+ - z).$$

Hence

$$0 \leq \frac{dc(\omega(t))}{dt} \leq z^+ - z,$$

and therefore

$$0 \leq c(\omega^+) - c(\omega) = \int_0^1 \frac{dc(\omega(t))}{dt} \, dt \leq z^+ - z. \quad \text{Q.E.D.}$$

LEMMA 4.6. *Let* $z^+ = z + \alpha/\sqrt{m}[c(x^+) - z]$, $\sqrt{m} > \alpha > 0$, *and* $x^+ \in$ int P_{z^+}. *If* $x^+ \in E(\omega, \delta)$, *then we have*

$$f(x^+, z^+) \leq f(x^+, z) + \frac{m\alpha}{\sqrt{m} - \alpha} \left(\frac{\delta}{\sqrt{m}} + \frac{\kappa\delta^2}{2m} \right) + \frac{\alpha^2 (1 + \frac{\delta}{\sqrt{m}} + \frac{\kappa\delta^2}{2m})^2}{1 - \frac{\alpha}{\sqrt{m}}(1 + \frac{\delta}{\sqrt{m}} + \frac{\kappa\delta^2}{2m})}.$$

PROOF. We may write

$$(4.6.1) \quad f(x^+, z^+) = f(x^+, z) + f(\omega, z^+) + m \ln \frac{[c(x^+) - z][c(\omega) - z^+]}{[c(x^+) - z^+][c(\omega) - z]}.$$

Since

$$\frac{[c(x^+) - z][c(\omega) - z^+]}{[c(x^+) - z^+][c(\omega) - z]} = 1 + \frac{(z^+ - z)[c(\omega) - c(x^+)]}{[c(x^+) - z^+][c(\omega) - z]}$$

$$= 1 + \frac{\alpha[c(x^+) - z][c(\omega) - c(x^+)]}{\sqrt{m}[c(x^+) - z^+][c(\omega) - z]}$$

$$= 1 + \frac{\alpha}{\sqrt{m} - \alpha} \frac{c(\omega) - c(x^+)}{c(\omega) - z},$$

by using (3.2.2) we have

(4.6.2)
$$\frac{[c(x^+) - z][c(\omega) - z^+]}{[c(x^+) - z^+][c(\omega) - z]} \leq 1 + \frac{\alpha}{\sqrt{m} - \alpha}\left(\frac{\delta}{\sqrt{m}} + \frac{\kappa\delta^2}{2m}\right).$$

We now bound $[c(\omega^+) - z^+]/[c(\omega) - z^+]$ from above to get a bound on the value of $f(\omega, z^+)$. Note that

$$\frac{c(\omega^+) - z^+}{c(\omega) - z^+} = \frac{c(\omega^+) - z}{c(\omega) - z}\left[1 + \frac{(z^+ - z)[c(\omega^+) - c(\omega)]}{[c(\omega^+) - z][c(\omega) - z^+]}\right].$$

Now by using Lemma 4.5 in the above equation we have
(4.6.3)
$$\frac{c(\omega^+) - z^+}{c(\omega) - z^+} \leq \frac{c(\omega^+) - z}{c(\omega) - z} + \frac{(z^+ - z)^2}{[c(\omega) - z][c(\omega) - z^+]}$$

$$= \frac{c(\omega^+) - z}{c(\omega) - z} + \frac{\alpha^2[c(x^+) - z]^2}{m[c(\omega) - z][c(\omega) - z^+]}$$

$$= \frac{c(\omega^+) - z}{c(\omega) - z} + \frac{\alpha^2}{m}\left[\frac{c(x^+) - c(\omega)}{c(\omega) - z} + 1\right]\left[\frac{c(x^+) - z}{c(\omega) - z - \frac{\alpha}{\sqrt{m}}[c(x^+) - z]}\right]$$

$$= \frac{c(\omega^+) - z}{c(\omega) - z} + \frac{\alpha^2}{m}\left[\frac{c(x^+) - c(\omega)}{c(\omega) - z} + 1\right]^2\left[\frac{1}{1 - \frac{\alpha[c(x^+)-z]}{\sqrt{m}[c(\omega)-z]}}\right]$$

$$\leq \frac{c(\omega^+) - z}{c(\omega) - z} + \frac{\frac{\alpha^2}{m}(1 + \frac{\delta}{\sqrt{m}} + \frac{\kappa\delta^2}{2m})^2}{1 - \frac{\alpha}{\sqrt{m}}(1 + \frac{\delta}{\sqrt{m}} + \frac{\kappa\delta^2}{2m})} \quad \text{(using (3.2.2))}.$$

The inequality (4.6.4) below follows by using (4.6.3). Now,
(4.6.4)
$$f(\omega, z^+) = \sum_{i=1}^{m} \ln\frac{c_i(\omega^+)}{c_i(\omega)} + m\ln\frac{c(\omega^+) - z^+}{c(\omega) - z^+}$$

$$\leq \sum_{i=1}^{m}\left[\frac{c_i(\omega^+)}{c_i(\omega)} - 1\right] + m\left[\frac{c(\omega^+) - z^+}{c(\omega) - z^+} - 1\right] \quad \text{(using } \ln(1 + t) \leq t)$$

$$\leq \sum_{i=1}^{m}\frac{c_i(\omega^+) - c_i(\omega)}{c_i(\omega)} + m\left[\frac{c(\omega^+) - z}{c(\omega) - z} + \frac{\frac{\alpha^2}{m}(1 + \frac{\delta}{\sqrt{m}} + \frac{\kappa\delta^2}{2m})^2}{1 - \frac{\alpha}{\sqrt{m}}(1 + \frac{\delta}{\sqrt{m}} + \frac{\kappa\delta^2}{2m})} - 1\right]$$

$$\leq \sum_{i=1}^{2m}\frac{\nabla c_i(\omega)^T(\omega^+ - \omega)}{c_i(\omega)} + \frac{\alpha^2(1 + \frac{\delta}{\sqrt{m}} + \frac{\kappa\delta^2}{2m})^2}{1 - \frac{\alpha}{\sqrt{m}}(1 + \frac{\delta}{\sqrt{m}} + \frac{\kappa\delta^2}{2m})}$$

$$\text{(since } c_i \text{ are concave functions)}$$

$$= \frac{\alpha^2(1 + \frac{\delta}{\sqrt{m}} + \frac{\kappa\delta^2}{2m})^2}{1 - \frac{\alpha}{\sqrt{m}}(1 + \frac{\delta}{\sqrt{m}} + \frac{\kappa\delta^2}{2m})} \quad \text{(since } \nabla f(\omega, z) = 0).$$

The proof of Lemma 4.6 is complete by combining (4.6.1), (4.6.2), and (4.6.4). Q.E.D.

Now we are ready to prove the following convergence theorem.

THEOREM 4.7. *Let z^* be the optimal objective function value of* (CP). *If $\varepsilon = 10^{-3}\kappa^{-5}$, $\alpha = 10^{-4}\kappa^{-5}$, and $\beta = 10^{-2}\kappa^{-4}$ are taken at each iteration of Algorithm 2.1, then the algorithm is well defined. Furthermore, at iteration k we have*

$$\frac{z^* - z^{k+1}}{z^* - z^k} \leq 1 - \frac{4.5 * 10^{-5}}{\kappa^5 \sqrt{m}}.$$

PROOF. We first show that

$$f(x^k, z^k) \leq \varepsilon = 10^{-3}\kappa^{-5}$$

by induction. This inequality is valid for $k = 0$ after the initialization step of Algorithm 2.1. Now we suppose that $f(x^k, z^k) \leq 10^{-3}\kappa^{-5}$. Plugging in $f(x^k, z^k) < 10^{-3}\kappa^{-5}$ and $\delta = 10^{-1}\kappa^{-2}$ in Lemma 4.1 we know that $x^k \in E(\omega^k, \delta)$, where $\delta = 10^{-1}\kappa^{-2}$ (see the proof of Corollary 4.4). Since $x^k \in E(\omega^k, 10^{-1}\kappa^{-2})$, for the choice of $\beta = 10^{-2}\kappa^{-4}$ in Corollary 4.4, we get

$$f(x^{k+1}, z^k) \leq 10^{-3}\kappa^{-5} - 10^{-4}\kappa^{-7}.$$

Because $\alpha = 10^{-4}\kappa^{-5}$ and $m \geq 1$, we have

$$\frac{m\alpha}{\sqrt{m} - \alpha}\left(\frac{\delta}{\sqrt{m}} + \frac{\kappa\delta^2}{2m}\right) + \frac{\alpha^2(1 + \frac{\delta}{\sqrt{m}} + \frac{\kappa\delta^2}{2m})^2}{1 - \frac{\alpha}{\sqrt{m}}(1 + \frac{\delta}{\sqrt{m}} + \frac{\kappa\delta^2}{2m})} \leq 10^{-4}\kappa^{-7}.$$

Lemma 4.6 now implies that $f(x^{k+1}, z^{k+1}) \leq 10^{-3}\kappa^{-5}$. This completes the induction. Corollary 3.3 and Lemma 4.1 then ensure that all $x^k \in E(x^k, \delta) \subset \text{int } P_{z^k}$, thus the algorithm is well defined for the given choices of parameters. Now since by using (3.2.2) and Lemma 3.4,

$$c(x^{k+1}) - z^k \geq \left(1 - \frac{\delta}{\sqrt{m}} - \frac{\kappa\delta^2}{2m}\right)[c(\omega^k) - z^k]$$

$$> 0.9[c(\omega^k) - z^k] \geq 0.45(z^* - z^k),$$

we have

$$z^* - z^{k+1} = z^* - z^k - \frac{\alpha}{\sqrt{m}}[c(x^{k+1}) - z^k] \leq \left(1 - \frac{4.5 * 10^{-5}}{\kappa^5 \sqrt{m}}\right)(z^* - z^k).$$

The proof of Theorem 4.7 is now complete. Q.E.D.

Using Theorem 4.7 we can easily show that, given $\underline{z}$, starting from a point satisfying the assumption in Algorithm 2.1, a solution satisfying $z^* - c(x^k) \leq \nu$ can be obtained in a time polynomial in n, m, and the curvature constant κ, and $|\ln[\nu/(z^* - \underline{z})]|$.

5. Conclusions

We have shown that the trajectory of centers can be followed by using Newton's method for a class of smooth convex programming problems. Our analysis indicates that the performance of the method depends on curvature of the objective and constraint functions near the trajectory.

Acknowledgment

We thank an anonymous referee for pointing out mistakes in an earlier draft of this paper and for providing valuable suggestions to improve its organization. We are also grateful to Professor Michael J. Todd for carefully reading this paper.

References

1. D. A. Bayer and J. C. Lagarias, *The nonlinear geometry of linear programming, I. Affine and projective scaling trajectories, II. Legendre transform coordinates and central trajectories*, Trans. Amer. Math. Soc. **314** (1989), 499–581.

2. P. Huard, *Resolution of mathematical programming with nonlinear constraints by the method of centers*, Nonlinear Programming (J. Abadie, ed.), North-Holland, Amsterdam, 1967, pp. 207–219.

3. F. Jarre, *On the convergence of the method of analytic centers when applied to convex quadratic programs*, manuscript, Institut für Angewandte Mathematik und Statistik, Universität Würzburg, Am Hubland, 8700 Würzburg, West Germany, 1989, Math. Programming (to appear).

4. ____, *The method of analytic centers for smooth convex programs*, Grottenthaler verlag Bamberg, 1989.

5. F. Jarre, G. Sonnevend, and J. Stoer, *An implementation of the method of analytic centers*, Lecture Notes in Control and Inform. Sci. (A. Benoussan and J. L. Lions, eds.), vol. 111, 1988.

6. M. Kojima, S. Mizuno, and T. Noma, *A new continuation method for complementary problems with uniform P-functions*, Math. Programming **43** (1989), 107–113.

7. N. Karmarkar, *A new polynomial-time algorithm for linear programming*, Combinatorica **4** (1984), 373–395.

8. G. P. McCormick, *Nonlinear programming*, Wiley, New York, 1983.

9. N. Meggido, *Pathways to the optimal set in linear programming*, Proc. 6th Math. Progr. Sympos. Japan, 1986, pp. 1–35.

10. S. Mehrotra and J. Sun, *An algorithm for convex quadratic programming that requires $O(n^{3.5}L)$ arithmetic operations*, Math. Oper. Res. **15** (1990), 342–363.

11. ____, *A method of analytic centers for quadratically constrained convex quadratic programs*, SIAM J. Numer. Anal. **28** (1991).

12. ____, *On computing the center of a convex quadratically constrained set*, Math. Programming (to appear).

13. J. E. Nesterov and A. S. Nemirovsky, *A general approach to polynomial-time algorithms design for convex programming*, Central Economical and Mathematical Institute, USSR Academy of Science, report at the 13th Internat. Sympos. Math. Programming, Tokyo, 1988.

14. J. M. Ortega and W. C. Rheinboldt, *Iterative solution of nonlinear equations in several variables*, Academic Press, New York, 1970.

15. J. Renegar, *A polynomial-time algorithm, based on Newton's method for linear programming*, Math. Programming **40** (1988), 59–93.

16. R. T. Rockafellar, *Convex analysis*, Princeton Univ. Press, New Jersey, 1970.

17. G. Sonnevend, *A new method for solving a set of linear (convex) inequalities and its applications for identification and optimization*, working paper, Department of Numerical Analysis, Institute of Mathematics, Eotvos University, 1088, Budapest, Muzeum Korut 6-8, 1985.

18. P. M. Vaidya, *An algorithm for solving linear programming which requires* $O(((m+n)n^2 + (m+n)^{1.5}n)L)$ *arithmetic operations*, Proc. 19th ACM Annual Sympos. on Theory of Computing, 1987, pp. 29–38.

DEPARTMENT OF INDUSTRIAL ENGINEERING AND MANAGEMENT SCIENCES, NORTHWESTERN UNIVERSITY, EVANSTON, ILLINOIS 60208

E-mail address, Sanjay Mehrotra: mehrotra@iems.nwu.edu

Contemporary Mathematics
Volume **114**, 1990

A Modified Kantorovich Inequality for the Convergence of Newton's Method

A. A. GOLDSTEIN

ABSTRACT. A sufficient condition is given guaranteeing quadratic convergence for Newton's method to find a solution of $F(x) = 0$. It assumed a Lipschitz-type condition for the inverse Jacobian $(DF)^{-1}(x)$ in the Newton's method search direction.

1. Introduction

Newton's method is now seen to play an important role in interior-point linear programming methods. The original Karmarkar algorithm has an interpretation as a global Newton method (see Bayer and Lagarias [1]), and other interior methods use Newton steps. The analyses of Renegar [12] and Renegar and Shub [13] described how to follow a central trajectory using Newton's method. Such trajectories were studied by Megiddo [10] and Bayer and Lagarias [2] and [3]. More recently Nesterov and Nemirovsky [11] have defined a large class of functions called self-concordant functions for which a variant of Newton's method has good global convergence properties. They showed that self-concordant functions can be found to solve many optimization problems including linear programming.

It is well known that Newton's method for finding a zero of a function $F(x)$ generally converges quadratically in a close neighborhood of a zero. It is of great interest to determine as large a neighborhood as possible in which quadratic convergence is guaranteed. Many results of this kind have been given, starting from Kantorovich [9], and including Smale's conditions at one point [15]. These results all require implicitly or explicitly some global constraint on the behavior of the function on an open set. Here we give a new variant of a Kantorovich-type inequality requiring only a Lipschitz-type condition in the direction of the Newton step for all points in an open set.

1980 *Mathematics Subject Classification* (1985 *Revision*). Primary 49D15.
Supported in part by Grants NIH RR012431 and NPS LMC-M4E1.

2. Sufficient conditions for convergence

For perspective and comparison, this section reviews two key results from the large literature on Newton's method.

We are given two real Banach spaces X and Y and an open convex subset U of X. A function F is a given mapping U into Y. We assume that F if Fréchet differentiable on U.

Newton's method for solving the equation $F(x) = 0$ has the following form. Assume that the Fréchet derivative $DF(x)$ has an inverse throughout U, denoted by $(DF)^{-1}(x)$. The Newton's method iterate of x is $x - s(x)$ where

$$s(x) = (DF)^{-1}(x)F(x)$$

is called the *Newton step*. Given a starting point x_0 in U, the *Newton sequence* is

$$x_{k+1} = x_k - (DF)^{-1}(x_k)F(x_k).$$

The totality of line segments joining the iterates x_k, $k = 1, 2, 3, \ldots$, will be called the *Newtonian trajectory*. The length of the Newton step $\eta(x)$, given by

$$\eta(x) = \|s(x)\| = \|(DF)^{-1}(x)F(x)\|,$$

is an important quantity in estimating convergence of Newton's method.

The sequence $\{x_k\}$ generated may not converge to a limit. Kantorovich [8] found sufficient conditions to guarantee that the point x_0 will start a Newton sequence that converges at a quadratic rate; these conditions depend on the existence of a global bound for $\|D^2F(x)\|$ on a convergence region S. Later on, Kantorovich and Akilov [9] weakened this to requiring a global bound on $\sup\{\|(Df)^{-1}(x_0)D^2F(u)\| : u \in S\}$ on a convergence region S. This was further relaxed to requiring only a Lipschitz condition on DF. The following result of this type is taken from Deuflhard and Heindl [4]. It uses an amalgamation of the optimal estimate of convergence by Gragg and Tapia [6], with the Kantorovich-Akilov theorem.

THEOREM 1.1 (Kantorovich-Akilov, Gragg-Tapia, and Deuflhard-Heindl). *Let* $F: X \to Y$ *be Fréchet differentiable. Suppose that one has a starting point* x_0 *with* $DF(x_0)$ *invertible and constants* $\eta(x_0) = \|DF^{-1}(x_0)F(x_0)\|$ *and* ω_0 *such that*

(1) *Lipschitz condition at* x_0.

$$\|(DF)^{-1}(x_0)(DF(y) - DF(x))\| \le \omega_0 \|x - y\| \quad \text{for all } x, y \text{ in } U$$

(2) *Proximal root condition.*

$$\hat{h}_0 = \eta(x_0)\omega_0 < \tfrac{1}{2}.$$

(3) *The set*

$$S = \left\{ x : \|x - x_0\| \leq \rho^- = \left(\frac{1 - \sqrt{1 - 2\hat{h}_0}}{\hat{h}_0} \right) \eta(x_0) \right\} \subset U.$$

Then

(i) *The Newton sequence* $\{x_n\}$ *is well defined, remains in* S, *and converges to a root* x^* *of* F.

(ii) *Let* $\rho^+ = \frac{1 + \sqrt{(1 - 2\hat{h}_0)}}{\omega_0}$ *and set* $\Theta = \frac{\rho^-}{\rho^+}$. *Then*

$$\|x_k - x^*\| \leq \left(\frac{2\sqrt{1 - 2\hat{h}_0}}{\hat{h}_0} \frac{\Theta^{2^k}}{1 - \Theta^{2^k}} \right) \eta(x_0).$$

Note that if A is a bounded invertible linear operator between Banach spaces Y and Z and if $G = AF$ then $DG^{-1}(x)G(x) = DF^{-1}(x)F(x)$. Thus the Newton sequences for F and G are the same. This invariance is present in the hypotheses of Theorem 1.1.

We now turn to the convergence criterion of Smale.

THEOREM 1.2 (Smale [15]). *Assume that* F *is an analytic map between real Banach spaces* X *and* Y; *that is, the Fréchet derivatives* $DF^k(x)$ *exist for all* $x \in X$ *and* $k = 1, 2, 3, \ldots$ *. Given* $x_0 \in X$, *assume that the inverse* $(DF)^{-1}(x_0)$ *exists. Set*

$$\eta(x_0) = \|(DF)^{-1}(x_0)F(x_0)\|$$

and

$$\hat{\gamma}(x_0) = \sup \left\{ \left\| \frac{1}{k!}(DF)^{-1}(x_0)DF^k(x_0) \right\|^{1/(k-1)} : k \geq 2 \right\}.$$

Suppose that

$$\eta(x_0)\hat{\gamma}(x_0) < .130707.$$

Then the Newton sequence is well defined and converges to a root x^* *of* F. *Furthermore*

(i) $\|x_{k+1} - x_k\| \leq 2 \left(\frac{1}{2}\right)^{2^k} \eta(x_0).$

(ii) $\|x_k - x^*\| \leq \frac{7}{4} \left(\frac{1}{2}\right)^{2^{k-1}} \eta(x_0).$

Smale's theorem has the merit that it requires only information available at the starting point. In particular, it is checkable for the case when F is a polynomial mapping. Its hypotheses are also invariant under bounded invertible linear transformations.

Smale's theorem implicitly supplies global information on the behavior of function $F(x)$ near x_0, which is inferable from the bound $\hat{\gamma}(x_0)$ by analytic continuation. Rheinboldt [14] has shown that Smale's theorem is a consequence of the Kantorovich-Akilov theorem [9], the precursor of Theorem 1.1.

3. Another variant of the Kantorovich inequality

Consider the set of proximal points given by Theorem 1.1, namely $\{x_0 \in U : \eta_0(x_0)\omega_0\} \leq \frac{1}{2}$. If ω_0 can be decreased this set grows larger. Our main result reduces ω_0 by replacing the Lipschitz condition on U defining ω_0 with a Lipschitz condition only in the direction of the Newton step $s(x) = (DF)^{-1}(x)F(x)$.

We first show that the existence of DF and a growth condition on it for a Newton method step suffices to obtain convergence of the Newtonian iterates. Recall that the Newton step length $\eta(x) = \|s(x)\|$.

LEMMA 2.1. *Let F be a map between real Banach spaces X and Y. Assume F is Fréchet differentiable on X and that the derivative DF is invertible. Assume that all the Newton iterates $\{x_n\}$ of x_0 exist and that*

$$\gamma(x_0) = \sup\{\|(DF)^{-1}(x_n)[DF(x_n) - DF(x_n - ts(x_n))]\|$$
$$: t \in (0, 1) \text{ and } n = 0, 1, 2, 3, \ldots\} \leq \tfrac{1}{3}.$$

Then the sequence $\{x_n\}$ converges to a limit x^* with $F(x^*) = 0$, and $\{x_n\}$ and x^* lie in

$$S = \{x \in U : \|x - x_0\| \leq 2\eta(x_0)\}.$$

Moreover

(i) $\eta(x_n) \leq (\tfrac{1}{2})^n \eta(x_0)$,
(ii) $\|x_n - x^*\| \leq (\tfrac{1}{2})^{n-1} \eta(x_0)$,
(iii) $\|F(x_n)\| \leq (\tfrac{2}{3})^n \|DF(x_0)\| \eta(x_0)$.

PROOF. Let $\gamma_n = \gamma(x_n)$ and $\eta_n = \eta(x_n)$ for the Newtonian iterates $\{x_n\}$. Suppose A is a bounded linear operator mapping X into itself, and let I denote the identity mapping. By a well-known consequence of the contraction mapping principle of Banach, if $\|A\| < 1$, then $(I - A)^{-1}$ exists and

$$\|(I - A)^{-1}\| \leq (1 - \|A\|)^{-1}.$$

By hypothesis, x_1 is well defined and belongs to S. Let

$$H(x_0, x_1) = (DF)^{-1}(x_0)DF(x_1) = I - (DF)^{-1}(x_0)(DF(x_0) - DF(x_1)).$$

The bounded linear operator $H = H(x_0, x_1)$ maps X into itself. Since $\|(DF)^{-1}(x_0)(DF(x_0) - DF(x_1))\| \leq \gamma_0 \leq \tfrac{1}{3}$, we have the estimate $\|(H)^{-1}\| = (1 - \gamma)^{-1} \leq \tfrac{3}{2}$. Also, $\|H\| \leq \tfrac{4}{3}$. Observe that $DF(x_0)H = DF(x_1)$ and $(H)^{-1}(DF)^{-1}(x_0) = (DF)^{-1}(x_1)$.

Let f be defined by the formula $f(x) = x - (DF)^{-1}(x_0)F(x)$. Since $f(x_0) = x_1$ and $f(x_1) = x_1 - (DF)^{-1}(x_0)F(x_1)$,

$$(DF)^{-1}(x_0)F(x_1) = f(x_0) - f(x_1).$$

By the generalized mean value theorem of Graves [7],

$$\|f(x_1) - f(x_0)\| \le \sup\{\|f'(\xi)\| : t \in (0,1) \quad \text{and} \quad \xi = tx_0 + (1-t)x_1\}\|x_1 - x_0\|.$$

But

$$\begin{aligned}
\sup\|f'(\xi)\| &= \sup\|I - (DF)^{-1}(x_0)DF(\xi)\| \\
&= \sup\|(Df)^{-1}(x_0)(DF(x_0) - DF(\xi))\| = \gamma_1 \le \gamma.
\end{aligned}$$

Thus
(2.1)
$$\eta_1 = \|(DF)^{-1}(x_1)F(x_1)\| = \|H^{-1}(DF)^{-1}(x_0)F(x_1)\| \le \gamma_1 \|H^{-1}\|\, \|x_1 - x_0\|,$$

whence

$$\eta_1 \le \frac{\gamma_1}{1 - \gamma_1}\eta_0 \le \frac{\eta_0}{2}.$$

Assume that $x_2, x_3, \ldots, x_n$ also belong to S. Then

(2.2)
$$\eta_n \le \frac{\gamma_1}{1 - \gamma_1}\frac{\gamma_2}{1 - \gamma_2} \cdots \frac{\gamma_n}{1 - \gamma_n}\eta_0 \le \left(\tfrac{1}{2}\right)^n \eta_0$$

and

$$\|DF(x_n)\| \le \left(\tfrac{4}{3}\right)^n \|DF(x_0)\|.$$

To complete the induction we show that $x_{n+1} \in S$. The triangle inequality gives

$$\|x_{n+1} - x_0\| \le \eta_0 + \eta_1 + \cdots + \eta_n \le \eta_0\left(1 + \tfrac{1}{2} + \tfrac{1}{4} + \cdots + \frac{1}{2^n}\right) \le 2\eta_0.$$

Similarly we have

$$\|x_{n+p+1} - x_n\| \le \eta_n + \eta_{n+1} + \cdots + \eta_{n+p} < 2\eta_n.$$

Since $\{\eta_n\}$ converges to 0, the sequence $\{x_n\}$ is Cauchy with limit, say x^*, and we get claim (ii) above. We also have (iii) since

$$F(x_n) = -DF(x_n)(x_{n+1} - x_n).$$

Thus,

$$\|F(x_n)\| \le \left(\tfrac{2}{3}\right)^n \|DF(x_0)\|\eta_0. \quad \square$$

Lemma 2.1 can be used to prove the quadratic convergence of Newton's method under the stronger assumption of Lipschitz continuity of DF, provided the bound $\gamma(x_0) < \tfrac{1}{3}$ holds. Indeed, if $\|(DF)^{-1}(x)\|$ is bounded on $\{x_n\}$ then there exists a number L such that

$$\gamma_{n+1} \le L\eta_n.$$

By the above Lemma η_n tends to 0. Thus γ_n also tends to 0, and by (2.2) we see that the convergence of η_n is superlinear.

The quantity $\gamma(x_0)$ of Lemma 2.1 depends on all the Newton iterates of x_0 and hence is difficult to bound. For our Kantorovich-type theorem we use a weaker quantity K, which codifies a growth bound in the Newton direction for all points in a region U closed under Newton method iteration. Our main result is as follows.

THEOREM 2.2. *Let F be a map between real Banach spaces X and Y. Assume that F is Fréchet differentiable on X, and that DF is invertible. Suppose there is an open set $U \subseteq X$ such that*

(1) *If $x \in U$ then the Newton step $x + s(x) \in U$.*
(2) *The quantity*

$$(2.3) \qquad K = \sup_{x \in U} \left\{ \frac{\|DF^{-1}(x)(DF(x) - tDF(x + s(x)))\|}{\|DF^{-1}(x)F(x)\|} : 0 \le t \le 1 \right\}$$

is finite

Assume that $x_0 \in U$ satisfies

$$\eta(x_0)K \le \tfrac{1}{3}$$

where $\eta(x) = \|S(x)\|$. Then the Newton iterates $\{x_n\}$ converge to a root x^* of F, and all iterates x_n and x^* lie inside the region

$$S = \{x \in U : \|x - x_0\| < 1.61\eta(x_0)\}.$$

Furthermore, for $n \ge 1$ and $\eta_n = \eta(x_n)$ we have

(i) $\dfrac{\eta_n}{\eta_0} \le \left(\tfrac{2}{3}\right)^n \left(\tfrac{3}{2}h_0\right)^{2^n - 1}$,
(ii) $\dfrac{\|x_n - x^*\|}{\|x_1 - x_0\|} < 1.61\dfrac{\eta_n}{\eta_0}$,
(iii) $\dfrac{\|F(x_n)\|}{\eta_0\|DF(x_0)\|} < 1.61\dfrac{\eta_n}{\eta_0}$,
(iv) $\|(DF)^{-1}(x_n)\| < 1.87\|(DF)^{-1}(x_0)\|$.

PROOF. We will actually prove a stronger result, which depends only on data on the Newtonian trajectory. For any starting point $x_0 \in U$, all the Newton iterates $\{x_n : n \ge 1\}$ are defined and lie in U, and we define

$$(2.4) \quad K(x_0) = \sup_{n \ge 0} \left\{ \frac{\|DF^{-1}(x_n)[DF(x_n) - DF(x_n - ts(x_n))]\|}{\|DF^{-1}(x_n)F(x_n)\|} : 0 \le t \le 1 \right\}.$$

It is clear that $K(x_0) \le K$ for any x_0. We prove that all the conclusions of the theorem hold assuming that

$$\eta(x_0)K(x_0) \le \tfrac{1}{3}.$$

In what follows let $\beta(x) = \|(DF)^{-1}(x)\|$ and $\beta_n = \beta(x_n)$. Also set $h_n = \eta(x_n)K(x_n)$ and since $K(x_n) \le K(x_0)$ one has $h_n \le \eta_n K(x_0)$.

Observe first that if $h_0 = \eta_0 K(x_0) \leq \frac{1}{3}$ then $\|(DF)^{-1}(x_0)(DF(x_0) - DF(\xi))\| \leq \frac{1}{3}$, for $\xi = tx_0 + (1-t)x_1$, $0 \leq t \leq 1$. Thus we may employ the proof of Lemma 2.1 up to the formula (2.1) for η_1. Since γ is bounded by h_0 we get

$$\eta_1 \leq \|H^{-1}\| h_0 \|x_1 - x_0\|.$$

We have $\|H^{-1}\| \leq (1-h_0)^{-1}$, and

$$\eta_1 \leq h_0(1-h_0)^{-1}\|s(x_0)\| = h_0(1-h_0)^{-1}\eta_0.$$

Next we have

$$\beta_1 = \|(DF)^{-1}(x_1)\| \leq \|(DF)^{-1}(x_0)\| \|H(x_0, x_1)\|$$
$$\leq (1-h_0)^{-1}\beta_0,$$

and also

$$h_1 \leq \eta_1 K(x_0) \leq h_0(1-h_0)^{-1}\eta_0 K(x_0) \leq h_0^2(1-h_0)^{-1}.$$

Note that $h_1 \leq \frac{1}{6}$ since $h_0 \leq \frac{1}{3}$.

Now we define $\hat{\eta}_n$, $\hat{\beta}_n$, $\hat{h}_n$ recursively by $\hat{\eta}_0 = \eta_0$, $\hat{\beta}_0 = \beta_0$, and $\hat{h}_0 = \frac{1}{3}$ and the recursion formulae

$$\hat{\eta}_n = \hat{\eta}_{n-1}\hat{h}_{n-1}(1-\hat{h}_{n-1})^{-1},$$
$$\hat{\beta}_n = \hat{\beta}_{n-1}(1-\hat{h}_{n-1})^{-1},$$
$$\hat{h}_n = \hat{h}_{n-1}^2(1-\hat{h}_{n-1})^{-1}.$$

We now prove by induction on n that

(2.5) $$\eta_n \leq \hat{\eta}_n, \qquad \beta_n \leq \hat{\beta}_n, \qquad h_n \leq \hat{h}_n.$$

It is trivially true for $n = 0$, and the inequalities above show that the $n = 0$ case implies (2.5) holds for $n = 1$. These inequalities complete the induction step from n to $n+1$ by taking x_n as x_0, provided we first establish that

$$\eta(x_n)K(x_n) \leq \frac{1}{3}$$

holds. And it does, since $K(x_n) \leq K(x_0)$ and $\eta_n \leq \hat{\eta}_n \leq \hat{\eta}_0 = \eta_0$ by the induction hypothesis. Thus (2.5) is proved.

Now $\{\hat{h}_n\}$ takes the values

$$\frac{1}{3}, \qquad \frac{1}{6}, \qquad \frac{1}{30}, \qquad \frac{1}{870}, \qquad \frac{1}{756030}, \cdots$$

Using

$$\hat{\beta}_{n+1} = \hat{\beta}_0 \prod_0^n \left(\frac{1}{1-\hat{h}_i}\right)$$

we establish the bound

(2.6)
$$\hat{\beta}_{n+1} \leq \hat{\beta}_0 \left(\frac{3 \cdot 6 \cdot 30 \cdot 870 \cdot 756030 \cdots 1}{2 \cdot 5 \cdot 29 \cdot 869 \cdot 756029 \cdots (1-h_n)}\right) \leq 1.82\beta_0 \quad \text{for } n \geq 1.$$

To check this inequality let $\theta = \prod_{i=0}^{3} 1/(1 - \hat{h}_i) < 1.8642$ and $\sigma = \prod_{i=4}^{\infty} 1/(1 - \hat{h}_i)$. Then

$$\log \sigma = -\log(1 - \hat{h}_4) - \log(1 - \hat{h}_5) \cdots < \hat{h}_4 + \hat{h}_5 + \cdots$$

and

$$\hat{h}_{n+3} \leq \frac{\hat{h}_{n+2}^2}{1 - (870)^{-1}}$$

$$= K(x_0)\hat{h}_{n+2}^2 \leq K(x_0)^{2^n - 1}\hat{h}_3^{2^n} < (K(x_0)\hat{h}_3)^{2^n} \quad \text{for } n = 1, 2, 3, \ldots.$$

Here $K(x_0)\hat{h}_3 = \frac{1}{869}$. Then $1 + \hat{h}_4 + \hat{h}_5 \cdots$ is bounded by a geometric series with ratio $(K(x_0)\hat{h}_3)^2$. We find that $\sigma < 1.00000132$.

Next we establish the bound

$$(2.7) \qquad \|DF(x_{n+1})\| < 1.61\|DF(x_0)\| \quad \text{for } n = 1, 2, 3, \ldots.$$

By definition

$$(2.8) \qquad \hat{\eta}_{n+1} = \eta_0 \left(\frac{1}{2} \cdot \frac{1}{5} \cdot \frac{1}{29} \cdot \frac{1}{869} \cdots \frac{\hat{h}_n}{1 - \hat{h}_n} \right).$$

One has

$$\|DF(x_{n+1})\| = \|DF(x_0)\| \prod_{i=0}^{n-1} \|H(x_i, x_{i+1})\|,$$

and (2.4) implies that

$$\|H(x_i, x_{i+1})\| \leq \|I\| + \|DF^{-1}(x_i)(DF(x_i) - DF(x_{i+1}))\|$$
$$\leq 1 + \eta_i K(x_i).$$

Combining this with $\eta_i \leq \hat{\eta}_i$ yields

$$\|DF(x_{n+1})\| \leq \|DF(x_0)\| \prod_{i=0}^{n-1} (1 + \hat{\eta}_i K(x_0))$$

$$\leq \|DF(x_0)\| \prod_{i=0}^{n-1} \left(1 + \frac{1}{3}\frac{\hat{\eta}_i}{\eta_0}\right).$$

Now (2.7) follows from this using (2.8) and numerical calculations similar to that used in proving (2.6).

To check that all $x_n \in S$, we use the triangle inequality and (2.4) to obtain

$$\|x_{n+1} - x_n\| \leq \sum_{i=0}^{n} \|s(x_i)\| \leq \hat{\eta}_0 + \cdots + \hat{\eta}_n.$$

Together with (2.8) one has
(2.9)

$$\|x_{n+1} - x_0\| \leq \eta_0 \left(1 + \frac{1}{2} + \frac{1}{10} + \frac{1}{252010} + \cdots + \prod_{k=1}^{n} \frac{\hat{h}_K}{1 - \hat{h}_K}\right) \leq 1.61\eta_0,$$

the last inequality obtained by estimates like those in (2.6).

Arguing as above starting from x_n,

$$\|x_{n+p+1} - x_n\| \leq \hat{\eta}_n + \hat{\eta}_{n+1} + \cdots + \hat{\eta}_{n+p} < 1.61\hat{\eta}_n.$$

Since $\{\hat{\eta}_n\}$ converges to 0, the sequence $\{x_n\}$ is Cauchy with limit, say x^*. Since $F(x_n) = -DF(x_n)(x_{n+1} - x_n)$, $\{F(x_n)\}$ converges to 0, and $F(x^*) = 0$.

We now estimate the rate of convergence. We have, at worst, that $\hat{h}_n \leq \frac{2}{3}(\frac{3}{2}\hat{h}_0)^{2^n} \leq \frac{2}{3}(\frac{1}{2})^{2^n}$. The recursion formula for $\hat{\eta}_n$ implies that

$$\eta_n \leq \hat{h}_{n-1}\hat{h}_{n-2}\cdots\hat{h}_0\hat{\eta}_0(1 - \hat{h}_{n-1})^{-1}(1 - \hat{h}_{n-2})^{-1}\cdots(1 - \hat{h}_0)^{-1}.$$

Also $\eta_n \leq \hat{\eta}_n \leq (\frac{2}{3})^n(\frac{3}{2}\hat{h}_0)^{2^{n-1}}(\frac{3}{2}\hat{h}_0)^{2n-2}\cdots(\frac{3}{2}\hat{h}_0)\hat{\eta}_0$. Since $\sum_{k=1}^{n} 2^{n-k} = 2^n - 1$,

$$\frac{\eta_n}{\eta_0} \leq (\tfrac{2}{3})^n (\tfrac{3}{2}\hat{h}_0)^{2^n-1},$$

which proves (i). Since $\|F(x_n)\| = \|DF(x_n)\|\eta_n$, while

$$\|DF(x_n)\| < 1.61\|DF(x_0)\|,$$

we obtain the bounds (ii) and (iii). The bound (iv) is obtained from (2.5) and (2.6). □

The self-concordant functions of Nesterov and Nemirovsky [11] are such that a bound like (2.3) can be obtained in a region where $\|s(x)\|$ is sufficiently small. The definition of self-concordance controls the size of F in all directions, however, now just the Newton direction as in the definition (2.3).

Theorem 2.2 captures the notion using the quantity $K(x_0)$ that only the data on the actual Newtonian trajectory are relevant to the behavior of Newton's method. If the difficult-to-obtain quantity $K(x_0)$ in (2.4) or the global quantity K is not known, then we can at least use Theorem 2.2 to show that a given point x_0 is *not* in the region by observing that fast convergence of $\|F(x_n)\|$ and $\|s(x_n)\|$ to zero has not occurred. For values $\eta(x_0)/\varepsilon$ equal to 10^{10}, 10^{20}, 10^{40}, 10^{80}, Theorem 2.2 asserts that at most 6, 7, 8, and 9 Newton iterations, respectively, are required to get $\|F(x_n)\|/\|F(x_0)\| < \varepsilon$. If this inequality is violated, then x_0 is not in the region for such points and a global Newton method is needed; see [5].

Acknowledgment

Thanks are due to the referees and Brad Bell for criticisms of an earlier version.

References

1. D. A. Bayer and J. C. Lagarias, *Karmarkar's linear programming algorithm and Newton's method*, Math. Programming (to appear).

2. ____, *The nonlinear geometry of linear programming* I. *Affine and projective scaling trajectories*, Trans. Amer. Math. Soc. **314** (1989), 499–526.

3. ____, *The nonlinear geometry of linear programming* II. *Legendre transform coordinates and central trajectories*, Trans. Amer. Math. Soc. **314** (1989), 527–581.

4. P. Deuflhard and G. Heindl, *Affine invariant convergence theorems for Newton's method and extensions to related methods*, SIAM J. Numer. Anal. **16** (1979), 1–10.

5. A. A. Goldstein, *A global Newton method* (to appear).

6. W. B. Gragg and R. A. Tapia, *Optimal error bounds for the Newton-Kantorovich theorem*, SIAM J. Numer. Anal. **11** (1974), 10–13.

7. L. M. Graves, *Taylor's theorem in general analysis*, Trans. Amer. Math. Soc. **29** (1927), 163–177.

8. L. V. Kantorovich, *Functional analysis and applied mathematics*, Uspekhi Mat. Nauk **3** 89–185; English transl., Rep. 1509, National Bureau of Standards, Washington, DC, 1952.

9. L. V. Kantorovich and G. P. Akilov, *Functional analysis in normed spaces*, Fizmatgiz, Moscow, 1959; English transl., Pergamon Press, Oxford, 1964.

10. N. Megiddo, *Pathways to the optimal set in linear programming*, Progress in Mathematical Programming (N. Megiddo, ed.), Springer-Verlag, Berlin and New York, 1989, pp. 131–158.

11. Ju. Nesterov and A. S. Nemirovsky, *Self-concordant functions and polynomial-time methods in convex programming*, USSR Academy of Sciences, Moscow, 1989.

12. J. Renegar, *A polynomial-time algorithm, based on Newton's method for linear programming*, Math. Programming **40** (1988), 59–93.

13. J. Renegar and M. Shub, *Simplified complexity analysis for Newton LP methods*, preprint.

14. W. C. Rheinboldt, *On a theorem of S. Smale about Newton's method for analytic mappings*, Appl. Math. Letters **1** (1988), 69–72.

15. S. Smale, *Newton's method estimates from data at one point*, The Merging of Disciplines (R. Ewing, K. Gross, and C. Martin, eds.), Springer-Verlag, Berlin and New York, 1986, pp. 185–196.

16. ____, *Algorithms for solving equations*, Invited Address, Proceedings of the International Congress of Mathematicians, vol. 1, Amer. Math. Soc., Providence, R. I., 1986, pp. 172–195.

DEPARTMENT OF MATHEMATICS, UNIVERSITY OF WASHINGTON, SEATTLE, WASHINGTON 98195

E-mail address: gold@math.washington.edu

- 5 -
Integer Programming
and
Multi-Objective Programming

Contemporary Mathematics
Volume **114**, 1990

An Interior-Point Approach
to NP-complete Problems—Part I

NARENDRA KARMARKAR

ABSTRACT. In this paper, we extend the interior-point approach used in Karmarkar's linear programming algorithm to solve NP-complete problems formulated as quadratic optimization problems over polytopes. We associate a potential function with the quadratic objective function and create a sequence of interior points to minimize the potential function. At each point in the sequence we optimize a local quadratic approximation to the potential function, over a search region that is similar to the search region used in linear programming.

We then define a class of functions more general than convex functions having pathwise connected level sets and prove that functions in this class do not have any spurious local minima (i.e., local minima that are not global minima). We prove that a particular family of potential functions used in our approach belongs to this class of functions.

1. Introduction

To represent NP-complete problems, we study the following integer programming problem:

IP PROBLEM. Find a feasible solution $\underline{x} \in \mathbf{R}^n$ such that

$$A\underline{x} \leq \underline{b} \quad \text{and} \quad x_i \in \{-1, 1\} \quad \forall i = 1, \ldots, n$$

where

$$A: \text{ is an } m \times n \text{ real matrix and } \underline{b} \in \mathbf{R}^m.$$

The more common form of integer programming, which requires each variable x_i to take 0–1 values, can be easily converted to the above form by the substitution $x_i = (1 + x_i')/2$, so that x_i' take ± 1 values.

Consider the following linear programming relaxation of the integer programming problem by allowing each variable x_i to take values in the closed interval $[-1, 1]$.

1980 *Mathematics Subject Classification* (1985 *Revision*). Primary 90C10; Secondary 65K05.

LP PROBLEM. Find $\underline{x} \in \mathbf{R}^n$ such that

$$A\underline{x} \leq \underline{b},$$
$$-1 \leq x_i \leq 1.$$

Solution of the LP relaxation can be useful in solving the IP problem in several ways:

1. If the LP problem is infeasible, then one can infer that the IP problem is also infeasible.
2. In a branch-and-bound type of method for solving IP [4], if the LP relaxation of the reduced problem obtained after partial assignment of ± 1 values to a subset of the variables is infeasible, then that partial solution cannot be extended to a feasible solution of the integer programming problem.
3. If a vertex of the polytope associated with the LP problem happens to have ± 1 coordinates, then it is also a solution to the integer programming problem. Many unwanted fractional solutions (including zero-valued solutions in case of ± 1 formulation) can be ruled out by augmenting the linear program with additional inequalities. An approach based on polyhedral combinatorics seeks to generate several such families of inequalities [9].

Further selectivity in the type of solution to the linear programming problem can be gained by using a quadratic objective function, $\sum_{i=1}^n x_i^2$, which prefers integral solutions over fractional ones, thus making it unnecessary to explicitly rule out fractional solutions with small values of the quadratic objective and achieving economy in the number of inequalities needed to obtain integral solution. In this paper, we elaborate on an interior-point approach based on quadratic objective function. In order to test our approach, we choose problem classes such as finding maximum independent sets in dense random graphs and finding the minimum width of Steiner triple systems, in which fractional solutions to the linear programming relaxation are abundant and it is very hard to find integral solutions. In 0–1 formulations of these problems, most coordinates have values $1/2$ and $1/3$, respectively. For both problems, lower bounds on combinatorial algorithms have been proved in restricted models of computation that grow like e^{kn} and $e^{k'\sqrt{n}}$, respectively [2], [1], [3]. For the latter problem, the model used is more general because it allows linear programming relaxation to be solved at each node of the branch-and-bound method. Hence these problem classes are good test cases to see additional selectivity gained by quadratic objective function using the interior-point approach. We have also tested our approach on the satisfiability problem in mathematical logic. Reports on experimental results on these problem classes can be found in [5], [11], [13].

However, many real-life applications of integer programming are much simpler. Our method has already been applied to developing a global router for VLSI design [10]. The method described in this paper is part of an ongoing research program in interior-point approach. Rather than describing a single algorithm, we have described a general approach for constructing interior-point algorithms for difficult combinatorial problems. We believe that efficient algorithms for many difficult combinatorial problems can be constructed on the basis of a common principle similar to what we have outlined in this paper, and these algorithms will differ mainly in the way additional structure of the particular problem class is exploited to gain further computational efficiency. Detailed description of the algorithm at the level of pseudocode as applied to any specific problem class will be included in the paper describing experimental results on that particular problem class.

Now we relate the following quadratic programming problem to the integer programming problem.

QP PROBLEM. Maximize

$$\sum_{i=1}^{n} x_i^2, \qquad \underline{x} \in \mathbf{R}^n$$

subject to

$$A\underline{x} \leq \underline{b},$$
$$-1 \leq x_i \leq 1.$$

Note that for any feasible solution to the quadratic programming problem, $|x_i| \leq 1 \Rightarrow \sum_{i=1}^{n} x_i^2 \leq n$. Hence the maximum value of objective function is at most n. If x is a feasible solution to IP, then it is a feasible solution to QP and $x_i \in \{-1, 1\} \Rightarrow \sum_{i=1}^{n} x_i^2 = n$; hence, it is an optimal solution to QP with optimal value n.

Conversely any optimal solution to QP with objective value n is a feasible solutions to the IP problem.

The problem transformation given above is by itself of little value since the new problem is also NP-complete and hopelessly difficult computationally, if one were to employ an extreme-point approach, as in the simplex method, to solve it. What makes the problem transformation computationally useful is the recent insight into the geometry of linear programming polytopes gained by interior-point approach such as Karmarkar's algorithm [6]. As a by-product of this approach we discovered a method of making good spherical approximation to a polytope, around any strictly interior point in the following sense.

Given any point in the strict interior of the polytope, there is a projective transformation of the space, so that in the transformed space it is possible to draw two balls, circumscribing and inscribing the polytope, both centered at the image of the given point and having a "small" ratio of radii. In the worst case the ratio can be guaranteed to be no more than n, but in the average case it is even smaller.

Secondly, optimizing a quadratic function over a ball or an ellipsoid is a computationally easy problem, even if the function is not convex [7].

2. Outline of the continuous minimization approach

For a feasible problem, the optimal value of $\sum_i x_i^2$ over the feasible region is n. Using this information, we can obtain the two forms of objective function we have used most often, called "spherical form" and "conical form," both of which have optimal value zero:

$$f(\underline{x}) = \sqrt{n - \sum_i x_i^2} \qquad \text{(spherical form)},$$

$$f(\underline{x}) = \sqrt{n + \delta} - \sqrt{\sum_i x_i^2 + \delta} \qquad \text{(conical form)}.$$

Here δ is some positive constant, used to avoid nondifferentiability of the square-root function when $\sum_i x_i^2 = 0$.

In the case of the projective form of the problem given later, it is easy to make $f(\underline{x})$ homogeneous by using the relation $\sum_i x_i = 1$. For example, the spherical form can be written as

$$f(\underline{x}) = \sqrt{n \cdot \left(\sum_i x_i\right)^2 - \sum_i x_i^2}.$$

If we want to impose equality constraints of the type $A\underline{x} = \underline{b}$ on the solution, their effect can be incorporated into the quadratic objective function $f(\underline{x})$ by changing it to $f(\underline{x}) + \|A\underline{x} - \underline{b}\|^2$. In the case of a feasible problem, the optimal value continues to be zero after this modification. We assume that such a modification has been done in stating the standard problem forms below, although equivalent problem formulations with explicitly imposed equality constraints are also possible.

From the objective function, we construct the associated potential function, just as in the linear programming.

Let $s_i(\underline{x})$, $i = 1, 2, \ldots, N$ denote the slack variables that are linear functions of $\underline{x}$. These include slack variables from upper- and lower-bound constraints $-1 \le x_i \le 1$ as well as slack variables from other inequality constraints $A\underline{x} \le \underline{b}$.

Then the potential function $g(\underline{x})$ associated with an objective function $f(\underline{x})$ is given by

$$g(\underline{x}) = \frac{(f(\underline{x}))^N}{\displaystyle\prod_{i=1}^{N} s_i(\underline{x})}.$$

A more general class of potential functions is based on nonnegative weights $\{w_i\}$:

$$g_w(\underline{x}) = \frac{(f(\underline{x}))^{\left(\sum_i w_i\right)}}{\displaystyle\prod_{i=1}^{N} s_i^{w_i}(\underline{x})}.$$

Suppose the potential function is reexpressed as a function of nonnegatively constrained variables $s_1, s_2, \ldots, s_N$.

Then the simplest projective form of the problem is

minimize $g(s_1, s_2, \ldots, s_N)$

subject to $s_i \geq 0$, $i = 1, \ldots, N$ and $\displaystyle\sum_{i=1}^{N} s_i = 1$

and the affine form is given by

minimize $g(s_1, s_2, \ldots, s_N)$

subject to $s_i \geq 0$, $i = 1, \ldots, N$.

Now we outline the general approach used, which is quite similar to the interior-point algorithms for linear programming:

1. Obtain an initial strictly interior solution to be used as the starting point. For many combinatorial problems, there is usually a trivial way of getting such a starting point. A more general method is to use an interior-point linear programming algorithm to obtain the starting point. In this case the starting point is also the center of the polytope, in the sense of maximizing the product of the distances from hyperplanes defining the polytope.

2. At each major iteration of the algorithm, transform the problem by projective or affine transformation to put the current point at the center, and construct the inscribed ball. Apply the corresponding transformation to the potential function.

3. Construct a Taylor series expansion of the transformed potential function and retain terms up to the quadratic term. (If only expansion up to the linear term is used, the resulting method can be rather easily incorporated into a linear programming system by augmenting it with a subroutine that supplies the gradient of the potential function to be used at each iteration in place of a fixed linear objective function. The resulting method can still be effective on simple real-world problems, although in general it will not perform as well as the quadratic method on harder problems.)

4. Optimize the quadratic approximation over the inscribed ball. In general, it is possible that the new point computed this way is worse (in terms of potential function value) than the current point due to the third- or higher-order terms that we ignored. In such a case the

radius of the search region must be reduced repeatedly until the optimization based on a quadratic approximation improves the current solution or the radius of the search region is smaller than a certain tolerance ε. In the latter situation, we regard the current point as a local minimum. From a theoretical point of view, we consider a point to be local minimum if the gradient of the potential function is zero and all eigenvalues of the hessian are nonnegative. Discussion of the treatment of local minimum is given in a latter section.

Minimization of a local approximation to the function over a search region is analogous to a "trust region" approach. However, there is a major difference. Whereas in a trust region approach, the choice of trust region is based on *local* considerations, in our approach the choice of inscribed ellipsoid as a search region is based on the fact that such a region provides a provably good *global* approximation to the polytope, as shown in the proof of polynomial-time behavior of the interior-point approach to linear programming [6].

5. Assuming that a local minimum is not reached, apply the inverse of the centering transformation to obtain the next point in the sequence. Then round the solution to the nearest ± 1 solution, or apply other rounding techniques described in a later section. If the rounded point satisfies all the constraints, the procedure terminates; otherwise repeat steps 2 to 5, from the new (unrounded) interior point.

3. Analysis of connectivity properties

3.1. Local minima in continuous minimization procedures. Throughout this section, f is a function from an open subset $U \subseteq \mathbf{R}^n$ to $\mathbf{R}$.

DEFINITION. A point $\underline{x} \in U$ is called a strict *local minimum* of the function f if there exists a neighborhood N of $\underline{x}$, $N \subseteq U$, such that $f(\underline{y}) > f(\underline{x})$ for all $\underline{y} \in N$, $\underline{y} \neq \underline{x}$.

It is well known that a first-order method such as the steepest descent method and a second-order method such as Newton's method for function minimization converge to a strict local minimum when started sufficiently close to it, for a $\mathbf{C}^2$ function.

DEFINITION. A point $\underline{x} \in U$ is called a *global minimum* of the function f over U if for all $y \in U$,

$$f(\underline{y}) \geq f(\underline{x}).$$

DEFINITION. A point $\underline{x} \in U$ is called a *spurious local minimum* of the function f over U if it is a strict local minimum of f but not a global minimum of f over U, and the function f has a global minimum over U.

The steepest descent method and Newton's method can converge to a spurious local minimum and thus fail to find a global minimum.

3.2. Importance and limitations of convexity. If the function $f(\underline{x})$ being minimized is convex, then we can rule out the possibility of spurious local minima a priori. This property makes the class of convex functions rather important in the study of function minimization methods and has attracted due attention from many researchers. Convex functions (with further suitable restrictions) can be minimized in polynomial time [12].

This leads to a natural question: Is convexity the characteristic property that separates the class of efficiently solvable minimization problems from the rest? Is there a broader class of functions for which existence of spurious local minima can be ruled out a priori?

Here we give at least two reasons why a broader class is needed:

1. Even if the given function $f(\underline{x})$ is not convex, it may be possible to define a Riemannian metric on the domain U such that for every pair of points in U there is a unique geodesic joining them and the levels sets of f are geodesically convex in the sense that for every pair of points in a level set, the geodesic joining them lies entirely in the same level set.

2. If the given function is convex and we apply a nonlinear transformation to the space that is a diffeomorphism, the transformed function need not be convex, but the transformed problem may still be easy in spite of nonconvexity. In fact, if there is a system of continuous trajectories defined in the original space that allow us to construct an algorithm for the global minimization problem, the image of these trajectories under the transformation gives us a way of solving the transformed problem. This argument suggests that the class of efficiently solvable continuous optimization problems should be larger than the class of convex problems, and the property characterizing the broader class should be topologically invariant.

In the next subsection we identify a class of functions, more general than convex, in terms of the connectivity of the level sets of the function.

In the subsection 3.4, we give a formulation of NP-complete problems as a continuous optimization problem belonging to this class.

3.3. Importance of connectivity.

DEFINITION. A subset $X \subseteq R^n$ is said to be pathwise connected if for every pair of points $\underline{a}, \underline{b} \in X$, there exists a continuous path in X joining the two points; i.e., there exists a continuous function $\underline{g}(t): [0, 1] \to X$ such that $\underline{g}(0) = \underline{a}$ and $\underline{g}(1) = \underline{b}$.

DEFINITION. Let $f(\underline{x})$ be a function from an open set $U \subseteq R^n$ to R. The level set of f corresponding to a level $\alpha \in R$ is denoted by $L_f(\alpha)$ and is defined as

$$L_f(\alpha) = \{\underline{x} \in U | f(\underline{x}) \le \alpha\}.$$

In the following theorem we explain the importance of the class of functions having connected level sets.

THEOREM. *Let $f(\underline{x})$ be a function from an open set $U \subseteq R^n$ to R. Suppose the level sets $L_f(\alpha)$ of f are pathwise connected for all values of α. Then the function f cannot have a spurious local minimum.*

PROOF. If possible let $\underline{x}_1$ be a spurious local minimum of f over U having value $\alpha = f(x_1)$ and let $\underline{x}_2$ be a global minimum. Then the level set $L_f(\alpha)$ contains both $\underline{x}_1$ and $\underline{x}_2$ but cannot be pathwise connected, since we can choose a neighborhood N of $\underline{x}_1$ such that $\underline{x}_2 \notin N$ and for all $\underline{y} \in N$, $f(\underline{y}) > f(\underline{x}_1)$, thus obtaining a contradiction.

3.4. Connectivity properties for NP-complete problems. In case of minimization of convex quadratic functions over a polytope, the level sets are connected and there is a polynomial-time algorithm for solving the problem. In contrast, minimization of a concave quadratic function over a polytope is NP-complete, and the level sets of such a function can have an exponential number of connected components; e.g., let the polytope be defined by

$$-1 \le x_i \le 1 \qquad i = 1, \ldots, n.$$

Let the concave function be

$$f(\underline{x}) = n - \sum_{i=1}^{n} x_i^2$$

and take the value of level α such that $0 < \alpha < 1$. Since each vertex of the polytope is in this level set and is disconnected from all other vertices, the number of connected components is exponential.

However, in our approach we associate a "potential function" with the objective function. The level sets of potential function appear to have far fewer connected components than those of the objective function. The following theorem offers a partial explanation of this phenomenon by showing that for a particular class of potential functions and the homogeneous form of the problem, there is only one connected component. The problem is defined as follows:

PROBLEM. Given a system of homogeneous linear equations $A\underline{x} = 0$, where $A: m \times n$ matrix of real numbers and $\underline{x} \in R^n$, find a solution $\underline{x}$ with each component

$$x_i = +1 \quad \text{or} \quad -1 \qquad i = 1, \ldots, n.$$

Note that there is no loss of generality in assuming the homogeneity. This problem too is NP-complete.

The simplest form of the potential function to which the theorem applies is

$$f(\underline{x}) = \frac{\left[n - \sum_{i=1}^{n} x_i^2 \right]^n}{\prod_{i=1}^{n} (1 - x_i^2)}.$$

A more general form of the potential function to which the theorem applies is defined in terms of a set of weights $\{w_i | i = 1, \ldots, n\}$ such that $w_i \geq 0$ and at least one $w_i > 0$:

$$f(\underline{x}) = \frac{\left[\sum_{i=1}^{n} w_i (1 - x_i^2) \right]^{\left(\sum_{i=1}^{n} w_i \right)}}{\prod_{i=1}^{n} (1 - x_i^2)^{w_i}}.$$

The level set corresponding to a level α is

$$L(\alpha) = \{\underline{x} | A\underline{x} = 0, -1 \leq x_i \leq 1, f(\underline{x}) \leq \alpha\}.$$

THEOREM. *The level set $L(\alpha)$ is pathwise connected for each α.*

PROOF. Let $\underline{x}$ be any feasible point; i.e.,

$$A\underline{x} = 0, \qquad -1 \leq x_i \leq 1.$$

By definition of level sets, $\underline{x} \in L(f(\underline{x}))$. Note that $-\underline{x}$ is also feasible and $f(-\underline{x}) = f(\underline{x})$, hence $-\underline{x} \in L(f(\underline{x}))$. We first show that the entire straight-line segment connecting $\underline{x}$ to $-\underline{x}$ is in $L(f(\underline{x}))$. Let $\underline{u} = \alpha\underline{x}$ be any point on this segment, so that $|\alpha| \leq 1$. Then $A\underline{u} = 0$ and $-1 \leq u_i \leq 1$, hence $\underline{u}$ is feasible. To show that $f(\underline{u}) \leq f(\underline{x})$ we define

$$g(\alpha) = \ln(f(\alpha\underline{x}))$$

$$= \left(\sum_{i=1}^{n} w_i \right) \ln \left[\sum_{i=1}^{n} w_i (1 - \alpha^2 x_i^2) \right] - \sum_{i=1}^{n} w_i \ln(1 - \alpha^2 x_i^2).$$

Hence

$$g'(\alpha) = -\frac{2\alpha \cdot \left[\sum_{i=1}^{n} w_i \right] \left[\sum_{i=1}^{n} w_i x_i^2 \right]}{\sum_{i=1}^{n} w_i (1 - \alpha^2 x_i^2)} + 2\alpha \cdot \sum_{i=1}^{n} \frac{w_i x_i^2}{1 - \alpha^2 x_i^2}.$$

It is easy to rewrite $g'(\alpha)$ in the following form:

$$g'(\alpha) = \frac{1}{\alpha} \cdot \frac{1}{\sum_{i=1}^{n} w_i (1 - \alpha^2 x_i^2)} \sum_{i,j=1}^{n} w_i w_j \left[\sqrt{\frac{1 - \alpha^2 x_i^2}{1 - \alpha^2 x_j^2}} - \sqrt{\frac{1 - \alpha^2 x_j^2}{1 - \alpha^2 x_i^2}} \right]^2.$$

From the expression on the right-hand side it is clear that $g'(\alpha) \geq 0$ when $\alpha > 0$ and $g'(\alpha) \leq 0$ when $\alpha < 0$.

Therefore

$$g(\alpha) \leq g(1) \quad \text{for all } |\alpha| \leq 1.$$

Hence

$$f(\alpha\underline{x}) \leq f(\underline{x}) \quad \text{for } |\alpha| \leq 1.$$

Connectivity of $L(f(\underline{x}))$ follows, since we have shown that each point in the level set can be connected to the origin (which belongs to the level set) by a continuous path, in fact by a straight line.

The above theorem by itself is not enough to allow construction of an algorithm with good theoretical properties, for reasons having to do with the nature of the minimum level set. If $\underline{x}$ is a feasible solution with $x_i = \pm 1$, so is $-\underline{x}$. If there are k such distinct pairs of solutions, denoted by $(\underline{x}^{(1)}, -\underline{x}^{(1)}), (\underline{x}^{(2)}, -\underline{x}^{(2)}), \ldots, (\underline{x}^{(k)} - \underline{x}^{(k)})$, the minimum level set consists of a union of k straight-line segments, where the ith segment is obtained by joining $\underline{x}^{(i)}$ to $-\underline{x}^{(i)}$. If we find, by means of a minimization procedure, any point in the minimum level set different from the origin, it is easy to obtain the corresponding pair of feasible solutions with ± 1 coordinates. But it is difficult to construct a method that is theoretically guaranteed to avoid the origin. Further elaboration of this point is given in Part II of this paper. On the other hand, practical algorithms based on the concepts described in this paper appear to be succeeding, for reasons that are more complex than what we can explain with the current theoretical understanding of the method. In our empirical study, we have worked with problems having multiple connected components. It is not enough to analyze just the number of connected components, but the ratios of their volumes is also important in the practical success of the approach. The above theorem merely provides an example of a situation in which the level sets of the potential have drastically fewer connected components than the level sets of the original objective function.

4. Local minima

In this section, we describe some simple techniques for treating local minima. While many further improvements are possible in these techniques, they are not so critical in the case of problem classes in which the number of local minima encountered is small or zero. Thus many real-world problems falling into this category can already be solved without the need for any sophisticated methods for resolving local minima. As an example, we mention the "global" wire-routing problem in VLSI design [10].

When a local minimum is encountered, we first round the solution to a ± 1 solution by one of these techniques:

1. Round to the nearest ± 1 point; i.e.,

$$
\begin{aligned}
\text{set } x_i &= 1 \quad \text{if } x_i \geq 0; \\
\text{set } x_i &= -1 \quad \text{if } x_i < 0.
\end{aligned}
$$

2. When the starting point maximizes the product of slack variables (which is the case when linear programming is used to obtain the starting point), rounding can be based on coordinate-wise comparison of the current solution point with the starting point.

    Let $\underline{x}^{(0)}$ denote the starting point and $\underline{x}$ the current solution.

Then
$$\begin{aligned}\text{set } x_i &= 1 \quad \text{if } x_i \geq x_i^{(0)};\\ \text{set } x_i &= -1 \quad \text{if } x_i < x_i^{(0)}.\end{aligned}$$

3. By a technique specific to the combinatorial problem solved, e.g., in the maximum independent set problem, round the largest x_i to $+1$; i.e., include the ith node in the independent set, exclude its neighbors from the graph, and repeat.

Let $\underline{w}$ be the rounded solution and suppose it does not satisfy all the constraints. Note that

$$\underline{w}^T \underline{x} = n \quad \text{for } \underline{x} = \underline{w}$$

and

$$\underline{w}^T \underline{x} \leq n - 2$$

for all $\underline{x}$ such that $x_i \in \{-1, 1\}$ and $\underline{w} \neq \underline{x}$. Thus we can exclude $\underline{w}$ by means of the inequality $\underline{w}^T \underline{x} \leq n - 2$ without converting any other feasible solution to the integer programming problem into an infeasible solution. We augment the problem with this inequality and restart the process from the center of the new polytope. Alternatively, it is also possible to make the current solution feasible w.r.t. the new inequality by means of a few steps of an interior-point linear programming algorithm, taking the normal vector to the newly added hyperplane as an objective function. Note that the new inequality does not necessarily eliminate the local minimum in the interior of the polytope, but it removes the point obtained by rounding the local minimum from the feasible region and also modifies the potential function.

REFERENCES

1. D. Avis, *A note on some computationally difficult set covering problems*, Math. Programming **18** (1980), 138–145.
2. V. Chvátal, *Determining the stability number of a graph*, SIAM J. Comput. **6** (1977), 643–662.
3. D. Fulkerson, G. Nemhauser, and L. Trotter, Jr., *Two computational difficult set covering problems that arise in computing the 1-width of incidence matrices of Steiner triple systems*, Math. Programming **2** (1974), 72–81.
4. R. Garfinkel and G. Nemhauser, *Integer programming*, John Wiley & Sons, New York, 1972.
5. N. Karmarkar, K. G. Ramakrishnan, and M. G. C. Resende, *An interior-point approach to the maximum independent set problem in dense random graphs*, in preparation.
6. N. Karmarkar, *A new polynomial-time algorithm for linear programming*, Combinatorica **4** (1984), 373–395.
7. J. J. Moŕe and D. Sorensen, *Computing a trust regions step*, SIAM J. Sci. Statist. Comput. **4** (1983), 553–572.
8. M. Morse and S. Cairns, *Critical point theory in global analysis and differential topology*, Academic Press, 1969.
9. M. Padberg and M. Grötschel, *Polyhedral computations*, E. Lawler, J. K. Lenstra, A. H. G. Rinnooy and D. Kan, Shmoys (eds.), *The traveling salesman problem*, John Wiley & Sons, New York, 1985.
10. R. Pai, N. Karmarkar and S. S. S. P. Rao, *A global router based on Karmarkar's interior-point method*, CSE report, Indian Institute of Technology, Bombay, April, 1988.

11. A. Kamath, N. Karmarkar, K. G. Ramakrishnam, and M. G. C. Resende, *Computational experience with an interior-point approach to the satisfiability problem*, To appear in Ann. Oper. Res., special issue on Computational Methods in Global Optimisation (P. M. Pardalos and J. B. Rosen, eds.), 1990.

12. P. Vaidya, *A new polynomial-time algorithm for minimization of convex functions over convex sets*, in preparation.

13. N. Karmarkar, M. G. C. Resende, and K. G. Ramakrishnan, *An interior-point algorithm to solve computationally difficult set covering problems*, to appear in Math. Programming, special issue on Interior Point Methods, Theory and Practice (C. Roos and J. Ph. Vial, eds.), 1991.

AT&T BELL LABORATORIES, MURRAY HILL, NEW JERSEY 07974

Contemporary Mathematics
Volume **114**, 1990

Solving Matching Problems Using Karmarkar's Algorithm

JOHN E. MITCHELL AND MICHAEL J. TODD

ABSTRACT. We describe a cutting-plane algorithm for solving matching problems. The primal projective standard-form variant of Karmarkar's algorithm for linear programming is applied to the duals of a sequence of linear programming relaxations of the matching problem. Computational performance of the algorithm is described. We indicate why we believe our algorithm holds some long-term promise, although the current implementation is not competitive with simplex-based cutting plane approaches.

1. Introduction

The matching problem is one of the fundamental problems of combinatorial optimization. It was shown to be polynomially solvable by Edmonds [6] in a classic paper in which he introduced and formalized the concept of a good algorithm. Edmonds [5] found a polyhedral description for the matching problem, giving a complete and nonredundant description of the facets of this polyhedron. The number of facets is exponential in the number of vertices of the graph, but Edmonds was able to use the structure of the set of facets to obtain an algorithm that runs in time polynomial in the number of vertices.

Grötschel and Holland [10] developed a cutting-plane algorithm for the perfect matching problem that uses the polyhedral description characterized by Edmonds. Their algorithm employs the simplex method to solve the successive linear programming relaxations that they obtain, and it appeared to

1980 *Mathematics Subject Classification* (1985 *Revision*). Primary 90C05, 05C70.

Computational facilities provided by the Cornell Computational Optimization Project and by the Cornell National Supercomputer Facility. The Cornell National Supercomputer Facility is a resource of the Center for Theory and Simulation in Science and Engineering at Cornell University, which is funded in part by the National Science Foundation, New York State, and the IBM Corporation. This research of both authors was partially supported by the U.S. Army Research Office through the Mathematical Sciences Institute of Cornell University.

Research partially supported by NSF Grants ECS-8602534 and DMS-8706133 and by ONR Contract N00014-87-K-0212.

run (at the time of publication) in time comparable to that taken by the best combinatorial codes. Subsequently, the combinatorial codes have been improved by incorporating some of the time-saving heuristics suggested by Grötschel and Holland (see [3], [4]).

In this report we describe an algorithm for the matching problem. This algorithm is also a cutting-plane algorithm, and it uses and extends several of the heuristics of Grötschel and Holland [10]. However, rather than using the simplex method, it uses Karmarkar's algorithm (see Karmarkar [13]) in deciding how to proceed while working on each successive linear programming relaxation of the original integer programming problem. Karmarkar's method is a polynomial algorithm to solve a single linear programming problem. Our algorithm does not share this theoretical property; we were attracted to using an interior point method in a cutting-plane context by the very favorable results quoted for such methods. (See, for example, [1], [17], [14].)

The performance of Karmarkar's algorithm on linear programming problems differs in several ways from the performance of the simplex method, and many of these differences have to be taken into account when designing a cutting-plane algorithm. It has been noted that interior point methods perform poorly when started from nonoptimal vertices; therefore, we do not solve each relaxation to optimality, but we attempt to add cuts as soon as possible. Because the amount of work required in each iteration is considerably higher for an interior point method than for simplex (with far fewer iterations needed), the cost of the separation routines relative to the cost of an iteration is far lower, so calling the separation routines whenever possible does not drastically increase the cost of an iteration. When one uses the simplex method, the dual simplex algorithm is used to solve a relaxation after cutting planes are added, so primal feasibility is only achieved when optimality is achieved. On the contrary, in our method we always maintain both primal and dual feasibility, so it is possible to call the separation routines whenever the solution to the relaxation is updated.

Since we call the separation routines before solving the current relaxation, we do not require nearly the number of iterations of Karmarkar's algorithm that might be anticipated. In addition, because of the nature of the cuts generated when solving the matching problem, we found that our algorithm usually examined fewer relaxations than a simplex-based cutting-plane algorithm. (See §4.)

The cutting-plane procedure described in this paper can be used to solve other combinatorial optimization problems. In particular, we have also developed an algorithm for solving linear ordering problems. (For a description of a cutting-plane algorithm based on the simplex method for the linear ordering problem, the reader is referred to [11].) This paper is intended to be an extended abstract of [15], [16]. More details of the algorithm, together with more motivation of various stages of it, can be found in those papers.

We now define the matching problem. A *graph* $G = [V, E]$ consists of a finite, nonempty set of *vertices* V together with a set of *edges* E, where each element e of E is defined as an unordered pair of vertices i and j in V—we write $e = ij$; these two vertices are then *adjacent* and are called the endvertices of e. If $G = [V, E]$ is a graph and $W \subseteq V$, the set of edges of the subgraph induced by W is defined to be $E(W) := \{ij \in E : i, j \in W\}$, and the cut induced by W is defined to be $\delta(W) := \{ij \in E : i \in W, j \in V \setminus W\}$. We write $\delta(v)$ for $\delta(\{v\})$, the set of edges incident to vertex v. A set of edges $M \subseteq E$ with no two edges sharing a common endvertex is called a *matching*. Let $n = |V|$. A matching of cardinality $n/2$ is called a *perfect matching*. Let $w: E \to \mathbf{R}$ be a weight function on the edges of the graph G. For a matching M, the weight of M is given by

$$w(M) := \sum_{e \in M} w(e).$$

Then the matching problem is to find a maximum weight matching in the graph G.

2. Formulation as a linear programming problem

We now recall Edmonds's polyhedral description of the matching problem.

Let A be a subset of the edges of the graph $G = [V, E]$. Define $m = |E|$, and consider the components of $\mathbf{R}^m$ indexed by E. The *incidence vector* $y(A)$ of A is the $\{0, 1\}$-vector in $\mathbf{R}^m$ given by

$$y_e(A) = \begin{cases} 1 & \text{if } e \in A; \\ 0 & \text{otherwise}. \end{cases}$$

The *matching polytope* $P_M(G) \subseteq \mathbf{R}^m$ of the graph G is defined to be the convex hull of the set of incidence vectors of matchings of the graph. Thus $P_M(G)$ can be written

$$P_M(G) := \text{conv}\{y(M) \in \mathbf{R}^m : M \subseteq E \text{ is a matching}\}.$$

There is a bijection between the vertices of $P_M(G)$ and the matchings M of the graph G.

It was shown by Edmonds that

THEOREM 1. *For every graph* $G = [V, E]$, $P_M(G)$ *is the solution set of the following system of inequalities*:

(1) $y(\delta(v)) \leq 1$ *for all* $v \in V$,

(2) $y(E(W)) \leq \frac{1}{2}(|W| - 1)$ *for all* $W \subseteq V$, $|W|$ *odd*,

(3) $y(e) \geq 0$ *for all* $e \in E$.

We refer to (1) as *degree* constraints, to (2) as *odd set* constraints, and to (3) as *nonnegativity* constraints. (The odd set constraints are sometimes called blossom constraints.) Thus, the matching problem can be expressed as

$$\max\{w^T y : y \in \mathbf{R}^m, y \text{ satisfies } (1), (2), (3)\}.$$

3. The algorithm

We solve linear programming relaxations of the matching problem using the primal projective standard-form variant of Karmarkar's algorithm, which is due to several authors, including Anstreicher [2], Gonzaga [9], Gay [7], Jensen and Steger [20], and Ye and Kojima [22]. The algorithm is applied to the dual of the relaxation.

Our initial linear programming relaxation of the matching problem is

$$\max_{\text{s.t.}} z$$

$$(D_0) \qquad \begin{aligned} y(\delta(v)) &\leq 1 & \forall v \in V \\ -y_e &\leq 0 & \forall e \in E \end{aligned}$$

$$-w^T y + z \leq 0.$$

Notice that the dual problem to (D_0) is of the appropriate form for the projective standard-form variant of Karmarkar's algorithm. We refer to the current relaxation of the matching problem as the *dual* linear programming problem, and we refer to its dual as the *primal* linear programming problem. Thus, (D_0) is our initial dual problem. As we proceed, we modify (D_0) by adding cutting planes of the form (2).

Our algorithm is outlined in Figure 1. We now describe in detail how the algorithm works.

Box 1. In order to save computational time, our initial linear programming relaxation only involves a subset $\overline{E} \subseteq E$ of the edges. This subset contains the shortest edges in the graph. In the results quoted later, the subset contains the 7 shortest edges adjacent to each vertex in the 60 vertex problems and the 10 shortest edges adjacent to each vertex in the 202 vertex problems.

Box 2. We use a greedy procedure to find a matching M. Let y^I be the incidence vector of M. Then $z^I := w^T y^I$ is a lower bound on the optimal value of

$$\max\{z : z \leq w^T y, y \in P_M(G)\}$$

and also on any relaxation of this linear program. We take (y^I, z^I) to be our initial dual solution. The point y^I is our best integral dual point; y is the best continuous dual point. Every primal constraint contains a surplus variable together with two terms corresponding to potentials on the vertices. Therefore, it is straightforward to find an interior primal solution if the potentials chosen are large enough. We take such a point as our initial primal solution.

Box 3. We use the Todd-Burrell procedure [21] to update the current dual solution (y, z).

Box 4. If $y \neq y^I$ then we round y to find a matching $\overline{M}$ whose value we compare with z^I. Rounding y involves taking all edges with value of at

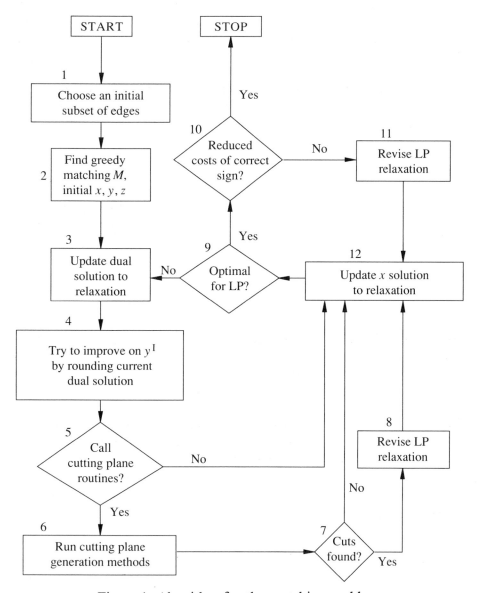

Figure 1. Algorithm for the matching problem.

least 0.5, pairing up remaining vertices randomly, and then using 2-opt to locally optimize the matching. (For a description of 2-opt, see, for example, [19].) Let $y(\overline{M})$ be the incidence vector of $\overline{M}$. Set $z^{\overline{M}} := w^T y(\overline{M})$. If $z^{\overline{M}} > z^I$ then we update y^I to $y(\overline{M})$ and z^I to $z^{\overline{M}}$. If $z^{\overline{M}} > z$ then we set $y = y(\overline{M})$ and $z = z^{\overline{M}}$.

Box 5. We call the cutting-plane routines if $y \neq y^I$.

Box 6. Our separation routines are designed to find cuts that separate y

from $P_M(G)$, and are an extension of those of Grötschel and Holland [10]. We extend their routines by looking for connected components of odd cardinality in several different subgraphs of the original graph. These subgraphs are obtained by deleting all edges with dual value less than some threshold. We did not implement the Padberg-Rao procedure [18], which is guaranteed to find a violated odd set constraint, if one exists. Therefore, our separation heuristics may fail.

Box 7. If cutting planes were found in Box 6 then we go to Box 8; otherwise we go to Box 12.

Box 8. We update our relaxation by adding the constraints found in Box 6. The dual point (y, z) is not feasible after we add the cutting planes, so we update z to z^I and y to y^I.

Box 9. We check whether the duality gap is sufficiently small. If it is then we move to Box 10 and check reduced costs; if it is not then we return to Box 3.

Box 10. Our initial formulation does not use all of the edges of the graph. Therefore, it is necessary to check the reduced costs of the unused edges in order to confirm optimality of the solution. If some edge has negative reduced cost then we move to Box 11. Otherwise we STOP.

Box 11. We choose some subset of the edges with negative reduced cost and add these variables to our formulation. We then construct a Phase I problem and an interior feasible solution to this problem. Each added primal constraint contains a slack variable, so our Phase I problem contains one artificial variable, and then we can obtain an interior feasible solution by choosing the slack variables and the artificial variable appropriately. After revising our linear program, we move to Box 12.

Box 12. The primal solution x is updated. If we come to this box from either Box 5 or Box 7 or Box 11, we have an interior primal feasible point and we use the projective standard-form variant of Karmarkar's algorithm to calculate a new point. If we enter this box from Box 8, the current primal feasible point is not an interior point because the additional variables have value zero. We find a strictly positive feasible point by trying directly to increase these variables in a rescaling of our current relaxation.

We have the following theorem regarding the performance of our algorithm.

THEOREM 2. *The algorithm converges. In addition, provided the separation routines do not fail, the algorithm converges to the optimal solution to the matching problem.*

The theorem holds because there are only a finite number of cutting planes, and the current dual solution (y, z) always satisfies all the cuts added so far, so each cut is added to the formulation at most once.

While Karmarkar's algorithm applied to a single linear program runs in polynomial time, it is not known if our algorithm runs in polynomial time. The reason is that there seems to be no way to obtain a polynomial bound for the number of stages. The analysis of Karmarkar's method is based on a guaranteed decrease in a potential function at each iteration, and the potential functions at different stages appear to be unrelated. (It should be noted that a cutting-plane algorithm based upon the ellipsoid algorithm does run in polynomial time, provided it has a polynomial separation oracle. See Grötschel et al. [12].)

4. Results and summary

In Table 1, we give the results for eight 60-node problems and eight 202-node problems that were solved using our algorithm.

Table 1. Several 60- and 202-node problems

Problem	Optimal value	Iterations	Time	Cuts	Stages	Gap
60i	2.5692	18	23.6	15	3	5
60ii	2.7601	12	12.6	13	1	5
60iii	2.8377	14	12.0	18	3	3
60iv	2.7939	15	19.7	7	2	8
60v	2.2073	10	8.6	2	1	5
60vi	2.5609	21	31.9	22	3	7
60vii	2.7714	31	47.6	29	5	5
60viii	2.6279	36†	31.4	21	4	7
202i	4.7793	39	267.8	54	6	6
202ii	4.8593	35	278.1	69	7	5
202iii	4.6887	24	162.2	41	4	6
202iv	4.4972	74	536.4	66	10	4
202v	4.6585	27	198.3	49	3	7
202vi	4.4679	28†	180.4	30	2	5
202vii	4.4000	31*	253.3	65	6	5
202viii	4.7317	55†	427.7	57	7	6

* This run finished with a nonintegral solution, i.e., the separation heuristics failed.

† This run required the addition of one extra edge that had reduced cost of the wrong sign, so an additional linear programming relaxation of the original problem was generated. Therefore, for this run, Box 11 was visited once.

The columns of Table 1 contain

- Optimal value: the optimal value of the perfect matching problem.

- Iterations: the number of iterations of Karmarkar's algorithm needed to solve the problem.
- Time: the run time in seconds.
- Cuts: the number of cutting planes added.
- Stages: the number of stages of adding cutting planes, i.e., the number of linear programming relaxations generated during the algorithm.
- Gap: the number of iterations after first adding cuts in order to decide whether more cuts may be necessary. This can be expressed as the number of iterations of Karmarkar's algorithm between the time cutting planes are first added and either (a) the next time the separation routines are called or (b) the time the problem is solved, whichever is less.

These problems were generated as follows:

- Generate k points uniformly in the unit square.
- Look at the complete graph on those vertices.
- Take the Euclidean distance between two vertices as the length of the corresponding edge.

We solved the perfect matching problem on these graphs, which involves finding the perfect matching of minimum weight. The perfect matching problem can be transformed to a matching problem by adjusting the edge weights appropriately. These runs were made on an IBM 3090, and the code was written in double precision FORTRAN. It should be noted that the run times quoted are only accurate to within approximately 10%. For more detailed results, the reader is referred to [16].

We believe that these results give some justification for optimism regarding the performance of our algorithm. This optimism is based on consideration of both the number of iterations and the number of stages of adding cutting planes, that is, the number of linear programming relaxations generated during the algorithm.

Using Karmarkar's algorithm to solve the final linear programming relaxation starting from scratch would take about half as many iterations as the number reported in the table. This is because we call the separation routines before optimality is reached, so the sequence of iterates does not approach the vertices of the polytope too closely, and thus the number of iterations in each stage is reasonably small.

The number of stages of adding odd set constraints using our algorithm is smaller than they would be if a cutting plane algorithm based on the simplex method were used. This is for at least two reasons: First, because we are trying to cut off an interior point, we are able to find more violated inequalities and cut off more of the polytope at each stage. Second, for many problems, the odd set constraints found at one stage are strongly related to the constraints found at subsequent stages. It is often necessary to add constraints that correspond to nested odd sets. Our separation heuristics are able to exploit this relationship because we do not solve the current relaxation to optimality.

It was not our intention to produce an algorithm that would be instantly competitive with combinatorial approaches to solving the matching problem; we were merely interested in investigating the feasibility of a cutting-plane approach based on an interior point method. Therefore, we concentrated on producing an implementation that was numerically stable and robust. We calculate projections using a method based on a QR-factorization. (See, for example, [8].) This is a very numerically stable method, but it is not fast. In addition, we refactorize the constraint matrix whenever we add constraints or variables, rather than using information from the previous factorization. For these two reasons our run times are not good.

REFERENCES

1. I. Adler, M. C. G. Resende, G. Veiga, and N. K. Karmarkar, *An implementation of Karmarkar's algorithm for linear programming*, Math Programming **44** (1989), 297–335.
2. K. M. Anstreicher, *A monotonic projection algorithm for fractional linear programming*, Algorithmica **1** (1986), 483–498.
3. U. Derigs, *Solving large-scale matching problems efficiently—A new primal approach*, Networks **16** (1986), 1–16.
4. U. Derigs and A. Metz, *On the use of optimal fractional matchings for solving the (integer) matching problem*, Computing **36** (1986), 263–270.
5. J. Edmonds, *Maximum matching and a polyhedron with* 0, 1 *vertices*, J. of Res. National Bureau of Standards **69B** (1965), 125–130.
6. ____, *Paths, trees, and flowers*, Canadian J. Math. **17** (1965), 449–467.
7. D. M. Gay, *A variant of Karmarkar's linear programming algorithm for problems in standard form*, Math. Programming **37** (1987), 81–90.
8. G. H. Golub and C. F. Van Loan, *Matrix Computations*, Johns Hopkins University Press, Baltimore, MD, 1983.
9. C. C. Gonzaga, *A conical projection algorithm for linear programming*, Math. Programming **43** (1989), 151–173.
10. M. Grötschel and O. Holland, *Solving matching problems with linear programming*, Math. Programming **33** (1985), 243–259.
11. M. Grötschel, M. Jünger, and G. Reinelt, *A cutting plane algorithm for the linear ordering problem*, Oper. Res. **32** (1984), 1195–1220.
12. M. Grötschel, L. Lovasz, and A. Schrijver, *The ellipsoid method and its consequences in combinatorial optimization*, Combinatorica **1** (1981), 169–197.
13. N. K. Karmarkar, *A new polynomial-time algorithm for linear programming*, Combinatorica **4** (1984), 373–395.
14. K. A. McShane, C. L. Monma, and D. Shanno, *An implementation of a primal-dual interior point method for linear programming*, ORSA Journal on Computing **1**(2) (1989), 70–83.
15. J. E. Mitchell and M. J. Todd, *Solving combinatorial optimization problems using Karmarkar's algorithm: Part* I: *Theory*, technical report, Mathematical Sciences, Rensselaer Polytechnic Institute, Troy, N.Y., 1989.
16. ____, *Solving combinatorial optimization problems using Karmarkar's algorithm: Part* II: *Computational results*, technical report, Mathematical Sciences, Rensselaer Polytechnic Institute, Troy, N.Y., 1989.
17. C. L. Monma and A. J. Morton, *Computational experience with a dual affine variant of Karmarkar's method for linear programming*, Oper. Res. Lett. **6**(6), (1987), 261–267.
18. M. W. Padberg and M. R. Rao, *Odd minimum cut-sets and b-matchings*, Math. of Oper. Res. **7** (1982), 67–80.
19. C. H. Papadimitriou and K. Steiglitz, *Combinatorial optimization: Algorithms and complexity*, Prentice-Hall, Englewood Cliffs, NJ, 1982.

20. A. E. Steger, *An extension of Karmarkar's algorithm for bounded linear programming problems*, Master's thesis, State University of New York at Stony Brook, NY, 1985.
21. M. J. Todd and B. P. Burrell, *An extension of Karmarkar's algorithm for linear programming using dual variables*, Algorithmica **1** (1986), 409–424.
22. Y. Ye and M. Kojima, *Recovering optimal dual solutions in Karmarkar's polynomial algorithm for linear programming*, Math. Programming **39(3)** (1987), 305–317.

DEPARTMENT OF MATHEMATICAL SCIENCES, RENSSELAER POLYTECHNIC INSTITUTE, TROY, NEW YORK 12180
E-mail address: Mitchell@turing.cs.rpi.edu

SCHOOL OF OPERATIONS RESEARCH AND INDUSTRIAL ENGINEERING, CORNELL UNIVERSITY, ITHACA, NEW YORK 14853
E-mail address: miketodd@cs.cornell.edu

Contemporary Mathematics
Volume **114**, 1990

Efficient Faces of Polytopes: Interior Point Algorithms, Parameterization of Algebraic Varieties, and Multiple Objective Optimization

S. S. ABHYANKAR, T. L. MORIN, AND T. TRAFALIS

ABSTRACT. This paper addresses the problem of computing the set of efficient faces of a bounded polyhedron in $\mathbf{R}^n$ that is defined by linear inequalities. It describes two algorithms. One algorithm is an interior point method that generalizes and extends the affine variants of Karmarkar's algorithm to multiple objective optimization. It finds an efficient (maximal) face in polynomial time. The other algorithm is based on techniques of algebraic geometry related to the parameterization of algebraic varieties in n-dimensional spaces. It approximates a portion of the set of efficient faces by an algebraic surface.

1. Introduction

Recent breakthroughs in linear programming (LP) have revolutionized the field. While it had long been known (Klee and Minty (1972)) that the simplex algorithm could require an exponential number of pivots, the computational complexity of LP remained an open question until 1979 when Khachian (1979) showed that LPs could be solved in polynomial time using the ellipsoid algorithm. More recently, Karmarkar (1984) and others (Adler, Resende, and Veiga (1986), Marsten, et al. (1988), McShane et al. (1988)) have demonstrated that interior point methods can be computationally effective.

Khachian and Karmarkar demonstrated that it is computationally more efficient to work with an algebraic hypersurface, e.g., an ellipsoid that approximates the convex polytope, than to work with the polytope itself, which may have an exponential number of extreme points. In multiple objective

1980 *Mathematics Subject Classification* (1985 *Revision*). Primary 90B50, 90C05, 52A25, 14A10.

This research was supported in part by a grant from the Office of Naval Research under University Research Initiative grant number N00014-86-K-0689, NSF grant DMS-88-16286, ONR grant N00014-88-K0402, and ARO contract DAAG 29-85-C0018 under Cornell MSI, at Purdue.

LP and nonlinear programming (NLP) problems one seeks the set of efficient faces, also called *Pareto-optimal faces* or nondominated faces (Benson and Morin (1977) and Yu (1974, 1985)) of a convex polytope. Since there can be an exponential number of efficient faces, this too is a natural setting to use interior point algorithms and algorithmic algebraic geometry (Abhyankar (1986)).

We apply interior methods to both linear and nonlinear multiple objective programming where the decision space is a convex polyhedron defined by linear inequalities. First, we develop an interior point algorithm that is a generalization of Renegar's algorithm (Renegar (1986)) to the multiple objective setting. The algorithm finds a point on the efficient frontier of a polytope, and hence an efficient (maximal) face, in polynomial time. The approach is geometrical and is a method of centers applied in multiple objective optimization problems. Second, we develop an algorithm that approximates the set of maximal faces (the efficient frontier or Pareto optimal set) of a polytope, with respect to an ordering given by a closed polyhedral cone, by constructing an increasing series of inscribed algebraic varieties. To do this we construct parameterizations of ellipsoids in n-space and investigate the existence and construction of parameterizations of quadrics without singularities (Abhyankar (1986)).

This paper is organized as follows. §1.1 discusses multiple objective optimization problems; §1.2 reviews the literature in MOLP. In §2.1 the MOLP problem is defined and notation developed. The analytical center of a polytope is discussed in §2.2, and an algorithm of centers for finding an efficient (maximal) face is developed in §2.3. A proof of convergence of the algorithm is presented in §2.4. The problem of finding the entire set of all efficient faces of a polytope is discussed in §3.1, and parameterizations of algebraic surfaces in n-dimensions and rational parameterizations of a general quadric hypersurface are presented in §3.2. In §3.3 a method of approximating the maximal facets of a polytope by algebraic surfaces is developed and illustrated by an example in §3.4. The parameterization of the set of efficient faces (the efficient frontier) of a polytope is discussed in §3.5. An example illustrating the connection of the algorithm of centers with dynamical systems is presented in §3.6. Applications in vector optimization are considered in §3.7.

1.1. The multiobjective optimization problem. Let X be a set of decisions, Y be the set of objectives, and assume that a function $f\colon X \to Y$ is given. The image set $f(X) = Q$ is called the set of *attainable objectives*.

We assume that Y is partially ordered by $<$. The partial order $<$ induces a partial order $<_f$ on the set X as follows:

$$x <_f x' \quad \text{iff} \quad f(x) < f(x').$$

MULTIOBJECTIVE OPTIMIZATION PROBLEM. *Compute the set of all maximal elements of X with respect to $<_f$.*

If X is an n-dimensional real Euclidean space $\mathbf{E}^n$, Y is an m-dimensional space $\mathbf{E}^m$, and the partial order on $\mathbf{E}^m$ is defined by

$$x < x' \quad \text{iff} \quad x_i \leq x_i', \qquad \text{for all } i = 1, \ldots, m,$$

with strict inequality for at least one i (componentwise strict partial order), then the maximal points of X under $<_f$ are called *efficient* (nondominated, Pareto-optimal) points.

The sets X and Y can also be infinite dimensional. However, for our purposes we shall assume that X and Y are subsets of finite dimensional Euclidean spaces. We will also assume that the partial order in Y is defined by a positive cone Λ as follows:

$$y \geq z \quad \text{iff} \quad y - z \in \Lambda \quad \text{and} \quad y > z \quad \text{iff} \quad y - z \in \Lambda - \{0\}.$$

The cone, therefore, describes the preferences of a decision maker. An example of a *canonical order* in $\mathbf{R}^p$ is given by the nonnegative orthant

$$\Lambda = \{y \in \mathbf{R}^p | y_1 \geq 0, \ldots, y_p \geq 0\}.$$

The *Pareto* maximal objective $\hat{y} = f(\hat{x}) \in Q$, the set $\hat{Q}$ of all *Pareto* maximal objectives, and the set of all maximal decisions $E(X, f, \Lambda)$ are then defined as follows:

$$\hat{Q} = \{\hat{y} \in Q | (\hat{y} + \Lambda) \cap Q = \{\hat{y}\}\}$$

and

$$E(X, f, \Lambda) = \{\hat{x} \mid f(\hat{x}) \text{ for some } \hat{y} \in \hat{Q}\}.$$

In the literature of multiple objective optimization the *Pareto maximal decisions* are called *efficient points*. Figure 1 shows an example in $\mathbf{R}^2$. In $\mathbf{R}^2$ with the usual componentwise vector ordering $\leq$,

$$\Lambda = \text{first quadrant}.$$

Then for any $x = (x_1, x_2) \in \mathbf{R}^2$, $x + \Lambda$ is x plus the set of all the points that dominate x. So if ABC is the polyhedron (polygon) of interest, then AC is the efficient face (edge).

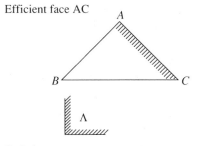

Efficient face AC

Ordering cone

FIGURE 1. Efficient face in $\mathbf{R}^2$

A particular case of interest is when X is a polytope in $\mathbf{R}^n$ and f is linear. This is called the *multiple objective linear program* (MOLP). Then Q is also a polytope, and $\hat{Q}$ can be described as a set of faces of the polytope Q. Note that Λ induces an ordering on $\mathbf{R}^n$ defined by the convex cone

$$\Lambda^{\#} = \{x \in \mathbf{R}^n | f(x) \in \Lambda\}.$$

It follows at once that $E(X, f, \Lambda) = E(X, I, \Lambda^{\#})$, where I is the identity mapping on X. Therefore, we will consider the following more general problem.

PROBLEM. *Given a polytope defined by linear inequalities, find the set of efficient faces with respect to an ordering cone Λ.*

This is a problem in the area of computational geometry (Preparata and Shamos (1985)). We seek efficient *faces* since this may include other than $(n-1)$-dimensional faces, i.e., facets. We use the adjective *efficient* to avoid confusion with the definition of a facet as a *maximal* face relative to set inclusion.

2. Simplex-based techniques in MOLP

The usual approach of solving the MOLP employs simplex-based methods (Yu and Zeleny (1975)). It is well known (Yu (1985)) that the set of efficient points of a polytope X is either the entire set of feasible solutions X or a part of or the whole boundary of X. If all the objective functions are linearly independent, then the set of efficient points lies on the boundary of X. Since X is a polyhedron, its boundary consists of faces (facets, edges, and points) that can be characterized by extreme points or extreme rays. If an interior point of a face or edge is efficient, then the whole face or edge, including its extreme points and edges, must also be efficient. The efficient set is connected (Yu (1985)). So for any points in the efficient set, one can move from one point to the other without leaving the set (efficient frontier).

The above results suggest a simplex-based strategy for generating the efficient frontier of X. Since a face is completely described by its extreme points, the primary goal of simplex-based methods is the determination of the efficient extreme points. The set of efficient extreme points can be constructed iteratively by beginning at an efficient extreme point and moving along one of the efficient edges emanating from it. The entire set of efficient extreme points is then constructed in a finite (but not necessarily polynomial) number of iterations. Once all the efficient extreme points have been generated, then one constructs all efficient (maximal) faces from the given set of efficient extreme points.

A variety of methods have been proposed to carry out the steps in this approach. For the problem of finding an initial efficient extreme point, most methods solve a linear programming problem. For more details see Ecker and Kouada (1975), Evans and Steuer (1973), and Zeleny (1974). There are several schemes for exploring adjacent efficient extreme points or edges. One approach proposed by Zeleny (1982), Yu and Zeleny (1975), Phillip (1972), and Evans and Steuer (1973) is based on checking all extreme points adjacent to the current efficient point for efficiency. For each adjacent extreme point, an LP problem is solved. Zeleny (1974) and Yu and Zeleny (1975) propose a simplex-based approach in which adjacent extreme points that need not be examined for efficiency can be avoided. Another approach (Gal (1976), Ecker and Kouada (1978)) is based on a weighting characterization. In this approach two vectors x_0 and x_1 are adjacent efficient extreme points iff (i) they are adjacent extreme points and (ii) they are both optimal solutions of the same scalarized linear programming problem $P(w)$ for some strictly positive vector w. The efficient faces may be identified by constructing appropriate undirected graphs or binary interaction matrices (Sage (1977)). Applications of this approach can be found in Zeleny (1974), Yu and Zeleny (1975), and Iserman (1977, 1979). Ideas about bookkeeping can be found in Gal (1976) and Ecker et al. (1980).

Extensive computer effort is needed to implement the above procedure, arising from the need to identify and store adjacent extreme points. Most applications require at least 150–800 constraints, 200–400 variables, and 2–6 objectives. Problems arise from the fact that we can have an exponential number of efficient extreme points. For example, Zeleny (1974, pp. 116–121) discussed two examples with 16 decision variables and 8 constraints. Both problems can have as many as 12,870 extreme points. Currently, multiple objective simplex methods cannot be successfully applied to large or even moderately sized problems.

Other approaches for MOLP include the *interval weights method* (Steuer (1976)) and the *noninferior set estimation method* (Cohon (1978)).

2.1. Interior point approach for MOLP. In this section we describe one algorithmic approach for the MOLP problem. We begin by first presenting some necessary notation and the statement of the problem.

Let P be a full-dimensional bounded polytope in $\mathbf{R}^n$ described by linear inequalities, i.e.,

$$P = \bigcap_{i=1}^{m} H_i, \quad \text{where } H_i = \{x \mid a_i^T x \geq b_i\},$$

and $\mathbf{R}^n$ is ordered by a constant closed polyhedral cone Λ. Recall that a set Λ is a cone if $a\Lambda = \Lambda$ for any $a > 0$. A polyhedral cone is a cone that is also a polyhedron. Thus, if Λ is a polyhedral cone then it can be represented by

$$\Lambda = \{x \mid \Lambda x \geq 0\},$$

where Λ is a matrix of proper dimension. From now on we will use the same notation for both the polyhedral cone and its defining matrix. Note also that we use P to denote a polytope, not the projective plane $\mathbf{P}$. Consider the following problem:

PROBLEM. *Find the set of points* $\widehat{P}$ *where*

$$\widehat{P} = \{\hat{x} : (\hat{x} + \Lambda) \cap P = \{\hat{x}\}\}.$$

This set is called the *efficient frontier* of P.

DEFINITION 1. A subset F of P is called a *face* of P if $F = P$, $\varnothing$ or if F is the intersection of P with a supporting hyperplane of P. A *facet* of P is an $(n - 1)$-dimensional face of P.

DEFINITION 2. A face F of P is called *efficient* (*maximal*) iff F is a subset of the efficient frontier of P.

REMARK. We will examine two aspects of the above problem.

  (i) Finding a point of $\widehat{P}$. This yields a point on the efficient frontier of P that can be used for the initialization process of the simplex-based techniques in MOLP reviewed in §1.2. The concept of an analytical center of a bounded polyhedron (Sonnevend (1985), Bayer and Lagarias (1989)) is fundamental to our approach.
  (ii) Finding a global approximation of $\widehat{P}$. Here we use techniques of algorithmic algebraic geometry.

2.2. A logarithmic barrier function and the center of a polytope. Consider a polytope P described by a finite number of inequalities

$$a_i^T x \geq b_i, \qquad i = 1, \ldots, m.$$

Let $H_i = \{x | a_i^T x \geq b_i\}$ and consider the Euclidean distance of x from the hyperplane H_i

$$d(x, H_i) = |a_i|^{-1} |a_i^T x - b_i|.$$

Define the logarithmic barrier function $f_P : \text{int}(P) \rightarrow \mathbf{R}$, where P is the polytope defined by the given linear inequalities, by

$$f_P(x) = \sum_{i=1}^{m} \ln d(x, H_i).$$

Our objective is to maximize the function f_p. The maximizer of this function will be a candidate for a center of the given polytope P. When P is bounded it is a unique point called the (analytical) center (Sonnevend (1985)).

An interesting property of the analytical center is that it does not depend on the particular ℓ^p metric for finite dimensional spaces.

PROPOSITION 1. *Let* $\| \ \|_p$ *be the usual* ℓ_p *norm in* $\mathbf{R}^n$. *We assume that* $1 < p < \infty$. *Then the defining equation for the center is invariant with respect to the particular* $\| \ \|_p$ *norm.*

PROOF. Consider the following finite optimization problem:

$$\min \sum_{j=1}^{n} |x_j - y_j|^p$$

subject to $a^T y = b$.

Since $\|\ \|_p^p$ is a convex function the problem has a solution y^*. Then y^* is characterized by the Lagrange multiplier conditions

$$-p|x_j - y_j^*|^{p-2}(x_j - y_j^*) + \lambda a_j = 0.$$

Hence

(1)
$$p|x_j - y_j^*|^{p-2}(x_j - y_j^*) = \lambda a_j$$

and therefore

$$|x_j - y_j^*|^p = (a_j x_j - a_j y_j^*)\frac{\lambda}{p}.$$

Summing both sides of the above equation from $j = 1$ to n and taking absolute values we have

$$\|x - y^*\|_p^p = |a^T x - b|\frac{|\lambda|}{p}.$$

Taking logarithms of both sides yields

$$p \ln \|x - y^*\|_p = \ln |a^T x - b| + \ln \frac{|\lambda|}{p}.$$

Hence, for the given bounded polytope P,

(2)
$$\sum_{i=1}^m \ln d(x, H_i)_p = \left(\frac{1}{p}\right) \sum_{i=1}^m \ln |a_i^T x - b_i| + \left(\frac{1}{p}\right) \sum_{i=1}^m \ln \frac{|\lambda_i|}{p}$$

where $d(x, H_i)_p$ describes the distance of x from the hyperplane H_i with respect to the ℓ^p metric in $\mathbf{R}^n$, and λ_i is the Lagrange multiplier of the ith constraint. By Equation 1 we have

$$p\frac{|x_j - y_j^*|^p}{(x_j - y_j^*)} = \lambda a_j.$$

Squaring both sides yields

$$p^2 \frac{(x_j - y_j^*)^{2p}}{(x_j - y_j^*)^2} = \lambda^2 a_j^2.$$

Hence

$$p^2 (x_j - y_j^*)^{2p-2} = \lambda^2 a_j^2,$$

$$(x_j - y_j^*)^{2p-2} = \frac{\lambda^2}{p^2} a_j^2,$$

$$(x_j - y_j^*) = \left(\frac{\lambda^2}{p^2}\right)^{1/(2p-2)} a_j^{2/(2p-2)},$$

$$(a_j x_j - a_j y_j^*) = \left(\frac{\lambda^2}{p^2}\right)^{1/(2p-2)} a_j^{p/(p-1)}.$$

Summing from $j = 1$ to n and taking absolute values we have

$$|a_i^T x - b| = \left(\frac{|\lambda|^2}{p^2}\right)^{1/(2p-2)} \left|\sum_{j=1}^m a_j^{p/(p-1)}\right|,$$

$$\frac{|a_i^T x - b|^{p-1}}{|\sum_{j=1}^n a_j^{p/(p-1)}|^{p-1}} = \frac{|\lambda|}{p}.$$

Taking logarithms of both sides gives

$$(3) \qquad (p-1)\left[\ln(|a_i^T x - b|) - \ln\left(\left|\sum_{j=1}^n a_j^{p/(p-1)}\right|\right)\right] = \ln\left(\frac{|\lambda|}{p}\right).$$

By (3) and (2) and upon differentiation we obtain

$$\sum_{i=1}^m \frac{a_i^T}{a_i^T x - b_i} = 0.$$

The above equation characterizes the analytical center of P. Hence, the defining equation of the analytical center does not depend on the particular ℓ^p metric, $1 < p < \infty$. $\quad\square$

2.3. Algorithm of centers for finding an efficient face. In this section we describe briefly the method of centers (Huard (1967)) applied to vector optimization. A similar approach was followed by Renegar (1986) in the single objective linear case. It results in a polynomial algorithm for linear vector optimization (Morin and Trafalis (1989)).

Let x^k be an interior point of P and consider the intersection P_k of $x^k + \Lambda$ and P. Next find the center x^{k+1} of P_k and start again with x^{k+1} instead of x^k. That is, take a sequence of points $\{x^k\}$ that is in the general case infinite. We will show that this sequence converges to a *Pareto* optimal solution. More formally,

ALGORITHM (i) [*Method of centers*].
0. *Set* $k = 0$.
1. *Let* x^k *be the interior point of* P. *Consider the intersection* P_k *of* $x^k + \Lambda$ *and* P. *Find the center* x^{k+1} *of* P_k.
2. If $|x^{k+1} - x^k| < \varepsilon$, **then** *stop.* **Else** *return to* 1, *with* $k \leftarrow k + 1$.

It can be shown (Bayer and Lagarias (1989)) that the above sequence of points lies on a real algebraic curve.

In order to find the center of a set we generally have to solve a nonlinear programming problem. If a polyhedron is defined by a set of linear inequalities then methods of computing its center exists (Vaidya (1987)).

POTENTIAL FUNCTION. Let P be the given polyhedron (which we assume to be *full dimensional and bounded*), Λ be the ordering cone, and x^k be a point in $\text{int}(P)$. Define

$$P_k = \{x^k + \Lambda\} \cap P,$$

where

$$P = \bigcap_{i=1}^{m} H_i,$$

$$H_i = \{x | a_i^T x \geq b_i\},$$

and

$$\Lambda = \{x | \Lambda_s^T x \geq 0, \ s = 1, \ldots, p\}.$$

It follows at once that P_k is a bounded polytope. Next define the potential function

$$f_{P_k} : \text{int}(P_k) \rightarrow \mathbf{R},$$

such that

$$f_{P_k}(x) = \sum_{i=1}^{m} \ln(a_i^T x - b_i) + \sum_{s=1}^{p} \ln(\Lambda_s^T x - \Lambda_s^T x^k).$$

Our objective is to maximize $f_{P_k}(x)$. It also is easy to prove (Bayer and Lagarias (1989)) that

PROPOSITION 2. f_{P_k} is strictly concave on $\text{int}(P_k)$.

Hence, it has a unique maximizer x_{k+1}, and x_{k+1} is a solution of

$$df_{P_k}(x) = 0,$$

or

$$\sum_{i=1}^{m} \frac{a_i^T}{a_i^T x - b_i} + \sum_{s=1}^{p} \frac{\Lambda_s^T}{\Lambda_s^T x - \Lambda_s^T x^k} = 0.$$

This is a nonlinear system of equations that can be solved by Newton's method (Sonnevend (1985), Vaidya (1987)).

2.4. Convergence of the algorithm. Next we discuss convergence of the algorithm developed in §2.3. We will use the following notation:

$$\Lambda(\lambda) = \{x | \Lambda x \geq \Lambda \lambda\},$$
$$P(\lambda) = P \cap \Lambda(\lambda),$$
$$I(\lambda) = \text{int}(P(\lambda)).$$

We will also use the following lemma for the proof of Theorem 2.

LEMMA 1 (Trafalis (1989)). *Let λ^* be an efficient point, λ an interior point in P, and ζ be the analytical center of $P(\lambda)$. Then*

$$\frac{1}{p+m} \|\Lambda_s^T(\lambda^* - \lambda)\| \leq \|\Lambda_s^T(\zeta - \lambda)\| \leq \|\Lambda_s^T\| \cdot \|\zeta - \lambda\|$$

$$\Rightarrow \|\zeta - \lambda\| \geq \frac{1}{(p+m)\|\Lambda_s\|} \|\Lambda_s^T(\lambda^* - \lambda)\|.$$

PROOF. See Trafalis (1989). □

We will make use of the following convergence theorem for multiple objective optimization.

THEOREM 1 (Pascoletti and Serafini (1982)). *Let X be any subset of P containing the set of efficient points. Suppose*

(a) *there exists a continuous function $\varepsilon \colon \mathbf{R}^n \to \mathbf{R}$ such that $\varepsilon(x) > 0$ for $x \notin X$,*

(b) *$\{x^k\}$ is contained in P,*

(c) *$x^{k+1} - x^k \in \Lambda$,*

(d) *$\|x^{k+1} - x^k\| \geq \varepsilon(x_k)$.*

Then every subsequence of x^k, $k = 1, \ldots$ converges to some point $x \in X$.

PROOF. See Theorem 2.2 (Pascoletti and Serafini (1982)). □

THEOREM 2. *Every subsequence of the sequence x^k, $k = 1, \ldots$ constructed by algorithm* (i) *converges to an efficient point.*

PROOF. Simply invoke Lemma 1 and Theorem 1. That is, select ε as follows:

$$\varepsilon \colon \mathbf{R}^n \to \mathbf{R} \quad \text{with } \varepsilon(\lambda) = \frac{1}{(p + m) \max_{1 \leq s \leq p} \|\Lambda_s^T\|} \cdot \min_{1 \leq s \leq p} \|\Lambda_s^T (\lambda^* - \lambda)\|.$$

We also consider X to be the set of efficient points. Therefore, invoking Lemma 1 and Theorem 1 yields the desired result. □

PROPOSITION 3. *x^k is an increasing sequence with respect to the cone ordering.*

PROOF. The proof is evident, since $x^{k+1} \in I(x^k)$. □

Algorithm (i) has several advantages over Karmarkar's for multiple objective optimization. The usual approach of solving multiple objective optimization problems is to scalarize the vector problem and solve the resulting single objective optimization problem. Suppose we have to solve the linear vector optimization problem of the form

$$\Lambda - \max Cx,$$
$$\text{subject to } Ax = b, \qquad x \geq 0,$$

where A is an $m \times n$ and C is a $p \times n$ matrix. Then we can solve the following equivalent scalarized linear programming problem

$$\max w^T Cx,$$
$$\text{subject to } Ax = b, \qquad x \geq 0,$$

where w is in $\text{int}(\Lambda^*)$. If we solve the scalarized problem by Karmarkar's method, then we find a single efficient extreme point in polynomial time. If instead the approach above is used, then we can find an efficient face in polynomial time (Morin and Trafalis (1989)). This results in a significant computational savings since finding an efficient face using Karmarkar's method requires at the very least k additional pivots (one for each of the k extreme points describing the efficient face). Furthermore, algorithm (i) also applies to *nonlinear* multiple objective optimization.

3.1. Efficient faces of polytopes. Now we consider the problem of finding the set of efficient faces of a polytope. We develop an algorithm that approximates portions of the set of efficient faces of a full-dimensional bounded polytope P in $\mathbf{R}^n$ described by linear inequalities, where $\mathbf{R}^n$ is ordered by a pointed closed convex cone Λ.

We solve this problem by constructing a sequence of algebraic surfaces $\{S_k\}$ that in the limit as $k \to \infty$ approaches a portion of the efficient frontier of P. The general surface S_k can be considered the loci of centers of the intersection of the ordering cone Λ and the polytope P, where the apex of Λ is moving on S_{k-1}. If we consider the initializing manifold S_0 to be part of the inscribed ellipsoid of P and use the concept of the analytical center as described in §2.2, a conceptual algorithm to do this is outlined as follows.

OUTLINE OF CONCEPTUAL ALGORITHM (ii).

0. *Set* $k = 0$.

1. *Find the center* x_0 *of* P *and consider the inscribed surface* S_k *(e.g., the ellipsoid* E_k *) with center* x_0 *. That is, solve the problem*

$$\max \sum_{i=1}^{m} \ln(a_i^T x_0 - b_i).$$

This reduces to the solution of the following equation:

$$\sum_{i=1}^{m} \frac{a_i^T}{a_i^T x_0 - b_i} = 0.$$

Thus,

$$E_k = x_0 + \frac{1}{m-1} \{z | \langle B^{-1} z, z \rangle \le 1\},$$

where

$$B = m(m-1)(L^+)^* L^+,$$

in which $L(e_i) = \frac{a_i^T}{a_i^T x_0 - b_i}$, L^+ *is the pseudo-inverse of* L, *and* e_i *is the ith unit vector. Parameterize* S_k; *i.e.,*

$$x(\lambda) \in S_k \subseteq \mathrm{int}(P).$$

2. *Take an arbitrary point* $x(\lambda)$ *on* S_k *and consider*

$$\{x(\lambda) + \Lambda\} \cap P = P(\lambda).$$

Calculate the next approximating surface

$$S_{k+1} = T(S_k),$$

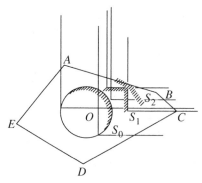

ABC efficient frontier

 Ordering cone

FIGURE 2. Illustration of algorithm (ii)

where T *is a nonlinear rational transformation.*

3. If $\|S_{k+1} - S_k\| < \varepsilon$, **then** *stop.* **Else** *return to* 2 *with* $k \leftarrow k+1$.

This algorithm is conceptual because we have not specified the rational transformation T. Below we shall use a Newton-method type operator for T.

Figure 2 illustrates algorithm (ii) for a bounded convex polygon in $\mathbf{R}^2$ where the ordering cone is the positive orthant. In Figure 2 the curves S_1 and S_2 after two iterations of the algorithm are depicted.

Note that Step 1 of the algorithm calls for parameterizing S_0. The algorithm uses this parameterization of a ellipsoid H in n-space to approximate a portion of the set of maximal facets (efficient frontier) of the polytope. The proposed algorithm is a generalization of the interior point affine version of Karmarkar's algorithm for linear programming, in which Λ is a half-space. However, our algorithm applies to both linear and nonlinear vector optimization, where the set of feasible solutions is a full-dimensional polytope in n-dimensions. In the nonlinear case the ordering cone in the decision space may not be constant. The portion of the efficient frontier obtained by algorithm (ii) depends on the geometry of the polyhedral cone Λ. Since the algorithm is dependent on parameterizations of algebraic hypersurfaces, we discuss these next.

3.2. Parameterization: Algorithmic algebraic geometry. The reason why we deal with ellipsoids rather than polytopes is that an ellipsoid has a single parameterization, while polytopes require a separate calculation for each facet. Recall from algebraic geometry (Abhyankar (1986)) that the standard form of an ellipsoid in n-space is given by

$$\frac{X_1^2}{a_1^2} + \frac{X_2^2}{a_2^2} + \cdots + \frac{X_n^2}{a_n^2} = 1$$

More generally, we can parameterize any quadric H in n-space (Abhyankar (1986)), Abhyankar and Bajaj (1987a, b, and c)) where

$$H : f_2(X_1, X_2, \ldots, X_n) = 0,$$

in which f_2 is a quadratic polynomial; i.e.,

$$f_2(X_1, \ldots, X_n) = a + \sum a_i X_i + \sum a_{ij} X_i X_j.$$

For example, the quadric (which we assume nonsingular)

$$H : f_2(X_1, \ldots, X_n) = a + \sum a_i X_i + \sum a_{ij} X_i X_j = 0,$$

can be parameterized as follows:

EXAMPLE. Parameterization of a quadric hypersurface.

Step 1. Take a point $\alpha = (\alpha_1, \ldots, \alpha_n)$ on H. Change coordinates:

(I) $$X_i = Y_i + \alpha_i \quad \text{for } 1 \leq i \leq n,$$

$$g(Y_1, \ldots, Y_n) = f(X_1, \ldots, X_n)$$
$$= \sum b_i Y_i + \sum b_{ij} Y_i Y_j.$$

Note that there is no constant term in g.

Step 2. At least one of the b_i must be nonzero. Say $b_n \neq 0$. Change coordinates:

$$Z_n = b_1 Y_1 + \cdots + b_n Y_n,$$

and

$$Z_i = Y_i, \quad \text{for } 1 \leq i \leq n - 1.$$

That is,

(II) $$Y_n = \frac{Z_n}{b_n} - \frac{b_1 Z_1}{b_n} - \cdots - \frac{b_{n-1} Z_{n-1}}{b_n},$$

$$Y_i = Z_i, \quad \text{for } 1 \leq i \leq n - 1,$$

and

$$h(Z_1, \ldots, Z_n) = g(Y_1, \ldots, Y_n)$$
$$= Z_n - \sum_{i \leq j \leq n} c_{ij} Z_i Z_j.$$

Step 3. Make the following fractional linear transformation:

(III) $$Z_n = \frac{1}{W_n} \quad \text{and} \quad Z_i = \frac{W_i}{W_n}, \quad \text{for } 1 \leq i \leq n - 1,$$

$$h(Z_1, \ldots, Z_n) = \frac{1}{W_n} - \frac{c_{nn}}{W_n^2} - \sum_{i < n} \frac{c_{in} W_i}{W_n^2} - \sum_{i \leq j < n} \frac{c_{ij} W_i W_j}{W_n^2},$$

$$h^*(W_1, \ldots, W_n) = W_n^2 h(Z_1, \ldots, Z_n)$$

(IV) $$= W_n - Q(W_1, \ldots, W_{n-1}) = 0,$$

where Q is the quadratic polynomial

$$Q(W_1, \ldots, W_{n-1}) = c_{nn} - \sum_{i<n} c_{in} W_i - \sum_{i \le j < n} c_{ij} W_i W_j.$$

Step 4. Let $T = (T_1, \ldots, T_{n-1})$, where $T_i = W_i$, for $1 \le i \le n - 1$. Use (IV) to get $W_n = Q(T)$. Then use (III) to get $Z_n = 1/Q(T)$ and $Z_i = T_i/Q(T)$, for $1 \le i \le n-1$. Use (II) to get $Y_i = L_i(T)/Q(T)$, for $1 \le i \le n$, where L_i is a linear function. Finally, use (I) to get $X_i = Q_i(T)/Q(T)$ for $1 \le i \le n$, where Q_i is a quadratic.

The rational parametric representation of a surface allows greater ease of transformation and shape control than the implicit form and is of most use in multiple objective optimization. The implicit form is useful for testing whether a point is above, on, or below the surface, where above and below is determined relative to the direction of the surface normal. It is crucial in applications to be able to go efficiently from one form to the other, especially when curves and surfaces of an object are automatically generated in one of the two representations (Boehm, Farin, and Kahmann (1984)).

In general, parameterization is more difficult than implicitization. For surfaces of degree higher than three, no rational parametric forms exist in general, although parameterizable subclasses can be identified. For low-degree curves and surfaces, Abhyankar and Bajaj (1987a, 1987b) have developed and implemented procedures for parameterizing implicit forms. The approach has been extended to parameterize planar curves of higher degree (Abhyankar and Bajaj (1987c)) and special space curves (Abhyankar and Bajaj (1986d)). These methods can be specialized to work over rational or real fields (both of characteristic 0).

For rational plane curves, i.e., curves having a rational parameterization, there is also a subclass of curves parameterizable using polynomials. Polynomial parameterization is related to whether the rational curve has one or more places at infinity. An algorithmic irreducibility criterion has been developed (Abhyankar and Bajaj (1987b)) for determining when a rational curve has one place at infinity and, thereby, determining when a rational curve also has a permissible polynomial parameterization.

3.3. Approximating the set of efficient faces of polytopes by algebraic surfaces. Consider the construction of the approximating surfaces in Step 2 of the algorithm of §3.2. The polytope $P(\lambda)$ can be described by the following inequalities:

$$a_i^T x \ge b_i, \qquad 1 \le i \le m,$$
$$\Lambda_s^T x \ge \Lambda_s^T x(\lambda), \qquad 1 \le s \le p,$$

where $x(\lambda) \in S^0$. The potential function for the problem depends on the parameter λ; i.e.,

$$f_P(x, \lambda) = \sum_{i=1}^{m} \ln(a_i^T x - b_i) + \sum_{s=1}^{p} \ln(\Lambda_s^T x - \Lambda_s^T x(\lambda)).$$

Maximizing this reduces to a solution of the following equation:

(4)
$$\sum_{i=1}^{m} \frac{a_i^T}{a_i^T x - b_i} + \sum_{s=1}^{p} \frac{\Lambda_s^T}{\Lambda_s^T x - \Lambda_s^T x(\lambda)} = 0.$$

The equation describes a portion of an algebraic surface as $x(\lambda)$ is moving on a part of S^0. If we consider the $(m + p) \times n$ matrix $\overline{A}$ such that

$$\overline{A} = \begin{bmatrix} a_1 \\ \vdots \\ a_m \\ \Lambda_1 \\ \vdots \\ \Lambda_p \end{bmatrix}$$

and

$$\Delta(x) = \mathrm{diag}(a_1 x - b_1, \ldots, a_m x - b_m, \Lambda_1(x - x(\lambda)), \ldots, \Lambda_p(x - x(\lambda))),$$

then Equation (4) can be written as

$$F(x) = \overline{A}^T \Delta(x)^{-1} e = 0.$$

How can this surface be described? If the above equation can be solved analytically with respect to $x(\lambda)$, then since $x(\lambda)$ satisfies the equation of S^{j-1}, we have an explicit formula for S^j. Otherwise, we can use interpolation techniques, e.g., spline surfaces (Farin (1987)). We solve the above equation by using Newton's method (Morin and Trafalis (1989)). Specifically, we construct a sequence of surfaces X^j in int(P) where X^j is defined in terms of x^j and x^j is obtained by applying one iteration of Newton's method beginning at x^{j-1} in an attempt to maximize $f_P(x, \lambda^j)$ as λ^j is moving on S^j, where

$$\lambda^j = \delta x^{j-1} + (1 - \delta)\lambda^{j-1},$$

and $0 < \delta < 1$.

CLAIM. If δ is chosen such that $0 < \delta < 1/(42\sqrt{p + m})$, then for every j, S^j is a good approximation to a subset of efficient faces (Morin and Trafalis (1989)).

Next we describe the operator T. In our algorithmic approach, T is the Newton operator defined by

$$T(x) = x - [DF(x)]^{-1} F(x).$$

The algorithm is as follows:

ALGORITHM (ii) [*Approximating surface algorithm*].
Initialization. *Set* $S^0 = part\ of\ E_{in}$, $j = 1$ *(for a description of S^0 see* §3.5), *and*

$$X^0 = \{x : \overline{A}^T \Delta(x)^{-1} e = 0, \lambda \in S^0\}.$$

Step 1. *Let*

$$S^j = \delta X^{j-1} + (1-\delta)S^{j-1}.$$

Step 2. *Apply Newton's method in*

$$f_P(x, \lambda^j) = \sum_{i=1}^{m} \ln(a_i^T x - b_i) + \sum_{s=1}^{p} \ln(\Lambda_s^T x - \Lambda_s^T \lambda^j)$$

and let X^j denote the resulting surface. More specifically, let

$$X^j = \{T(x^{j-1})|x^{j-1} \in X^{j-1}\}.$$

Step 3. $j \leftarrow j + 1$ *and return to Step 1.*

The above algorithm yields a sequence of surfaces S^j that converges to a part of the efficient frontier of P (Morin and Trafalis (1989)).

3.4. Example: *Approximating surface algorithm.* Now we give a simple example in which we solve explicitly the equation of centers to illustrate our algorithm.

Consider the following problem in $\mathbf{R}^2$ where the ordering cone is the positive orthant. Find the *Pareto* optimal points for the polytope P defined by the following linear inequalities:

$$2x_1 + \frac{2}{\sqrt{3}}x_2 \le 1,$$

$$-2x_1 + \frac{2}{\sqrt{3}}x_2 \le 1,$$

and

$$x_2 \ge 0.$$

Applying our algorithm yields
Step 1. *Find the center x_0 of P and consider the inscribed ellipsoid S_0. For sake of simplicity, let S_0 be an inscribed circle, which has center $(0, \sqrt{3}/6)$ and radius $\sqrt{3}/6$.*
Step 2. *Parametrize S_0 as follows:*

$$x_1(t) = \frac{\sqrt{3}}{6} \cos t$$

and

$$x_2(t) = \frac{\sqrt{3}}{6} + \frac{\sqrt{3}}{6} \sin t.$$

Step 3. *Take a point $x(t)$ on the efficient frontier of S_0 and consider the set*

$$\{x(t) + \Lambda\} \cap P = P(t).$$

Now $P(t)$ is a new polytope. Find the center $\hat{x}(t)$ of $P(t)$, where $P(t)$ is described by the following inequalities:

$$2x_1 + \frac{2}{\sqrt{3}}x_2 \leq 1,$$

$$x_1 \geq \frac{\sqrt{3}}{6}\cos t,$$

and

$$x_2 \geq \frac{\sqrt{3}}{6} + \frac{\sqrt{3}}{6}\sin t.$$

In order to find the center of $P(t)$ we consider the function

$$f_{P(t)}: \operatorname{int}(P(t)) \to \mathbf{R},$$

where

$$f_{P(t)}(x) = \ln\left(1 - 2x_1 = \frac{2}{\sqrt{3}}x_2\right)$$

$$+ \ln\left(x_1 - \frac{\sqrt{3}}{6}\cos t\right) + \ln\left(x_2 - \frac{\sqrt{3}}{6} - \frac{\sqrt{3}}{6}\sin t\right).$$

Setting the differential $df_{P(t)}$ equal to zero results in the following system of equations:

$$-\frac{2}{1 - 2x_1 - \frac{2}{\sqrt{3}}x_2} + \frac{1}{x_1 - \frac{\sqrt{3}}{6}\cos t} = 0$$

and

$$-\frac{2}{\sqrt{3}}\frac{1}{1 - 2x_1 - \frac{2}{\sqrt{3}}x_2} + \frac{1}{x_2 - \frac{\sqrt{3}}{6}\frac{\sqrt{3}}{6}\sin t} = 0.$$

Solving the above system of equations for $x_1(t)$ and $x_2(t)$ yields

$$x(t) = \begin{bmatrix} x_1(t) \\ x_2(t) \end{bmatrix} = A\hat{x} + b$$

where

$$A = \begin{bmatrix} 2 & \frac{1}{\sqrt{3}} \\ \sqrt{3} & 2 \end{bmatrix},$$

$$b = \begin{bmatrix} -\frac{1}{2} \\ -\frac{\sqrt{3}}{2} \end{bmatrix},$$

Now consider the transformation

$$y = Ax + b.$$

The above transformation is affine. Its inverse corresponds to the transformation T of §3.1. In this particular example $T = A^{-1}$ and is affine. It *transforms ellipses into ellipses. After n iterations the algorithm will result in an algebraic curve of second degree with general formula*

$$[A^n x + (A^{n-1} + A^{n-2} + \cdots + A + I)b - x_0]^2 = 12.$$

3.5. Parametrizing the efficient frontier of a polytope. Sonnevend (1985) showed how to approximate a bounded polytope P by two homothetic ellipsoids. Specifically, he showed that if P is described by a set of linear inequalities as in §3.1, then we have

$$x_0 + \frac{1}{m-1} E_{\text{out}} \subset P \subset x_0 + E_{\text{out}},$$

$$E_{\text{out}} = \{z | \langle B^{-1} z, z \rangle \le 1\},$$

where

$$B = m(m-1)(L^+)^* L^+,$$

in which

$$L(e_i) = \frac{a_i^T}{A_i^T x_0 - b_i},$$

L^+ is the pseudo-inverse of L, and e_i is the ith unit vector.

We describe how to use the parameterization of the inscribed and circumscribed ellipsoids to compute the efficient frontier of the polytope P. Consider the line that passes from the center x_0 and crosses the inscribed ellipsoid at a point p on its efficient frontier. This line also crosses the efficient frontier of E_{out} at a point q. The point q can be completely described by the parameterization of E_{out}. Now translate Λ from x_0 to q until its intersection with P reduces to the empty set. Figure 3 illustrates the procedure for a convex polygon $ABCDE$ in the plane, where the ordering cone is the positive orthant. We move the positive orthant from the center O of both the inscribed and circumscribed circles along the ray Oq until we hit the efficient frontier ABC at the point r. The procedure can be described by the following linear programming problem

(LP)   maximize t

                  subject to $x - x_0 - tq \in \Lambda$ and $x \in P$.

(LP) is equivalent to the following:

                  maximize t

                  subject to $a_i^T x \ge b_i$, $1 \le i \le m$

and

$$\Lambda_s^T x \ge \Lambda_s^T x_0 + t\Lambda_s^T q, \qquad s = 1, \ldots, p.$$

If an optimal solution (t^*, x^*) exists, then x^* is an efficient point and

$$x^*(T) = x_0 + tq(T) \quad \text{where } T \text{ is the parameter for } E_{\text{out}}.$$

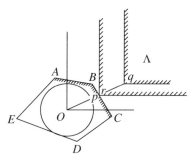

r is an efficient point
Efficient frontier = ABC

FIGURE 3.   Parametrizing the efficient frontier of a polygon.

In §3.3 we considered the initializing manifold S^0 to be part of the inscribed ellipsoid E_{in}. S^0 can be parameterized by all T for which the above (LP) problem has a solution.

3.6. Example: Trajectories and differential inclusions. Our algorithm can be related to dynamical systems. Bayer and Lagarias (1989) discussed this relationship for single objective LP problems, whose trajectories are solutions of ordinary differential equations. In our algorithm the trajectories can be considered solutions of differential inclusions (Clarke and Aubin (1977)). This is illustrated in the following example:

Find a point on the efficient frontier of the polytope P in $\mathbf{R}^2$ defined by the following system of linear inequalities:

$$-1 \leq x_1 \leq 1,$$
$$-1 \leq x_2 \leq 1.$$

The ordering cone is defined to be the positive orthant in $\mathbf{R}^2$. Geometrically it is easy to see that there is only one efficient point. More precisely, the point $(1, 1)$ is the only efficient point. Next, we describe algorithm (i) for finding a point on the efficient frontier of P. Let x^k be a point in the interior of P. Then P_k is defined as follows:

$$x_1 \geq x_1^k,$$
$$x_2 \geq x_2^k,$$
$$-1 \leq x_1 \leq 1,$$

and

$$-1 \leq x_2 \leq 1.$$

The center of P_k can be found as follows. First define

$$f_{P_k}: \text{int}(P_k) \rightarrow \mathbf{R},$$

where

$$f_{P_k}(x_1, x_2) = \ln(x_1 - x_1^k) + \ln(x_2 - x_2^k) + \ln(1 - x_1)$$
$$+ \ln(1 + x_1) + \ln(1 - x_2) + \ln(1 + x_2).$$

Then the center of P_k is defined as the solution of

$$df_{P_k}(x) = 0.$$

Solving this results in the following system of difference equations:

$$3(x_1^{k+1})^2 - 2x_1^k x_1^{k+1} - 1 = 0,$$
$$3(x_2^{k+1})^2 - 2x_2^k x_2^{k+1} - 1 = 0.$$

This system of difference equations describes a discretized trajectory of the point x_k. As k goes to infinity in the limit, these difference equations yield the following system of equations:

$$3x_1^2 - 2x_1^2 - 1 = 0,$$
$$3x_2^2 - 2x_2^2 - 1 = 0,$$

where x_1 and $x_2 \geq 0$. Hence, the solution is $(1, 1)$.

3.7. Applications in multiobjective optimization. Consider the following vector optimization problem:

$$\Lambda - \text{maximize } f(x)$$
$$\text{subject to } a_i^T x \geq b_i, \qquad i = 1, \ldots, m.$$

Assume that f is a differentiable function from $\mathbf{R}^n$ to $\mathbf{R}^k$ and that Λ is the ordering cone in the space of objectives.

Tamura (1974) considered the problem of constructing the polar cone of a polyhedral cone. This construction is also useful in our application, in constructing the cone of increasing directions. Consider first the polar cone Λ^* of Λ. That is, Λ^* is given by

$$\Lambda^* = \{x \in \mathbf{R}^m | x = \sum_{i=1}^{p} \alpha^i v_i, \alpha^i \geq 0\}.$$

The *cone of increasing directions* $Q(x)^>$ at x is given by

$$Q^> = \{y \in \mathbf{R}^n : F(x)^T y \geq 0\},$$

where

$$F(x) = [\nabla f(x)^T v_1, \ldots, \nabla f(x)^T v_p] \in \mathbf{R}^{n \times p},$$

and v_i are the edge vectors of Λ^*. From this point the procedure is the same as Algorithm (i) where $P = \{x | a_i^T x \geq b_i, i = 1, \ldots, m\}$ and the ordering cone is $Q^>$. If f is linear, then $Q^>$ is constant; otherwise $Q^>$ is a function of x.

Note that if $Q^>$ is constant, then the ideas of affine-scaling vector fields (Bayer and Lagarias (1989)) can be generalized by considering affine "cone fields" and affine scaling trajectories as solutions of differential inclusions. Specifically, consider the following MOLP problem.

$$\Lambda - \max Cx$$

$$\text{subject to } Ax = b, \qquad x \geq 0,$$

where AA^T is invertible and C is a $p \times n$ matrix. Define the following "cone field"

$$V_A(x, C) = X\pi_{(Ax)^L}(XC),$$

where C is the cone of increasing directions

$$C = \{y \in \mathbf{R}^n \mid Cy \geq 0\},$$

$\pi_{(Ax)^L}$ is the orthogonal projection operator onto A^L, and

$$X = \text{diag}(x_1, \ldots, x_n).$$

Then the affine scaling trajectory $T_A(x; C, A, b)$ in the vector optimization setting can be defined as the solution of the differential inclusion

$$\frac{d}{dt}x(t) \in V_A(x(t), C),$$

$$x(0) = x_0,$$

where x_0 is an interior point of the feasible polytope.

Acknowledgments

The authors are grateful to the referee for numerous useful comments, in particular an improved proof of Theorem 2. This research was supported in part by the Office of Naval Research under Purdue's University Research Initiative in Computational Combinatorics, ONR Contract N00014-86-K0689, NSF grant DMS-88-16286, ONR grant N00014-88-K0402, and ARO contract DAAG 29-85-C0018 under Cornell MSI, at Purdue.

References

S. S. Abhyankar, *Lectures on algorithmic algebraic geometry*. I, Technical Report CC-86-1, University Research Initiative in Computational Combinatorics, IIES, Purdue University, West Lafayette, Ind., 1986.

S. S. Abhyankar and C. Bajaj, *Automatic rational parametrization of curves and surfaces*. I: *Conics and conicoids*, Comput. Aided Design **19** (1987a), 11–14.

_____, *Automatic rational parametrization of curves and surfaces*. II: *Cubics and cubicoids*, Comput. Aided Design **19** (1987b), 499–502.

_____, *Automatic rational parametrization of curves and surfaces* III: *Algebraic plane curves*, Technical Report CC-82-2, University Research Initiative in Computational Combinatorics, IIES, Purdue University, West Lafayette, Ind., 1987c.

_____, *Automatic rational parametrization of curves and surfaces* IV: *Algebraic space curves*, Technical Report CSD-TR-703, Department of Computer Sciences, Purdue University, West Lafayette, Ind., 1987d.

I. Adler, M. G. Resende, and G. Veiga, *An implementation of Karmarkar's algorithm for linear programming*, Operations Research Center, Report 86-8, University of California, Berkeley, Calif., 1986.

J. B. Aubin, A. Cellina, and J. Noel, *Monotone trajectories of multivalued dynamical systems*, Ann. Mat. Pura Appl. **115** (1977), 99–117.

D. Bayer and J. Lagarias, *The nonlinear geometry of linear programming*. I, II, Trans. Amer. Math. Soc. **314** (1989), 499–526; 527–581.

H. P. Benson and T. L. Morin, *The vector optimization problem: Proper efficiency and stability*, SIAM J. Appl. Math. **32** (1977), 64–72.

W. Boehm, A. Farin, and J. Kahmann, *A survey of curve and surface methods in* CAGD, Comput. Aided Geom. Design **1** (1984), 1–60.

F. H. Clarke and J. P. Aubin, *Monotone invariant solutions to differential inclusions*, J. London Math. Soc. **16** (1977), 357–366.

J. L. Cohon, *Multiobjective programming and planning*, Academic Press, New York, 1978.

J. G. Ecker, N. S. Hegner, and I. A. Kouada, *Generating all maximal efficient faces for multiobjective linear programs*, J. Optim. Theory Appl. **30** (1980), 353–381.

J. G. Ecker and I. A. Kouada, *Finding efficient points for linear multiple objective programs*, Math. Programming **8** (1975), 375–377.

_____, *Finding all efficient extreme points for linear multiple objective programs*, Math. Programming **14** (1978), 249–261.

J. P. Evans and R. E. Steuer, *A revised simplex method for linear multiobjective programs*, Math. Programming **5** (1973), 54–72.

A. Farin, ed., *Geometric modeling: Algorithms and new trends*, SIAM, 1987.

T. Gal, *A general method for determining the set of all efficient solutions to a linear vector maximum problem*, Report No. 76/12, Institut für Wirtschaftswissenschaften, Aachen, Germany, 1976.

G. Hazen and T. L. Morin, *Steepest ascent algorithms for nonconical multiple objective programming*, J. Math. Anal. Appl. **100** (1984), 188–221.

P. Huard, *Resolution of mathematical programming with nonlinear constraints by the method of centers*, Nonlinear Programming (J. Abadie, ed.), North-Holland, Amsterdam, 1967.

M. Isermann, *The enumeration of the set of all efficient solutions for a linear multiple objective program*, Oper. Res. Quart. **28** (1977), 711–725.

_____, *The enumeration of the set of all efficient solutions for a linear multiple objective transportation problem*, Naval Res. Logist. Quart. **26** (1979), 123–139.

N. Karmarkar, *A new polynomial-time algorithm for linear programming*, Combinatorica **4** (1984), 373–395.

L. G. Khachian, *A polynomial algorithm for linear programming*, Dokl. Akad. Nauk SSSR **244** (1979), 1093–1096.

V. Klee and G. L. Minty, *How good is the simplex method*, Inequalities III (O. Shisha, ed.), Academic Press, New York, 1972, pp. 159–175.

S. R. Lay, *Convex sets and their applications*, John Wiley & Sons, New York, 1982.

R. E. Marsten, M. J. Saltzman, D. F. Shanno, G. S. Pierce, and J. F. Ballintijn, *Implementation of a dual affine interior point algorithm for linear programming*, Working Paper, University of Arizona, Tuscon, Ariz., 1988.

K. A. McShane, C. L. Monma, and D. Shanno, *An implementation of a primal-dual interior point method for linear programming*, Working Paper, 1988.

T. L. Morin and T. B. Trafalis, *A polynomial-time algorithm for finding an efficient face of a polyhedron*, Technical Report, Purdue University, West Lafayette, Ind., 1989.

A. Pascoletti and P. Serafini, *An iterative procedure for vector optimization*, J. Math. Anal. Appl. **89** (1982), 95–106.

F. P. Preparata and M. I. Shamos, *Computational geometry: An introduction*, Springer-Verlag, New York, 1985.

J. Renegar, *A polynomial-time algorithm, based on Newton's method, for linear programming*, Report 07118-86, Mathematical Sciences Research Institute, University of California, Berkeley, Calif., 1986.

A. P. Sage, *Methodology for large-scale systems*, McGraw-Hill, New York, 1977.

G. Sonnevend, *An analytical center for polyhedrons and new classes of global algorithms for linear (smooth convex) programming*, Proc. 12th IFIP Conference on System Modelling, Budapest, 1985.

G. Sonnevend and J. Stoer, *Global ellipsoidal approximations and homotopy methods for solving convex analytic programs*, preprint, 1988.

R. E. Steuer, *Multiple objective linear programming with interval criterion weights*, Management Sci. **23** (1976), 305–316.

K. Tamura, *A method for constructing the polar cone of a polyhedral cone, with applications to linear multicriteria decision problems*, J. Optim. Theory and Appl. **19** (1974), 547–564.

T. B. Trafalis, *Efficient faces of a polytope: Interior methods in multiple objective optimization*, Ph.D. dissertation, Purdue University, August, 1989.

R. M. Vaidya, *A locally well-behaved potential function and a simple Newton-type method for finding the center of a polytope*, Technical Report, AT&T Bell Laboratories, Murray Hill, N. J., 1987.

P. L. Yu, *Cone convexity, cone extreme points, and nondominated solutions in decision problems with multiobjectives*, J. Optim. Theory Appl. **14** (1974), 318–377.

P. L. Yu and M. Zeleny, *The set of all nondominated solutions in linear cases and a multicriteria simplex method*, J. Math. Anal. Appl. **49** (1975), 430–468.

P. L. Yu, *Multiple-criteria decision making*, Plenum Press, New York, 1985.

M. Zeleny, *Linear multiobjective programming*, Springer-Verlag, Berlin/Heidelberg, 1974.

____, *Multiple criteria decision making*, McGraw-Hill, New York, 1982.

DEPARTMENTS OF MATHEMATICS AND COMPUTER SCIENCE, SCHOOL OF INDUSTRIAL ENGINEERING, PURDUE UNIVERSITY, WEST LAFAYETTE, INDIANA 47907

E-mail address, S. S. Abhyankar, T. L. Morin, and T. Trafalis: iies@ecn.purdue.edu